Handbook of Experimental Pharmacology

Volume 180

Editor-in-Chief

K. Starke, Freiburg i. Br.

Editorial Board

S. Duckles, Irvine, CA
M. Eichelbaum, Stuttgart
D. Ganten, Berlin
F. Hofmann, München
C. Page, London
W. Rosenthal, Berlin
G. Rubanyi, San Diego, CA

Bone Marrow-Derived Progenitors

Contributors

M. Andreeff, T. Asahara, C. Beauséjour, A.J. Boyle,
B.H. Davis, I. Dimarakis, C. Flores, J. Galipeau,
M.Y. Gordon, N.A. Habib, B. Hall, J.M. Hare, L.-Y. Huw,
I. Kan, A. Karsan, K. Kauser, B. Larrivée, S.R. Larsen,
J.-P. Lévesque, N. Levičar, D.W. Losordo, F. Marini,
E. Melamed, D. Offen, H.-S. Qian, A. Quraishi,
J.E.J. Rasko, M.M. de Resende, K.H. Schuleri, J. Stagg,
D.A. Taylor, J. Tracey, I.G. Winkler, A. Zenovich

Editors
Katalin Kauser and Andreas-Michael Zeiher

Katalin Kauser MD, PhD, DSc
Director, Cardiovascular Disease
Boehringer Ingelheim
Pharmaceuticals, Inc.
175 Briar Ridge Road,
Ridgefield, CT 06877
USA
e-mail: kkauser
 @rdg.boehringer-ingelheim.com

Andreas-Michael Zeiher MD
Professor Molecular Cardiology
Department of Internal Medicine III
University of Frankfurt
Theodor-Stern-Kai 7
D-60590 Frankfurt a. Main

e-mail: zeiher@em.uni-frankfurt.de

With 18 Figures and 6 Tables

ISSN 0171-2004

ISBN 978-3-540-68975-1 Springer Berlin Heidelberg New York

This work is subject to copyright. All rights reserved, whether the whole or part of the material is concerned, specifically the rights of translation, reprinting, reuse of illustrations, recitation, broadcasting, reproduction on microfilm or in any other way, and storage in data banks. Duplication of this publication or parts thereof is permitted only under the provisions of the German Copyright Law of September 9, 1965, in its current version, and permission for use must always be obtained from Springer. Violations are liable for prosecution under the German Copyright Law.

Springer is a part of Springer Science+Business Media
springer.com

© Springer-Verlag Berlin Heidelberg 2007

The use of general descriptive names, registered names, trademarks, etc. in this publication does not imply, even in the absence of a specific statement, that such names are exempt from the relevant protective laws and regulations and therefore free for general use.

Product liability: The publishers cannot guarantee the accuracy of any information about dosage and application contained in this book. In every individual case the user must check such information by consulting the relevant literature.

Editor: Simon Rallison, London
Desk Editor: Susanne Dathe, Heidelberg
Cover design: *design & production* GmbH, Heidelberg, Germany
Typesetting and production: LE-TEX Jelonek, Schmidt & Vöckler GbR, Leipzig, Germany
Printed on acid-free paper   27/3100-YL - 5 4 3 2 1 0

Preface

There is excitement generated almost daily about the possible uses of stem cells to treat human disease. The ability of stem cells to acquire different desired phenotypes has opened the door for a new discipline: regenerative medicine. Much of the interest for this purpose is generated by embryonic stem cells, but their use is still controversial for moral as well as scientific reasons. Less controversial and readily available are the adult bone marrow-derived progenitors, including hematopoietic stem cells, endothelial progenitors, and mesenchymal stem cells, which are the subjects of this book. These cells can be isolated by simple procedures directly from the bone marrow or from peripheral blood after being stimulated, i.e., mobilized. By reaching sites of damage through the circulation or even after local administration, these cells can overcome the hurdles of delivery approaches that limit the success of gene therapy. Adult bone marrow-derived cells have been shown to regenerate diseased hepatocytes and contribute to neurons, blood vessels, and skeletal and cardiac muscle cells. The increasing amount of new data, sometimes with conflicting results, is making us appreciate the molecular complexity of cell differentiation and potential mechanisms of action involved in these cell-mediated processes. It is becoming increasingly important to understand the biology of these cells to potentially improve their therapeutic efficiency and to facilitate their proper therapeutic use. Examining the cell-mediated processes can ultimately lead to the discovery of pathways and molecular mechanisms of organ repair, which can be further utilized in drug development. With patients' growing attention to the most recent research developments, there is increasing medical need for a better understanding—developed through rationally designed, randomized clinical trials that will move these strategies quickly and carefully toward medical reality—to parallel the increased enthusiasm.

In this volume of the series *Handbook of Experimental Pharmacology* published by Springer, we hope to achieve the ambitious goal of providing a comprehensive overview of the currently available information related to the therapeutic utility of adult bone marrow-derived cells. Chapters in Part I focus on basic principles, including a general introduction to the different bone marrow-derived cell types, mechanisms contributing to their development and localization in the bone marrow niche, mechanisms leading to their mobilization, the current understanding about their immune plasticity, the effect of aging, and the potential enhancement of their survival or function using cell–gene combinations. Part II is dedicated to therapeutically relevant preclinical experiences and the most recent clinical experiences with these cells for cardiac diseases, neurodegenerative disorders, liver diseases, and diabetes. The potential role of bone marrow-derived cells in tumorigenesis and their potential contribution to tumor angiogenesis are also discussed. Although their exact role in cancer pathology remains to be better understood, harnessing the ability of these cells to deliver antitumor agents provides an additional therapeutic opportunity, which is introduced within the therapeutic section.

Each chapter is written or co-authored by accomplished scientists, leading experts in their field, ensuring the delivery of up-to-date information regarding our current understanding of bone marrow-derived progenitor cell biology and its applications to specific disease indications. The editors focused their efforts on providing a balanced overview of the recent developments in the field without major interference with the content and style of the individual chapters. In some instances reiteration of basic principles in the different chapters may appear redundant when looking at the volume as a whole, but it is necessary to allow each chapter to serve as a self-standing overview of the chosen principle.

The editors thank the authors of the chapters for their excellent contributions, and Springer for its highly professional work and timely publication of the book. We would like to express our specific gratitude to Susanne Dathe from Springer for her patience and guidance throughout the development of this book. We also appreciate the interest and support of the HEP Editorial Board, specifically acknowledging Gabor M. Rubanyi among the board members for his enthusiastic support and encouragement from the very beginning of this project.

Ridgefield and Frankfurt am Main,   Katalin Kauser
March 2007   Andreas-Michael Zeiher

List of Contents

Part I. Basic Principles

Mobilization of Bone Marrow-Derived Progenitors 3
  J.-P. Lévesque, I. G. Winkler, S. R. Larsen, J. E. J. Rasko

Role of Endothelial Nitric Oxide in Bone Marrow-Derived
Progenitor Cell Mobilization 37
  M. Monterio de Resende, L.-Y. Huw, H.-S. Qian, K. Kauser

Immune Plasticity of Bone Marrow-Derived Mesenchymal Stromal Cells 45
  J. Stagg, J. Galipeau

Bone Marrow-Derived Cells: The Influence of Aging
and Cellular Senescence 67
  C. Beauséjour

Involvement of Marrow-Derived Endothelial Cells in Vascularization . 89
  B. Larrivée, A. Karsan

Part II. Therapeutic Implication and Clinical Experience

Comparison of Intracardiac Cell Transplantation:
Autologous Skeletal Myoblasts Versus Bone Marrow Cells 117
  A. G. Zenovich, B. H. Davis, D. A. Taylor

Ischemic Tissue Repair by Autologous Bone Marrow-Derived Stem Cells:
Scientific Basis and Preclinical Data 167
  A. Quraishi, D. W. Losordo

Cell Therapy and Gene Therapy Using Endothelial Progenitor Cells
for Vascular Regeneration 181
  T. Asahara

Mesenchymal Stem Cells for Cardiac Regenerative Therapy 195
  K. H. Schuleri, A. J. Boyle, J. M. Hare

Autotransplantation of Bone Marrow-Derived Stem Cells
as a Therapy for Neurodegenerative Diseases 219
   I. Kan, E. Melamed, D. Offen

Stem Cells as a Treatment for Chronic Liver Disease and Diabetes . . . 243
   N. Levičar, I. Dimarakis, C. Flores, J. Tracey, M. Y. Gordon, N. A. Habib

The Participation of Mesenchymal Stem Cells
in Tumor Stroma Formation and Their Application
as Targeted-Gene Delivery Vehicles 263
   B. Hall, M. Andreeff, F. Marini

Subject Index 285

List of Contributors

Addresses given at the beginning of respective chapters

Andreeff, M. , 263
Asahara, T. , 181

Beauséjour, C. , 67
Boyle, A.J. , 195

Davis, B.H. , 117
Dimarakis, I. , 243

Flores, C. , 243

Galipeau, J. , 45
Gordon, M.Y. , 243

Habib, N.A. , 243
Hall, B. , 263
Hare, J.M. , 195
Huw, L.-Y. , 37

Kan, I. , 219
Karsan, A. , 89
Kauser, K. , 37

Larrivée, B. , 89
Larsen, S.R. , 3

Lévesque, J.-P. , 3
Levičar, N. , 243
Losordo, D.W. , 167

Marini, F. , 263
Melamed, E. , 219

Offen, D. , 219

Qian, H.-S. , 37
Quraishi, A. , 167

Rasko, J.E.J. , 3
de Resende, M.M. , 37

Schuleri, K.H. , 195
Stagg, J. , 45

Taylor, D.A. , 117
Tracey, J. , 243

Winkler, I.G. , 3

Zenovich, A.G. , 117

Part I
Basic Principles

Mobilization of Bone Marrow-Derived Progenitors

J.-P. Lévesque[1] (✉) · I. G. Winkler[1] · S. R. Larsen[2,3] · J. E. J. Rasko[2,3]

[1] Mater Medical Research Institute, University of Queensland, Biotherapy Program, Aubigny Place, Raymond Terrace, 4101 South Brisbane, Queensland, Australia
jplevesque@mmri.mater.org.au

[2] Centenary Institute of Cancer Medicine and Cell Biology, University of Sydney, New South Wales, Australia

[3] Cell and Molecular Therapies, Institute of Haematology, Sydney Cancer Center, Royal Prince Alfred Hospital, Camperdown, New South Wales, Australia

1	Introduction		4
2	Mobilization of Hematopoietic Stem and Progenitor Cells		5
2.1	Diversity of Mobilization Mechanisms and Resulting Kinetics		5
2.2	Stem and Progenitor Cells Reside in Separate Niches Within the Bone Marrow		6
2.3	Finding Your Home and Staying There: A Matter for Cell Adhesion Molecules and Chemokines		8
	2.3.1	Bone Marrow Hematopoietic Stem Cell Niches	8
	2.3.2	Bone Marrow-Committed Hematopoietic Progenitor Cell Niches	10
2.4	What Keeps HSPCs Within the BM?		11
2.5	What Makes HSPCs Leave the BM?		12
	2.5.1	Neutrophilia and/or Neutrophil Activation Are Essential for HSPC Mobilization	12
	2.5.2	Role of Neutrophil Proteases	12
	2.5.3	The Role of Bone-Forming Cells	14
2.6	Strategies to Increase HSPC Mobilization		15
	2.6.1	Why the Need to Enhance Mobilization	15
	2.6.2	Chemical Stabilization of G-CSF	17
	2.6.3	Stem Cell Factor	17
	2.6.4	Chemokines and Their Analogs: The AMD3100 Success Story	18
3	Mobilization of Bone Marrow Endothelial Progenitor Cells		18
3.1	Mobilization of Pro-angiogenic Progenitor Cells		18
3.2	What Are Endothelial Progenitor Cells? A Simple Question, a Complex Answer		19
3.3	Is Re-vascularization Due to EPC Mobilization?		21
4	Mobilization of Bone Marrow Mesenchymal Stem Cells		23
4.1	What Are Mesenchymal Stem Cells?		23
4.2	Mesenchymal Stem Cells Are Present in Circulating Human Blood During Gestation		24
4.3	Can MSCs Be Isolated from Adult Peripheral Blood and Can They Be Mobilized?		25
5	Conclusion		27
	References		28

Abstract Bone marrow (BM) is a source of various stem and progenitor cells in the adult, and it is able to regenerate a variety of tissues following transplantation. In the 1970s the first BM stem cells identified were hematopoietic stem cells (HSCs). HSCs have the potential to differentiate into all myeloid (including erythroid) and lymphoid cell lineages in vitro and reconstitute the entire hematopoietic and immune systems following transplantation in vivo. More recently, nonhematopoietic stem and progenitor cells have been identified that can differentiate into other cell types such as endothelial progenitor cells (EPCs), contributing to the neovascularization of tumors as well as ischemic tissues, and mesenchymal stem cells (MSCs), which are able to differentiate into many cells of ectodermal, endodermal, and mesodermal origins in vitro as well as in vivo. Following adequate stimulation, stem and progenitor cells can be forced out of the BM to circulate into the peripheral blood, a phenomenon called "mobilization." This chapter reviews the molecular mechanisms behind mobilization and how these have led to the various strategies employed to mobilize BM-derived stem and progenitor cells in experimental and clinical settings. Mobilization of HSCs will be reviewed first, as it has been best-explored—being used extensively in clinics to transplant large numbers of HSCs to rescue cancer patients requiring hematopoietic reconstitution—and provides a paradigm that can be generalized to the mobilization of other types of BM-derived stem and progenitor cells in order to repair other tissues.

Keywords Mobilization · Hematopoietic stem cells · Endothelial progenitor cells · Mesenchymal stem cells · Transplantation · Tissue repair

1
Introduction

In the adult, bone marrow (BM) is a source of various stem and progenitor cells that are able to regenerate a variety of tissues following transplantation. Schofield identified the first BM stem cells, which were hematopoietic stem cells (HSCs) (Schofield 1970). HSCs have the potential to differentiate into all myeloid (including erythroid) and lymphoid cell lineages in vitro and reconstitute the entire hematopoietic and immune systems following transplantation in vivo. More recently, nonhematopoietic stem and progenitor cells have been identified that can differentiate into other cell types such as endothelial progenitor cells (EPCs), contributing to the neovascularization of tumors as well as ischemic tissues (Asahara et al. 1999), and mesenchymal stem cells (MSCs), which are able to differentiate into many cells of ectodermal, endodermal, and mesodermal origins (such as adipocytes, chondrocytes, osteoblasts, hepatocytes, neurons, myocytes, and endothelial and epithelial cells) in vitro as well as in vivo (Sale and Storb 1983; Barry et al. 1999, 2001; Devine et al. 2001; Dennis and Charbord 2002).

In steady-state conditions in adult mammals, most HSCs, EPCs, and MSCs reside in the BM, with a few HSCs (Wright et al. 2001; Abkowitz et al. 2003) and EPCs (Lin et al. 2000) circulating in the peripheral blood; circulating MSCs are usually not detectable (Lazarus et al. 1997; Wexler et al. 2003). For this reason, the stem and progenitor cells used to be isolated by BM aspiration for subsequent transplantation into patients requiring immune and hematopoietic reconstitution or tissue repair. As multiple painful BM aspirations are necessary to obtain sufficient numbers of stem cells to reconstitute a patient, general anesthesia and hospitalization of the donor were required, a process posing a risk for the donor. Because of this limitation, alternative sources of transplantable stem cells have been investigated. Following stress, challenge, or stimulation of the BM compartment, a proportion of these BM stem and progenitor cells egress from the BM and circulate into the blood. This phenomenon is called "mobilization." Mobilized cells are harvested by apheresis from the peripheral blood so they can be concentrated, eventually enriched, and stored for transplantation. If sufficient numbers of stem cells can be mobilized, mobilization offers the advantage of a cost-effective, relatively safe procedure to collect transplantable stem cells. The advantage of mobilized stem cells over BM aspiration is illustrated by the speed at which mobilized hematopoietic stem and progenitors cells (HSPCs) have supplanted BM aspiration as the preferred source of HSPCs for transplantation. Since its first discovery in human patients in the 1980s, cellular support using mobilized HSPCs has now grown to the point where it is used with over 45,000 patients a year worldwide. The main reason why mobilization is preferred to BM aspiration for collection of HSPCs is increased safety and comfort for the donor, faster reconstitution, and greater disease-free survival in recipient (To et al. 1997; Korbling and Anderlini 2001).

The molecular mechanisms that drive BM stem cell mobilization and the reasons behind certain molecules eliciting that mobilization have remained a mystery for a long time. In recent years, systematic analysis of (1) the molecular mechanisms responsible for the retention of BM stem and progenitor cells at their specific niches and (2) how these mechanisms are perturbed during mobilization has shed some light on the processes behind mobilization. The identification of some of the mechanisms responsible for mobilization (although many remain to be identified) has led to the development of new molecules that will considerably improve HSPC mobilization in the clinic.

The aim of this chapter is to review the molecular mechanisms and how these have led to the various strategies employed to mobilize BM-derived stem and progenitor cells in experimental and clinical settings. Mobilization of HSPCs will be reviewed first as it is the best understood and provides a paradigm that can be generalized to the mobilization of other types of BM stem and progenitor cells.

2
Mobilization of Hematopoietic Stem and Progenitor Cells

2.1
Diversity of Mobilization Mechanisms and Resulting Kinetics

HSPC mobilization can occur in response to a variety of stimuli that can be grouped according to their molecular nature, kinetics, and efficiency to mobilize HSPC. For instance, intense and prolonged physical exercise causes a limited but nevertheless significant HSPC mobilization as observed in marathon runners (Barrett et al. 1978; Bonsignore et al. 2002). Neutrophil-activating chemokines such as interleukin (IL)-8/CXCL8 or Groβ/CXCL2 induce extremely rapid and brief HSPC mobilization within minutes, whereas hematopoietic growth factors induce more sustained mobilization within days. At the other end of the spectrum, cytotoxic drugs such as cyclophosphamide (CY) or 5′-fluorouracil (5-FU) induce HSPC mobilization within weeks during the recovery that follows myeloablation of the BM. Collectively, this broad range of kinetics and these levels of mobilization suggests that the different agents induce HSPC mobilization through different mechanisms (Table 1).

Currently, the agent most commonly used to elicit HSPC mobilization in the clinical setting is granulocyte-colony stimulating factor (G-CSF) used alone or in combination with myelosuppressive chemotherapy or stem cell factor (SCF) (To et al. 1994; Korbling and Anderlini 2001). The administration of G-CSF induces a 10- to 100-fold increase in the level of circulating HSPC in humans, primates, and mice. G-CSF-induced mobilization is time- and dose-dependent, involving a rapid neutrophilia (evident within hours) and a gradual increase in HSPC numbers in the blood, peaking between 4 and 7 days of G-CSF administration in humans. Mobilization with chemotherapeutic agents such as CY occurs during the recovery phase following the chemotherapy-induced neutropenia, that is, days 6–8 in mice, and days 10–14 in humans.

In order to understand how BM HSPCs are mobilized into the blood, it is first necessary to understand why they remain in the BM in steady-state conditions.

2.2
Stem and Progenitor Cells Reside in Separate Niches Within the Bone Marrow

Hematopoietic stem cells and lineage-restricted progenitor cells (HPCs) do not distribute randomly in the BM but are localized according to their differentiation stage. The majority of true HSCs are found at the bone–BM interface (endosteum) (Lord et al. 1975; Nilsson et al. 2001) in contact with osteoblasts (Zhang et al. 2003; Arai et al. 2004), whereas more committed progenitors accumulate in the central BM (Lord et al. 1975; Nilsson et al. 2001). The first experiments to illustrate this were performed by Brian Lord in the 1970s. The BM was dissected according to its proximity to the femur shaft. Most of the short-term reconstitution cells that colonize the spleen of lethally irradiated mice (colony forming units-spleen, CFU-S) were found close to the bone, while lineage-restricted HPCs that form colonies in vitro (CFU-C), but are unable to reconstitute hematopoiesis in vivo in lethally irradiated mice, accumulated away from the bone, in the central BM (Lord et al. 1975). As CFU-S can reconstitute hematopoiesis in vivo for a few weeks (unlike true HSCs that reconstitute life-long hematopoiesis), while CFU-C cannot, it was concluded that the most

Table 1 Classes of mobilizing agents

Class of agents	Time to maximum mobilization in humans and mice	Examples	Mechanism(s)
Chemokines and their analogs	15 min to few hours	CCL3, CXCL2, CXCL8	Direct granulocyte activation
		CXCL12, CXCR4 antagonists (AMD3100)	Blockade of CXCR4 function
Stress	1–2 h	Intense exercise, ACTH	Demargination, granulocyte activation?
Polyanions	1–2 h	Fucoidans	Release of cytokines and chemokines from intracellular and extracellular stores
		Dextran sulfate, polymethacrylic acid	ND
	5 days	Defibrotide	ND
Toxins	2–3 days	Pertussis toxin	Blocked CXCR4-mediated signaling
	ND	Bacterial endotoxins	ND
Antibodies	1–2 days	Anti-CD49d, anti-VCAM-1	Direct inhibition of VLA4:VCAM-1 interaction
Growth factors	4–6 days	G-CSF, pegylated G-CSF	Granulocyte expansion/activation
		KIT ligand/SCF	Synergism with G-CSF
		IL-3, GM-CSF, Flt-3 ligand, thrombopoietin, growth hormone, erythropoietin	ND
		VEGF, angiopoietin-1	Activation of Nos3, release of MMP-9
Myelosuppressive	1–3 weeks	CY, 5-FU	Granulocyte expansion, activation following BM suppression

G-CSF, granulocyte colony-stimulating factor; GM-CSF, granulocyte-macrophage colony-stimulating factor; MMP, matrix metalloproteinase; ND, not determined; SCF, stem cell factor; VCAM, vascular cell adhesion molecule; VEGF, vascular endothelial growth factor; VLA, very late activation antigen primitive HPCs and HSCs reside in proximity of the bone, whereas more lineage-restricted HPCs reside in the central marrow (Lord et al. 1975). This was recently confirmed in the mouse by sorting lineage-negative (Lin$^-$) wheat-germ agglutinin (WGA)^{dim} rhodamine 123 (Rho123)^{dull} cells or Lin$^-$ Sca-1$^+$ c-KIT$^+$ cells, which are both enriched in HSCs, and Lin$^-$ WGA^{dim} Rho123^{bright} cells, which are enriched in HPCs. HSC- and HPC-enriched populations were then labeled with the fluorescent cell tracker carboxylfluorescein diacetate succinimidyl ester (CFSE) and injected intravenously into nonirradiated syngeneic recipients. While HPCs distributed randomly in the whole BM, HSCs lodged preferentially at the bone–BM interface or endosteum (Nilsson et al. 2001, 2005). Further studies have confirmed that cells that display the antigen profile of mouse HSCs (Lin$^-$ Sca-1$^+$ c-KIT$^+$ Tie-2$^+$) are in close association with bone-forming osteoblasts that line the endosteum (Calvi et al. 2003; Zhang et al. 2003; Arai et al. 2004), whereas myeloid and lymphoid HPCs are in contact with fibroblastoid elements of the central marrow that express vascular cell adhesion molecule-1 (VCAM-1) and the chemokine CXCL12 (Tokoyoda et al. 2004). This differential distribution of stem and progenitor cells within the BM is in good accord with the notion that each type of stem and progenitor cell lodges in specific niches that provide specific signals adapted to the specific requirements of these cells (Schofield 1978; Fuchs et al. 2004; Tumbar et al. 2004; Kiel et al. 2005; Nagasawa 2006; Yin and Li 2006). These niche-specific signals control survival, quiescence, proliferation, and differentiation in order to maintain a steady stock of stem and progenitor cells while producing an adequate number of mature blood cells.

2.3
Finding Your Home and Staying There: A Matter for Cell Adhesion Molecules and Chemokines

2.3.1
Bone Marrow Hematopoietic Stem Cell Niches

In addition to the anatomical evidence for an HSC niche at the endosteum, there is strong genetic evidence that osteoblasts are crucial to BM hematopoiesis. For instance, transgenic mice with increased bone formation have an increased number of HSCs in the BM. Transgenic mice expressing a constitutively active mutant of the parathyroid hormone receptor 1 (PTHR1) in osteoblasts as well as mice carrying an inducible deletion of the bone morphogenic protein (BMP) receptor 1A (BMPR1A) gene, both have an increased number of osteoblasts, increased trabeculae, and a doubling in the number of HSCs per femur despite a markedly reduced BM cavity due to increased bone formation (Calvi et al. 2003; Zhang et al. 2003). Conversely, the ablation of osteoblasts results in a drop in HSC number. In transgenic mice carrying a herpesvirus thymidine kinase (HTK) suicide gene driven by an osteoblast-specific promoter, administration of ganciclovir specifically induces osteoblast death. Osteoblast ablation results in a rapid and strong reduction in BM cellularity and a tenfold reduction in the number of HSCs present in the BM (Visnjic et al. 2004). Importantly, HSPCs and hematopoiesis relocate to the spleen and liver of these conditionally osteoblast-ablated mice (Visnjic et al. 2004). Collectively, these data strongly suggest that osteoblasts are an essential component of the HSC niche in the BM.

To understand how bone formation and BM hematopoiesis are coupled, one has to consider the molecular interactions between osteoblasts and HSCs. Lodgment at the endosteal niche is driven by a calcium gradient (Adams et al. 2006) and an array of osteoblast-mediated adhesive interactions (Fig. 1). At the endosteum, the bone is in constant turnover with concomitant bone formation driven by osteoblasts and osteoclast-mediated bone degradation. This continuous bone degradation releases soluble Ca^{2+} in the BM fluid, thus forming a Ca^{2+} gradient. HSCs express a chemotactic Ca^{2+} receptor that senses this gradient and promotes their migration to and lodgment at the endosteal HSC niche (Adams et al. 2006). Similarly, adhesive molecules produced by osteoblasts such as osteopontin, N-cadherin, transmembrane c-KIT ligand stem cell factor (tm-SCF), and the polysaccharide hyaluronic acid keep HSCs at the endosteum (see Fig. 1). Deletion of any of these molecules or their receptors results in random distribution of HSCs in the BM following transplantation (Nilsson et al. 2003, 2005; Hosokawa et al. 2005; Stier et al. 2005; Adams et al. 2006). Interestingly, adhesive interactions mediated by osteopontin, hyaluronic acid, and N-cadherin also initiate signaling events in HSCs that together delay their proliferation (Nilsson et al. 2003, 2005; Hosokawa et al. 2005; Stier et al. 2005). Thus, these adhesive interactions not only mediate tight adhesion of HSCs to osteoblasts at the hematopoietic niche, but are also likely to regulate HSC self-renewal in vivo.

The dramatic effect of osteoblast ablation on hematopoiesis may be explained by the loss of osteoblast-mediated adhesive interactions as well as loss of the cytokines and chemokines they produce (Fig. 1). In particular, osteoblasts are the main source of (1) chemokine stromal cell-derived factor-1 (SDF-1)/CXCL12 in the BM (Semerad et al. 2005), (2) angiopoietin-1, the ligand for the tyrosine-kinase receptor Tie-2 which is expressed by HSCs (Arai et al. 2004), and (3) Jagged-1, a ligand for Notch1 also expressed at the surface of HSCs (Calvi et al. 2003). Importantly, these interactions all regulate the survival/quiescence/proliferation of HSCs (Carlesso et al. 1999; Lataillade et al. 2000; Cashman et al. 2002; Arai et al. 2004). Hence, in light of these recent findings, it is not surprising that elimination of these essential interactions by specific ablation of osteoblasts, or conversely enhancement of osteoblast-specific ligands through increased bone formation, suppresses or enhances BM hematopoiesis. Interestingly, ablation of osteoblasts results in the massive migration of HSPCs to the spleen and liver via the blood (Visnjic et al. 2004). Hence, suppression of bone formation results in HSPC mobilization.

It must be noted that an "endothelial HSC niche" has recently been suggested as some HSCs that express CD150 are found in direct contact with the many

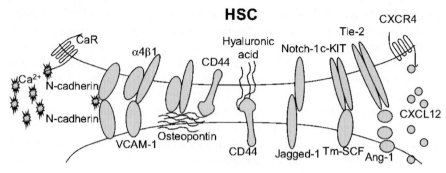

Fig. 1 Molecular interactions between HSCs and osteoblasts at the endosteal HSC niche. Attracted by CXCL12 secreted by osteoblasts and by soluble calcium released from bone degradation, HSCs expressing CXCR4 and calcium receptors lodge at the endosteum to establish direct cell–cell contact with osteoblasts through N-cadherin homotypic adhesive interactions, VCAM-1 interaction with α4β1 integrin, osteopontin with β1 integrins and CD44, hyaluronic acid with CD44. These cell–cell adhesive interactions allow Notch-1 to interact with its ligand Jagged-1, Tie-2 with angiopoietin-1 (Ang-1), and c-KIT with transmembrane SCF (tmSCF) expressed at the surface of osteoblasts

endothelial sinuses that irrigate the BM (Kiel et al. 2005). It is not clear, however, whether these HSCs that locate to endothelial sinuses represent cells in transit (as HSCs permanently leak into the circulation and home back to the BM) or whether they represent a separate HSC pool that is regulated differently. What is clear, however, is that these HSCs that lodge in specific endothelial niches must utilize a separate array of adhesive interactions from the HSCs found at the endosteum. Indeed, unlike osteoblasts, BM endothelial cells that compose this endothelial niche do not express osteopontin nor N-cadherin, but express instead high levels of platelet endothelial cell adhesion molecule-1 (PECAM-1/CD31), VCAM-1/CD106, P-selectin, and E-selectin (Jacobsen et al. 1996; Schweitzer et al. 1996; Sipkins et al. 2005) whose counter-receptors are all expressed at the surface of HSCs (Yong et al. 1998; Lévesque et al. 1999; Winkler et al. 2004). Similar to the adhesion molecules expressed in the endosteal niches, P- and E-selectin-mediated adhesion regulates HSPC survival, proliferation, and differentiation (Lévesque et al. 1999; Winkler et al. 2004; Eto et al. 2005). Thus, the two anatomically distinct HSC niches, endosteal and endothelial, must regulate HSC turnover differently.

2.3.2
Bone Marrow-Committed Hematopoietic Progenitor Cell Niches

Due to the number and diversity of HPCs (multipotential, myeloid, lymphoid, etc.) their niches are much less understood and defined. B lymphoid progenitor cell niches, however, have been characterized. In mice with a green fluorescent protein (GFP) complementary DNA (cDNA) knocked into the CXCL12 gene, GFP is specifically expressed by cells expressing CXCL12, a potent chemokine and survival factor for B cells as well as HSCs and HPCs. In these mice, the most primitive B cell progenitors (pre-pro B cells) are in direct contact with BM stromal cells located in the central BM that coexpress CXCL12 and VCAM-1. In contrast, the more mature pro-B cells are in contact with a distinct population of BM stromal cells that express IL-7 but not CXCL12 (Tokoyoda et al. 2004). Interestingly, $Sca-1^+ KIT^+$ cells, which include HSCs, multipotential HPCs, and most pluripotent colony-forming cells, were also found in direct contact with stromal cells that coexpress CXCL12 and VCAM-1 and are scattered throughout the BM (Tokoyoda et al. 2004).

Similarly, it has been found that megakaryocytes migrate to a specific niche proximal to BM endothelial sinuses, where they directly interact with BM endothelial cells via CXCL12 and VCAM-1. Furthermore, this direct interaction between megakaryocytes and BM endothelial cells was found to be necessary for platelet production (Avecilla et al. 2004).

From these observations, it has been speculated that the endosteal niche may maintain HSCs in a quiescent state, whereas the various niches scattered throughout the BM stroma may represent proliferative niches in which HSCs and HPCs proliferate to renew the pool of mature blood leukocytes as well as erythrocytes.

2.4
What Keeps HSPCs Within the BM?

The two interactions thought to be most important in retaining HSPCs within the BM are: (1) the adhesive interaction between VCAM-1 expressed by BM stromal cells and integrin α4β1 (very late activation antigen-4, VLA-4) expressed by HSPCs, and (2) the chemotactic interaction between the chemokine CXCL12 (or SDF-1) and its sole receptor CXCR4, also expressed by HSPCs. The conditional deletion of either the integrin α4 gene or the VCAM-1 gene (its ligand) results in a permanent and robust HSPC mobilization (Scott et al. 2003; Ulyanova et al. 2005) as does the systemic administration of CXCR4 antagonists such as AMD3100 (Liles et al. 2003; Broxmeyer et al. 2005) or the systemic delivery of CXCL12 via recombinant adenoviruses (Hattori et al. 2001b). Both VCAM-1 and CXCL12 are expressed by osteoblasts, BM stromal cells, and endothelial cells, which are all cellular components of the various HSC and HPC niches in the BM. As discussed further in the following section, it is therefore not surprising that, by targeting these two critical interactions, most mobilizing agents induce mobilization of HSCs as well as most HPCs (collectively termed as HSPCs).

2.5
What Makes HSPCs Leave the BM?

2.5.1
Neutrophilia and/or Neutrophil Activation Are Essential for HSPC Mobilization

Neutrophilia always precedes the HSPC mobilization induced by physical exercise (Barrett et al. 1978), ACTH (Barrett et al. 1978), endotoxin (Cline and Golde 1977), sulfated polysaccharides and polyanions (van der Ham et al. 1977), myelosuppressive chemotherapy (To et al. 1984, 1989), chemokines (Pruijt et al. 1999; King et al. 2001), or hematopoietic growth factors (Molineux et al. 1990, 1991, 1997; Sato et al. 1994; Glaspy et al. 1997; Torii et al. 1998), and the degree of neutrophilia is correlated to the level of mobilization (Roberts et al. 1997; Krieger et al. 1999).

The critical importance of neutrophils and neutrophilia to the mobilization of HSPCs is illustrated by the fact that mice that are made neutropenic by homozygous targeted deletion of the G-CSF receptor gene (G-$CSFR^{-/-}$ knockout mice), or by administration of specific antineutrophil antibodies, do not mobilize in response to G-CSF, CY, or IL-8 (Liu et al. 1997; Pruijt et al. 2002; Pelus et al. 2004). Thus, neutrophils are essential for HSPC mobilization.

2.5.2
Role of Neutrophil Proteases

Mobilization with G-CSF or CY induces the accumulation of neutrophils and their precursors in the BM (Lévesque et al. 2002) with the release of large amounts of neutrophil proteases such as neutrophil elastase (NE), cathepsin G (CG), and matrix metalloproteinase-9 (MMP-9) (Lévesque et al. 2001, 2002) directly in the BM fluid. These proteases selectively cleave and inactivate the adhesion molecules and chemokines necessary for the retention of HSPC within the BM, particularly VCAM-1 (Lévesque et al. 2001), CXCL12, and the CXCL12 receptor CXCR4 (Petit et al. 2002; Lévesque et al. 2003a), as well as the tyrosine-kinase receptor c-KIT (Lévesque et al. 2003b) and transmembrane c-KIT ligand SCF (Heissig et al. 2002) whose roles in BM retention and mobilization of HSPC has also been reported (Papayannopoulou et al. 1998; Nakamura et al. 2004). Thus, active proteases released by neutrophils during mobilization disrupt the adhesive and chemotactic interactions that are essential for retaining HSPC within the BM, resulting in their release into the circulation.

In support of this model is the fact that specific NE inhibitors reduce mobilization by 60%–70% in the mouse (Petit et al. 2002; Pelus et al. 2004), and the administration of the serine-protease inhibitor serpina1/α_1-antitrypsin (a physiological inhibitor of both NE and CG) completely blocks mobilization (van Pel et al. 2006). Similarly, anti-MMP-9 monoclonal antibodies that block MMP-9 proteolytic function inhibit mobilization induced by IL-8 (by 90%)

and G-CSF (by 40%) in the mouse (Pruijt et al. 1999; Pelus et al. 2004). In humans, raised plasma NE and MMP-9 concentrations correlate significantly with the level of HPC mobilization (Lévesque et al. 2001). A similar increase in blood NE, CG, MMP-9, and cleaved soluble VCAM-1 has been observed in idiopathic myelofibrosis and polycythemia vera patients, correlating with the level of constitutive HSPC mobilization that occurs with these conditions (Xu et al. 2005; Passamonti et al. 2006). It thus appears that mobilizing agents that induce the accumulation, activation, or accumulation and activation of neutrophils in the BM (e.g., G-CSF, cytotoxic, IL-8, Groβ) disrupt the proteolytic balance within the BM, maintenance of which is essential to the regulation of the BM microenvironment, homeostasis, and HSPC trafficking.

It must be noted, however, that body fluids are normally loaded with naturally occurring protease inhibitors to protect tissues from proteolytic damage. For instance, blood contains approximately 2 mg/ml α_2-macroglobulin, an inhibitor of a wide range of proteases, and approximately 1 mg/ml serpina1/α1-antitrypsin and serpina3/α1-antichymotrypsin, both of which are specific inhibitors of NE and CG. BM extracellular fluids are devoid of α_2-macroglobulin (which is produced by the liver and cannot diffuse through the endothelial cell barrier because of its large size) but contain large amounts of serpina1 and serpina3 (Winkler et al. 2005). Unlike blood serpins, which are produced by the liver, BM serpins are transcribed and produced within the BM to protect the BM stroma from neutrophil serine proteases (Winkler et al. 2005). We have recently demonstrated that during mobilization induced by either G-CSF or CY, the levels of these serpins drop dramatically within the BM (from a few mg/ml to below detection) boosting the levels of active neutrophil serine proteases with concomitant cleavage and inactivation of molecules essential for the retention of HSPCs (Winkler et al. 2005). Interestingly, serpina1 expression by the liver and hence in the plasma remains unchanged during mobilization (Winkler et al. 2005). That the downregulation of serpina1 and serpina3 is critical to mobilization is also supported by the recent finding that prior administration of human serpina1 into mice prevents HSPC mobilization in response to IL-8 (van Pel et al. 2006). Therefore, the downregulation of serpin expression is likely to be a permissive step enabling the accumulation of active neutrophil proteases in the BM extracellular fluid.

Despite this array of convergent observations, there remains a discrepancy between the results of studies using short-term, systemic administration of protease inhibitors or conditional gene deletions (many of which alter mobilization) and those using mice where the targeted gene has been deleted throughout development (where mobilization is generally not altered). For example, systemic administration of specific NE or CG inhibitors decreases mobilization (Petit et al. 2002; Pelus et al. 2004) while mice knocked out for NE and CG mobilize normally even in the presence of a soluble MMP inhibitor (Lévesque et al. 2004). Functional redundancy between proteases may be partly responsible for these conflicting results, as neutrophils express many other proteases that were not targeted or might have been overexpressed to compensate for the lack of NE and CG.

2.5.3
The Role of Bone-Forming Cells

Recently we have found that in addition to cleavage by neutrophil proteases, CXCL12 messenger RNA (mRNA) levels also drop in the BM of mobilized mice, suggesting that an alternative mechanism may also be involved (Semerad et al. 2005). By sorting various mouse BM cell populations, we found the $Lin^-CD45^-CD31^-CD51^+$ population that is enriched in osteoblasts most actively transcribed CXCL12 (Semerad et al. 2005). This is of particular interest as osteoblasts have been recently identified as an essential component of the HSC endosteal niche (Visnjic et al. 2001; Calvi et al. 2003; Zhang et al. 2003; Zhu and Emerson 2004).

A common side effect of G-CSF administration is bone pain, which affects 80% of mobilized donors (Vial and Descotes 1995). This may be due to the dramatic reduction in bone turnover that occurs with G-CSF administration (Takamatsu et al. 1998). Systemic administration of G-CSF rapidly inhibits osteoblast-mediated bone formation as well as increasing bone degradation by osteoclasts in both human and mouse (Takamatsu et al. 1998). Osteocalcin, a bone matrix protein specifically produced by osteoblasts, is a good indicator of bone formation and osteoblast activity. In humans, osteocalcin concentration in the plasma drops during HSPC mobilization, and this drop is significantly correlated with the number of CFU-GM mobilized in the peripheral blood (Takamatsu et al. 1998). Similarly in mice, the concentration of osteocalcin mRNA in the bone marrow drops approximately 50-fold during G-CSF administration (Semerad et al. 2005), while the number of osteoblasts lining the endosteum is reduced. The mechanisms by which osteoblasts are inhibited during HSPC mobilization are still poorly understood as none of the cytokines used to induce mobilization directly binds to osteoblasts or alters their function (Lévesque et al. 2005; Semerad et al. 2005; Katayama et al. 2006). It is not a direct effect of G-CSF as osteoblasts do not express the G-CSF receptor and do not respond to G-CSF in vitro (Semerad et al. 2005). It has recently emerged that sympathetic nerves that extend through the BM and bones may play an important role in the maintenance of osteoblast function and HSC mobilization (Katayama et al. 2006). Using mice knocked out for the uridine diphosphate-galactose ceramide galactosyltransferase (an enzyme necessary for myelin synthesis) gene or for the dopamine β hydroxylase (the enzyme converting dopamine to norepinephrine) gene, it has been shown that functional sympathetic nerves are necessary for HSPC mobilization and osteoblast inhibition in neonatal mice, but to a much lesser extent in adults (Katayama et al. 2006). As adrenergic nerves express G-CSF receptors and osteoblasts express $β_2$ adrenergic receptors, it is possible that G-CSF may directly act on these nerves eliciting osteoblast inhibition and HSPC mobilization. However, lethally irradiated $G\text{-}CSFR^{-/-}$ adult recipients reconstituted with wild-type BM cells (thus containing $G\text{-}CSFR^{-/-}$ nerves with $G\text{-}CSFR^{+/+}$ hematopoietic cells) mobilize normally in response to G-CSF or CY (Liu et al. 2000). Similarly, HSPC mobilization can be completely abrogated by neutrophil depletion (Liu et al. 1997, 2000; Pruijt et al. 2002; Pelus et al. 2004) despite apparently normal neuronal function. Taken together, it is therefore likely that the predominant mechanisms behind mobilization in adults are independent of neurons.

Collectively, these data indicate that mobilization follows the inhibition of bone-forming osteoblasts, which in turn would lead to a decrease in both CXCL12 production and number of HSPC endosteal niches. Therefore, HSPC mobilization may involve at least two underlying mechanisms. The first mechanism involves the release of active proteases by BM neutrophils that cleave and inactivate chemotactic and adhesive molecules that retain HSPC within the BM, particularly CXCL12, c-KIT, tmSCF, and VCAM-1. Since most HSPCs are in direct contact with BM stromal and endothelial cells expressing these molecules and express receptors for these molecules, the protease-mediated mechanism could be responsible for the mobilization of multipotential, myeloid and lymphoid progenitors located in the central BM (Fig. 2).

A second mechanism targets the osteoblasts that form the HSC niche by decreasing their number and function (Fig. 2). Ultimately, this inhibition not only results in decreased CXCL12 expression and release by osteoblasts, but also in a net decrease in the number of functional HSC niches at the endosteum. Thus, this mechanism could be involved in the mobilization of most primitive HSCs residing at the endosteum, by depleting the endosteal niches, and forcing their migration to more a central location within the BM where protease-dependant mechanisms could take the relay to force their egress into the peripheral blood (Fig. 2).

2.6
Strategies to Increase HSPC Mobilization

2.6.1
Why the Need to Enhance Mobilization

The dose of mobilized $CD34^+$ HSPCs infused correlates directly with the speed of recovery to acceptable levels of neutrophils and platelets and the overall survival of transplanted patients (To et al. 1986; Sheridan et al. 1994; Siena et al. 2000). A minimal threshold of $2\text{--}5\times10^6$ $CD34^+$ cells/kg is necessary for successful hematopoietic reconstitution (Demirer and Bensinger 1995; To et al. 1997; Siena et al. 2000). Furthermore, it now appears that the higher the dose of $CD34^+$ cells infused the shorter is the leukopenic period and the faster is the recovery.

A significant and well-documented problem with G-CSF is patient-to-patient variability in HSPC mobilization which extends over 2.5 orders of

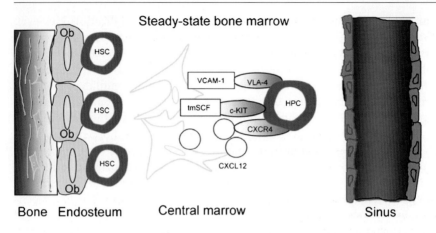

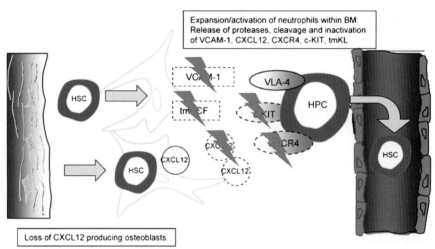

Fig. 2 A two-step model of HSPC mobilization. The inhibition of osteoblasts results in a loss of the HSC endosteal niche and decrease production of the chemokine CXCL12 causing the migration of HSCs away from the endosteum into the central marrow. In parallel, neutrophils expand and accumulate within the BM, releasing proteases. Because the expression of protease inhibitors is downregulated, neutrophil proteases accumulate in an active state, cleaving and inactivating molecules essential to HSPC retention in the BM such as VCAM-1, tmSCF, c-KIT, CXCL12, and CXCR4

magnitude (Roberts et al. 1995; Villalon et al. 2000). A consequence of this variability is the failure to harvest sufficient numbers of peripheral blood HSPC to reach minimal thresholds in 5%–10% of allogeneic healthy donors and up to 50%–60% of patients who have already undergone several courses of high-dose chemotherapy (mobilized for autologous transplantation) (Bensinger et al. 1995; Demirer and Bensinger 1995; Brown et al. 1997; Villalon et al. 2000). Thus, the discovery of more efficient mobilizing agents is paramount to improve transplantation outcome.

2.6.2
Chemical Stabilization of G-CSF

The commercially available form of human G-CSF used to mobilize HSPCs (Filgrastim or Neupogen, Amgen, Thousand Oaks, CA) in clinics is produced in recombinant form in *Escherichia coli* and is thus unglycosylated. Unglycosylated G-CSF is more sensitive to proteolytic degradation by both NE and CG (El Ouriaghli et al. 2003; Carter et al. 2004), the very same proteases that become active during mobilization. More recently, the recombinant human G-CSF produced in *E. coli* has been stabilized by attaching long chains of polyethylene glycol ("pegylation"). Because of its increased resistance to proteases and lower renal filtration, due its large size, this pegylated G-CSF (Pegfilgrastim or Neulasta, Amgen, Thousand Oaks, CA) persists at pharmacologically active concentrations in the plasma over 4 days following a single injection. As a consequence, a single injection of pegylated G-CSF is sufficient to induce robust mobilization in rodents and humans instead of a 4–5 day course of daily injections of nonpegylated G-CSF (de Haan et al. 2000; van Der Auwera et al. 2001). In addition, transplantations using cells mobilized with pegylated G-CSF or progenipoietin-1 (a fusion protein made of G-CSF and Flt-3 ligand) display (1) reduced graft-vs-host disease (GvHD) in fully mismatched murine models and (2) increased graft-vs-leukemia reaction (GvL) (Morris et al. 2004, 2005; Kiel et al. 2005). Because of its improved stability and its strong immunomodulatory functions, the use of pegylated G-CSF for HSPC mobilization is likely to increase rapidly and replace nonpegylated G-CSF.

2.6.3
Stem Cell Factor

In the last 15 years, a number of hematopoietic growth factors have been tried for their potential to mobilize HSPCs, with varying degrees of success. For instance, GM-CSF, IL-3, IL-1, and IL-6 showed little potential, while ligands of tyrosine kinase receptors such as SCF synergized when used with G-CSF (To et al. 1994; Roberts et al. 1997; Stiff et al. 2000; To et al. 2003). Of those, the KIT ligand/SCF is the only one that has made it to the bedside. Because of the strong synergistic effect of SCF with other hematopoietic growth factors, and particularly G-CSF, this combination of SCF together with G-CSF increases by a factor 5 to 20 the number of HSPCs mobilized by G-CSF alone. Despite its very potent effect on HSPC mobilization, SCF is no longer used the USA due to a high incidence of mast cell-mediated reactions despite antihistamine prophylaxis (Stiff et al. 2000). SCF, however, is still used in Canada, Europe, and Australia to boost mobilization in patients who have failed to respond adequately in response to G-CSF alone.

2.6.4
Chemokines and Their Analogs: The AMD3100 Success Story

As reviewed earlier, CXCR4 is central to the retention of HSCs and HPCs within the BM, and perturbations of the CXCR4:CXCL12 chemotactic interaction within the BM results in mobilization. Since exogenous CXCL12 is rapidly cleared from the circulation, injection of CXCL12 has a limited effect on mobilization. Therefore more stable CXCR4 agonists and antagonists were designed to induce mobilization. The first series of compounds comprises a shorter cyclic version of CXCL12 (Perez et al. 2004; Pelus et al. 2005). A second category of compounds comprises small synthetic nonpeptide molecules such as the bicyclam AMD3100 (Mozobil, AnorMED-Genzyme, Cambridge, MA). AMD3100 was designed as a CXCR4 antagonist to treat human immunodeficiency virus (HIV) infection, as CXCR4 is a coreceptor of HIV. Although AMD3100 was efficacious in blocking HIV entry into T cells in vitro, clinical trials were stopped because of its long-term toxicity and low effect on HIV viral load (Hendrix et al. 2000; Hendrix et al. 2004). However, when administered for a few days, AMD3100 induces potent HSPC mobilization and strongly synergizes with G-CSF, increasing mobilization by one to two logs over G-CSF alone (Liles et al. 2003; Devine et al. 2004; Broxmeyer et al. 2005; Flomenberg et al. 2005). There are currently eight clinical trials from phase I to phase III ongoing in the USA to further evaluate the safety and efficacy of AMD3100 to increase G-CSF-induced HSPC mobilization in autologous transplantations (multiple myeloma, non-Hodgkin's lymphoma) as well as in healthy donors for transplantation into patients affected with a variety of hematopoietic diseases. These diseases include myelodysplastic syndrome, multiple myeloma, non-Hodgkin's lymphoma, Hodgkin's disease, and acute myelogenous, acute lymphoblastic, chronic myelogenous, and chronic lymphocytic leukemia (http://www.clinicaltrials.gov).

3
Mobilization of Bone Marrow Endothelial Progenitor Cells

3.1
Mobilization of Pro-angiogenic Progenitor Cells

Following limb ischemia or myocardial infarct, rapid reperfusion and reestablishment of circulation is critical in order to limit the extent of tissue damage and necrosis. Since terminally differentiated endothelial cells isolated from the vascular lumen (i.e., from umbilical cord or from aorta) have an enormous capacity to proliferate and migrate in vitro, postnatal neovascularization was thought to involve the migration and proliferation of differentiated endothelial cells from neighboring preexisting vessels. In 1997, however, Asahara et al. reported the existence of a subpopulation of $CD34^+$ cells isolated from human peripheral blood that adhere to fibronectin, proliferate, and differentiate in culture into cells that display the hallmarks of endothelial cells: spindle-shaped morphology, formation of tube-like structures, incorporation of fluorescent acetylated low-density lipoproteins (AcLDL), expression of endothelial nitric oxide synthase (eNOS)/Nos3 and vascular endothelial growth factor receptor 2 (VEGFR2)/kinase insert-domain containing receptor (KDR). Furthermore, when injected intravenously into immunodeficient mice with a prior hind-limb ischemia, these cells colonized the neovascularized ischemic limb, suggesting that they directly contributed to the neovascularization process and as such were labeled "endothelial progenitor cells" or EPCs (Asahara et al. 1997). From this pioneering observation, a flurry of studies followed in order to determine whether these putative EPCs that circulate in steady-state blood could be mobilized to accelerate postischemic revascularization. Two years later, the same group showed that administration of human vascular endothelial growth factor 165 ($VEGF_{165}$) into mice mobilized cells displaying these EPC characteristics. Furthermore, VEGF administration resulted in a 50% increase of the neovascularization taking place in the cornea following corneal injury. When these experiments were repeated in chimera mice transplanted with BM cells from transgenic mice expressing LacZ under the Tie-2 promoter, administration of VEGF resulted in the contribution of $LacZ^+$ cells in the neovascularized cornea. $LacZ^+$ cells were very rare in the neovascularized cornea of control mice injected with bovine serum albumin, proving that these endothelial-like cells in neoformed vasculature were (1) coming from the BM and (2) mobilized into the circulation in response to VEGF (Asahara et al. 1999). Subsequently, it was shown that EPCs are mobilized by many of the cytokines that mobilize HSPCs such as VEGF-A, angiopoietin-1 (Hattori et al. 2001a), G-CSF (Orlic et al. 2001), GM-CSF (Takahashi et al. 1999), or erythropoietin (Heeschen et al. 2003). In these experiments, evidence of EPC mobilization was demonstrated using *Tie-2*LacZ BM transplantation chimera with enhanced neovascularization of damaged tissues and enhanced contribution of $LacZ^+$ bone marrow-derived cells in neoformed vessels following corneal injury, hind-limb ischemia, or acute myocardial ischemia.

3.2
What Are Endothelial Progenitor Cells? A Simple Question, a Complex Answer

That the phenotypic profile of HSCs and HPCs is now well-established in both humans and mice has greatly helped our understanding of HSPC mobilization. Transplantation of single-sorted cells and clonal analyses have resolved the identity of HSCs and most types of HPCs (Morrison and Weissman 1994; Kondo et al. 1997, 2003; Akashi et al. 2000). This knowledge has enabled the precise localization of HSCs and HPCs in a variety of tissues in adults and during development, follow their trafficking in vivo, or purify them to analyze their fundamental biological properties as well as their potential to reconstitute the various lineages of the hematopoietic and immune systems. In respect to EPCs, their proper identification and characterization has been a lot more challenging, and to date their precise nature remains still open to debate.

EPCs are still currently defined as cells that, upon culture, form colonies of adherent spindle-shaped cells displaying characteristics of endothelial cells such as the incorporation of AcLDL and binding of von Willebrand factor (vWF) and BS-1 lectin. While this functional definition is sufficient to enable the identification of molecules that accelerate neovascularization and/or enhance EPC mobilization in vivo, it gives little insight on the nature, ontology, and biology of the cells responsible for this effect, whether it is due to the mobilization of so-called EPCs and/or to comobilized hematopoietic cells secreting pro-angiogenic cytokines.

The task of precisely identifying and defining EPCs has been hampered not only by the lack of specific markers but also by the very nature of endothelial cells, since they share many markers with hematopoietic cells. This promiscuity of endothelial vs hematopoietic markers is likely to be due to the fact that during development they derive from a common cell, the hemangioblast. The task is also complicated by the fact that many hematopoietic cells are mobilized together with EPCs, and these hematopoietic cells secrete a wide range of proangiogenic cytokines and directly contribute to the repair of the ischemic tissue, particularly macrophages and granulocytes (Balsam et al. 2004; Minatoguchi et al. 2004; De Palma et al. 2005).

Mobilized EPCs express CD34, CD133, VEGFR2, and VE-cadherin in humans and are Lin^- c-KIT^+ Sca-1^+ $VEGFR2^+$ VE-$cadherin^+$ in mice (Rafii and Lyden 2003). However, this phenotype is shared with that of HSCs. Other frequently used markers are CD31 and CD146 (P1H12). While CD31 is expressed by many nucleated hematopoietic cells in the BM, CD146 is expressed by a subset including HSCs. Amazingly, CD45, a transmembrane phosphatase exclusively expressed by nucleated hematopoietic cells, is very rarely included in these studies. It is therefore difficult to assess whether mobilized EPCs are of hematopoietic or endothelial origin, and whether accelerated revascularization following ischemia is due to mobilized EPCs, mobilized hematopoietic cells, or both. To illustrate the importance of this still unresolved question, a subpopulation of $CD45^+$ $CD11b^+$ monocytes expressing Tie2 (as HSCs and EPCs do) play a critical role in tumor neovascularization and ischemia revascularization (De Palma et al. 2005). These Tie2-expressing monocytes express proangiogenic factors, and their ablation suppresses tumor neovascularization. Reciprocally, incorporation of Tie2-expressing monocytes into Matrigel plugs implanted under the skin promotes robust angiogenesis, suggesting that recruitment of these Tie2-expressing monocytes to the site of ischemic injury is sufficient to support revascularization (De Palma et al. 2005).

The hematopoietic nature of mobilized EPCs is suggested by a recent study showing that injection of monoclonal antibodies blocking α4-integrin function mobilize EPCs (Qin et al. 2006) in parallel to HSPCs (Papayannopoulou and Nakamoto 1993; Papayannopoulou et al. 1995). The ability of blood cells to form endothelial colonies (defined as mobilized EPCs) was contained within the $CD45^+$ α4-integrin$^+$ population, which comprises only hematopoietic cells (as CD45 is exclusively expressed by hematopoietic cells). Unfortunately, the potential of $CD45^-$ cells was not tested, and it is therefore impossible to conclude from this study whether the entire EPC activity is contained within the $CD45^+$ population.

The possibility that mobilized hematopoietic cells are essential to the revascularization of ischemic tissues is further suggested by the finding that ischemia-induced revascularization could be due to "hemangiocytes," a subpopulation of HPCs expressing VEGFR1 together with CXCR4 (Jin et al. 2006). Unlike endothelial cells, these hemangiocytes do express CD45 and CD11b (and are therefore of hematopoietic origin), but do not express VE-cadherin, vWF, E-selectin, or smooth muscle α-actin (Grunewald et al. 2006). The hemangiocytes are mobilized following VEGF-A administration and home to the site of ischemia due to the release of CXCL12 by fibroblasts surrounding the damage vessels. Once homed to ischemic vessels, the hemangiocytes may release proangiogenic paracrine factors that stimulate the proliferation of adjacent endothelial cells from the damaged vessel, or of mobilized EPCs that have been recruited the site of ischemia by a similar mechanism (Grunewald et al. 2006).

3.3
Is Re-vascularization Due to EPC Mobilization?

It is intriguing that many of the agents that mobilize HSPCs also induce revascularization post ischemia and are therefore presumed to mobilize EPCs (e.g., VEGF-A, placental growth factor, G-CSF, GM-CSF, erythropoietin, CXCL12, function-blocking anti-α4-integrin). This raises the question about whether revascularization is due to mobilized EPCs, and if so, is this function due to intrinsic properties of EPCs or rather BM leukocytes that comobilize with them, such as Tie2-expressing monocytes, granulocytes, and hemangiocytes (Balsam et al. 2004; Minatoguchi et al. 2004; Ingram et al. 2005; Jin et al. 2006; Kopp et al. 2006).

An answer to this difficult question has been provided by Shahin Rafii's group. The blood concentration of cytokines such as thrombopoietin/megakaryocyte growth and differentiation factor (TPO), soluble SCF, erythropoietin, and GM-CSF and of CXCL12 chemokine rises dramatically between 24 and 72 h following hind-limb ischemia. Furthermore, mice deficient for TPO, its receptor c-Mpl, G-CSF, or GM-CSF exhibit reduced ischemia-induced revascularization, demonstrating that in vivo revascularization is critically dependent on endogenous cytokines that also induce mobilization of HSPCs (Jin et al. 2006). Interestingly, $MMP\text{-}9^{-/-}$ mice, in which the shedding of tmSCF into soluble SCF is compromised, also exhibit impaired postischemia revascularization (Jin et al. 2006). TPO and soluble SCF induce the release of CXCL12 stored in platelet granules both in vitro and in vivo, as there is no rise in CXCL12 blood concentration following ischemia in thrombocytopenic $TPO^{-/-}$ or $c\text{-}mpl^{-/-}$ mice (Jin et al. 2006). Revascularization is critically dependant on CXCL12 as systemic delivery of CXCL12 following infection with recombinant adenoviruses expressing CXCL12 restores postischemic revascularization in $TPO^{-/-}$ or $c\text{-}mpl^{-/-}$ mice (Jin et al. 2006). Finally, these authors show that the rise in plasma CXCL12 mobilizes BM VEGFR1$^+$ CXCR4$^+$ hemangiocytes, which then home to damaged vessels in response to the release of CXCL12 and VEGF-A by the surrounding hypoxic tissue (Jin et al. 2006). In favor of this model, injection of purified hemangiocytes into $MMP\text{-}9^{-/-}$ ischemic mice (which have impaired hemangiocyte mobilization and revascularization) restores revascularization. Thus, accelerated revascularization following mobilization may not involve the mobilization of EPCs but rather that of hemangiocytes, which are nonendothelial.

The intimate link between acceleration of revascularization and mobilization of BM hematopoietic cells is further illustrated in mice deficient for eNOS/Nos3. $Nos3^{-/-}$ mice have impaired neovascularization following ischemia. This intracellular enzyme is expressed by endothelial cells and myocytes but not by BM hematopoietic cells. It produces nitric oxide (NO) by converting L-arginine into citrulline. Once released by endothelial cells, NO is a vasodilator that relaxes smooth muscle cells and increases vessel permeability, as well as promotes endothelial cell survival and proliferation. Increased Nos3 activity and NO release are characteristic of the ischemic response. Although Nos3 is not expressed by BM hematopoietic cells, the deletion of the Nos3 gene reduces BM hematopoietic cell survival and recovery in response to cytotoxic injury (particularly in response to 5-FU), reduces EPC and HSPC mobilization in response to VEGF, and impairs revascularization following hind-limb ischemia (Aicher et al. 2003). This effect is due to nonhematopoietic cells, as lethally irradiated $Nos3^{-/-}$ recipients reconstituted with wild-type BM cells have impaired mobilization and revascularization whereas wild-type recipients reconstituted with $Nos3^{-/-}$ BM cells mobilize normally in response to VEGF-A. The effect of Nos3 on mobilization and revascularization seems to involve MMP-9 because (1) MMP-9 release is reduced in the BM of $Nos3^{-/-}$ mice (Aicher et al. 2003), and (2) MMP-9 cleaves tmSCF into soluble SCF, a critical step of VEGF-induced and 5-FU-induced mobilization of EPCs and HSPCs (Heissig et al. 2002; see Fig. 3). Therefore, although not expressed in hematopoietic cells, Nos3 regulates the homeostasis of the BM stroma that regulates the fate of both EPCs and HPCs.

Therefore, from these studies, it is clear that a population of still not well defined EPCs is mobilized from the BM into the blood, and that this mobi-

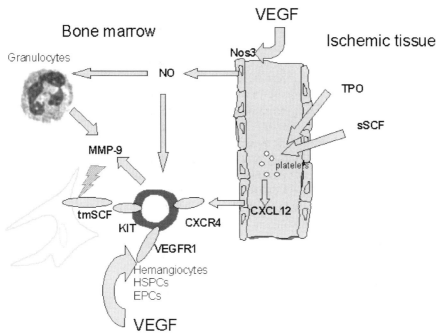

Fig. 3 Model of EPC mobilization. Ischemic tissues release TPO and soluble SCF (sSCF) that induce CXCL12 release from blood platelets. CXCL12 directly induces mobilization of HSPCs, and hemangiocytes that home to the ischemic tissue to release proangiogenic factors. Ischemia and VEGF activate Nos3 expressed by endothelial cells and BM stromal cells. NO is released, enhancing vascular permeability, blood flow, and MMP-9 release by neutrophils and HSPCs, which leads to the cleavage of tmSCF, releasing EPCs, HSPCs, and hemangiocytes into the circulation. By binding to VEGFR1 expressed on EPCs, HSPCs, and hemangiocytes, VEGF induces their mobilization

lization accelerates and facilitates revascularization following limb or cardiac ischemia. However, the accelerated revascularization involves many players, particularly mobilized hemangiocytes, Tie2 expressing monocytes, and EPCs, and the exact contribution of each of these players, particularly EPCs, remains to be fully understood.

4 Mobilization of Bone Marrow Mesenchymal Stem Cells

4.1 What Are Mesenchymal Stem Cells?

Mesenchymal stem cells may be loosely defined as nonhematopoietic, plastic-adherent precursor cells that can differentiate into specific connective tissues including osteoblasts (Haynesworth et al. 1992b), chondrocytes (Johnstone et al. 1998), and adipocytes (Pittenger et al. 1999). These cells are isolated by an initial step of adherence to plastic, followed by a long period of cell culture to eliminate hematopoietic cells that spontaneously adhere to plastic such as granulocytes, monocytes, and macrophages. Due to controversy about whether these cells fulfill the criteria for stem cell activity, they have been variously termed marrow stromal cells (Pochampally et al. 2004; Horwitz et al. 2005), mesenchymal progenitor cells (Lazarus et al. 1997), and stromal cell progenitors. Interestingly, a population of nonhematopoietic cells with similar properties to plastic-adherent MSCs (particularly clonogenicity, high proliferative potential, ability to differentiate into osteoblasts, chondrocytes and adipocytes) expresses the STRO-1 antigen, and can be directly isolated from human BM without the step of plastic adherence and cell culture (Simmons and Torok-Storb 1991; Gronthos et al. 1994). The STRO-1 antigen is also expressed by cultured plastic-adherent MSCs (Chen et al. 2006), suggesting that STRO-1 could be used to directly isolate MSCs from human tissues without prior plastic-adhesion and ex vivo expansion.

Phenotypically, MSCs express a number of nonspecific markers, and it is generally accepted that MSCs are devoid of hematopoietic and endothelial markers, including, CD11b, CD31, and CD45. Plastic-adherent MSCs express numerous surface adhesion molecules such as CD44, integrin $\alpha 4$/CD49d, and integrin $\beta 1$/CD29, and are typically positive for MHC I (Pittenger et al. 1999). Although STRO-1 is a marker of MSCs in humans (Simmons and Torok-Storb 1991; Gronthos et al. 1994; Bensidhoum et al. 2004; Chen et al. 2006), MSC populations are heterogeneous between species and within cultures, with variable expression of CD90/Thy-1, c-KIT/CD117, SH2 (CD105 or endoglin), SH3, and SH4 (CD73) being observed (Haynesworth et al. 1992a). A population of MSCs has also been shown to express CXCR4, hepatocyte growth factor receptor c-met, and certain MMPs such as membrane type 1 (MT1)-MMP (Wynn et al. 2004; Son et al. 2006). Additionally, these MSCs were strongly attracted by CXCL12 and hepatocyte growth factor gradients in vitro, providing evidence that such receptor–ligand interactions, along with MMPs, may be involved in trafficking of MSCs.

4.2
Mesenchymal Stem Cells Are Present in Circulating Human Blood During Gestation

MSCs are present in circulating human blood during the first trimester (Campagnoli et al. 2001; Gotherstrom et al. 2004; Chan et al. 2005). These cells are abundant from at least 7 weeks of gestation, and they persist until approximately 12 weeks gestation, at which time they decline and almost disappear from the fetal circulation. The phenotype of MSCs in fetal blood is similar to adult MSCs; however, they demonstrate greater multipotentiality in vitro, contain a higher frequency of CFU-fibroblasts (CFU-F), and appear to have more rapid growth profiles with maintenance of multipotentiality through a greater number of passages. This observation suggests that circulating mesenchymal cells present early in development may provide a mechanism by which the stromal compartments of hematopoietic and other tissues are populated. Although the presence of circulating MSCs decreases after 12 weeks gestation, there is conflicting evidence that a very low-frequency population of circulating multipotent cells may persist. Three groups have reported the isolation from umbilical cord blood of adherent cells with the morphology, immunophenotype, and in vitro differentiation characteristics of adult MSCs (Erices et al. 2000; Bieback et al. 2004; Lee et al. 2004; Yu et al. 2004). Erices et al. found that 25% of 31 umbilical cord blood harvests contained cells with these characteristics and that there was a correlation with early gestational age and their presence. Bieback et al. found MSC-like cells in 63% of 59 umbilical cord blood units whereas Lee et al. isolated similar cells in all of 11 units using negative immunoselection and limiting dilution. Three other groups have reported difficulty in isolating MSCs from full-term cord blood samples (Gutierrez-Rodriguez et al. 2000; Wexler et al. 2003; Yu et al. 2004).

4.3
Can MSCs Be Isolated from Adult Peripheral Blood and Can They Be Mobilized?

The issue of whether MSCs can be isolated from peripheral blood and indeed whether they can be mobilized has been controversial. Little is known about the physiological role of putative circulating MSCs in vivo. One possible role could be as a cellular resource for tissue repair; hence, there is enthusiasm for the use of these cells to treat conditions such as chronic heart failure and spinal cord injuries (Berry et al. 2006; Deng et al. 2006). If MSCs do indeed play a role in tissue repair then an open question remains whether these cells are recruited locally from sites of injury or from other sites via the peripheral blood.

There are at least seven reports on the detection of MSCs in the peripheral blood, and they are summarized in Table 2. Firm conclusions cannot be drawn as these studies are not comparable based on variations in the source of cells obtained, methods of culture, and techniques to identify MSCs in vitro. Fernandez et al. reported the detection of cells with a stromal cell phenotype in the peripheral blood of breast cancer patients after G-CSF administration (Fernandez et al. 1997); however, a follow-up report suggested that these cells were most likely to be of the monocytic lineage (Purton et al. 1998). Kuznetsov et al. sought to isolate MSCs from the peripheral blood of mice, rabbits, guinea pigs, and humans (Kuznetsov et al. 2001). Cells with some of the characteristics of MSCs were isolated from rodents; however, colonies were only rarely observed from human peripheral blood and these did not express characteristic MSC surface markers. Lazarus et al. did not detect MSCs from the peripheral blood of 11 patients mobilized with chemotherapy and G-CSF for autologous transplantation or from 3 healthy donors who received G-CSF alone to mo-

Table 2 Published reports on mobilization of MSCs

Source	Presence in blood	Comment	Reference
Breast cancer patients PBMC chemotherapy +G-CSF	Yes	Follow-up report suggests they are monocytes	Fernandez et al. 1997
Normal donors and cancer patients PBMC chemotherapy +G-CSF	No	10 ml PBMC cultured	Lazarus et al. 1997
Normal donors Blood buffy coats	Yes		Zvaifler et al. 2000
Normal donors, blood	Yes	Rare. Not typical MSC phenotype	Kuznetsov et al. 2001
Cord blood Normal donors, PBMC G-CSF	No	PBMC taken from bag washings	Wexler et al. 2003
Normal donors, PBMC G-CSF	Yes	Used fibrin microbeads	Kassis et al. 2006
Rats after chronic hypoxia, blood	Yes		Rochefort et al. 2006

PBMC, peripheral blood mononuclear cells

bilize hematopoietic stem cells for allogeneic transplantation (Lazarus et al. 1997). One limitation of this study was the fact that cells from only 10 ml of mobilized peripheral blood were cultured, a restriction consequent to ethical issues. A low level of circulating MSCs may have been missed due to this limitation. Wexler et al. compared human bone marrow, mobilized peripheral blood, and umbilical cord blood as sources of MSCs (Wexler et al. 2003). While MSC were routinely cultured from bone marrow at a mean frequency of 1 in 3.4×10^4 nucleated cells, they could not be identified in either cord blood or G-CSF-mobilized peripheral blood from normal donors. In these experiments, G-CSF-mobilized peripheral blood was obtained from washings of the collection bags, and the number of cells cultured was not stated thereby raising the possibility that a low frequency of MSCs may have been missed. Conversely, Zvaifler et al. cultured plastic-adherent cells with stromal cell morphology derived from 50 ml of buffy coat from more than 100 consecutive normal blood donors (Zvaifler et al. 2000). These cells demonstrated both osteogenic and adipogenic differentiation. Kassis et al. used a novel strategy to culture MSCs from G-CSF-mobilized peripheral blood from normal donors (Kassis et al. 2006). Fibrin microbeads provide a highly cross-linked, dense, stable, and partially denatured three-dimensional fibrin matrix and have high cell-binding properties (Bensaid et al. 2003). Rochefort et al. compared the efficacy of isolating MSCs from mobilized peripheral blood using conventional plastic-adherence with their own method using fibrin microbeads. They collected MSCs using fibrin microbeads in 8 out of 11 patients, but none of the MSCs was obtained using the conventional method. In rats, they reported on more than a 14-fold increase in peripheral blood MSCs after exposure to chronic hypoxia consistent with a role of MSCs in tissue repair (Rochefort et al. 2006).

To further investigate the potential for MSC mobilization, we (S.R.L. and J.E.J.R.) have tested various cytokines and cytokine combinations in a nonhuman primate model (manuscript in preparation). CFU-Fs were not observed in the peripheral blood of 8 baboons studied on two separate occasions in steady-state. We did not see any evidence for the mobilization of MSCs after G-CSF, pegylated G-CSF (supplied by Amgen), or G-CSF plus pegylated TPO/megakaryocyte growth and development factor (supplied by Kirin Brewery). However, we did observe CFU-F in the peripheral blood of 3 out of 5 animals receiving G-CSF plus SCF (supplied by Amgen). We understand that this observation may provide the first evidence of mobilization of MSCs after cytokine administration in primates. These cells expressed the same immunophenotype as MSCs cultured from BM in the same animals, and also demonstrated both adipogenic and osteogenic differentiation. It is possible that the presence of CFU-F in the peripheral blood may correlate with the degree of hematopoietic mobilization rather than the actual cytokine(s) administered, as all three animals in which peripheral CFU-F were observed also mobilized well for $CD34^+$ cells, colony forming cells, and severe combined immunodeficiency (SCID) repopulating cells.

In conclusion, despite conflicting evidence in the literature, it is likely that MSCs are present in the peripheral circulation at a very low frequency. Given appropriate stimuli it is likely that a small number of MSCs can be mobilized from the BM. Mobilization of MSCs into the peripheral blood may provide an alternative route for their harvest using leukapheresis to be applied in a burgeoning number of potential therapeutic applications.

5
Conclusion

BM is a privileged tissue where stem and progenitor cells reside that can be used for cellular therapy. Ten years of experimentation and clinical practice have clearly indicated that while transplantable HSPCs with the capacity to reconstitute the entire hematopoietic and immune systems are undoubtedly mobilized in large numbers into the circulation, the mobilization of nonhematopoietic stem and progenitor cells is more elusive. Although the revascularization of damaged tissue can be enhanced in vivo following mobilization, it is still unclear if this is due to the mobilization of EPCs, proangiogenic hematopoietic cells, or both types of cell. Finally, the mobilization of adult bone marrow MSCs in sufficient numbers for practical, clinical application remains a challenge. The identification of mechanisms responsible for HSPC mobilization has revealed some essential aspects of the biology of these cells and how critical integrin-mediated adhesive interactions and CXCR4-mediated chemotactic interactions are to maintain these cells in their specific niches within the BM. Interestingly, most mobilizing agents used to date disrupt these interactions either directly (anti-integrin monoclonal antibodies, CXCR4 antagonists) or indirectly via, for instance, neutrophils and their proteases. Since all stem and progenitor cells use a common repertoire of integrins and chemokine receptors (particularly CXCR4) to remain within the BM, it is not surprising that the mobilization of MSCs or EPCs is achieved when HSPCs are also mobilized. The fact that HSPCs are mobilized in greater numbers and by a wider number of agents than EPCs or MSCs may simply reflect the fact that HSPCs are small cells, are very motile, and have relatively low adhesiveness compared to EPCs or MSCs. These properties of HSPCs are likely to make their migration through the BM endothelial barrier into the circulation easier. This may also mean that because or their large size and high adhesiveness, the mobilization of MSCs in large numbers will remain challenging, while solid tissues will remain the main source of large numbers of transplantable MSCs.

Acknowledgements Jean-Pierre Lévesque is a Senior Research Fellow of the Queensland Cancer Found.

References

Abkowitz JL, Robinson AE, Kale S, et al (2003) Mobilization of hematopoietic stem cells during homeostasis and after cytokine exposure. Blood 102:1249–1253
Adams GB, Chabner KT, Alley IR, et al (2006) Stem cell engraftment at the endosteal niche is specified by the calcium-sensing receptor. Nature 439:599–603
Aicher A, Heeschen C, Mildner-Rihm C, et al (2003) Essential role of endothelial nitric oxide synthase for mobilization of stem and progenitor cells. Nat Med 9:1370–1376
Akashi K, Traver D, Miyamoto T, et al (2000) A clonogenic common myeloid progenitor that gives rise to all myeloid lineages. Nature 404:193–197
Arai F, Hirao A, Ohmura M, et al (2004) Tie2/angiopoietin-1 signaling regulates hematopoietic stem cell quiescence in the bone marrow niche. Cell 118:149–161
Asahara T, Murohara T, Sullivan A, et al (1997) Isolation of putative progenitor endothelial cells for angiogenesis. Science 275:964–966
Asahara T, Takahashi T, Masuda H, et al (1999) VEGF contributes to postnatal neovascularization by mobilizing bone marrow-derived endothelial progenitor cells. EMBO J 18:3964–3972
Avecilla ST, Hattori K, Heissig B, et al (2004) Chemokine-mediated interaction of hematopoietic progenitors with the bone marrow vascular niche is required for thrombopoiesis. Nat Med 10:64–71
Balsam LB, Wagers AJ, Christensen JL, et al (2004) Haematopoietic stem cells adopt mature haematopoietic fates in ischaemic myocardium. Nature 428:668–673
Barrett AJ, Longhurst P, Sneath P, et al (1978) Mobilization of CFU-C by exercise and ACTH induced stress in man. Exp Hematol 6:590–594

Barry F, Boynton R, Murphy M, et al (2001) The SH-3 and SH-4 antibodies recognize distinct epitopes on CD73 from human mesenchymal stem cells. Biochem Biophys Res Commun 289:519–524

Barry FP, Boynton RE, Haynesworth S, et al (1999) The monoclonal antibody SH-2, raised against human mesenchymal stem cells, recognizes an epitope on endoglin (CD105). Biochem Biophys Res Commun 265:134–139

Bensaid W, Triffitt JT, Blanchat C, et al (2003) A biodegradable fibrin scaffold for mesenchymal stem cell transplantation. Biomaterials 24:2497–2502

Bensidhoum M, Chapel A, Francois S, et al (2004) Homing of in vitro expanded Stro-1$^-$ or Stro-1$^+$ human mesenchymal stem cells into the NOD/SCID mouse and their role in supporting human CD34 cell engraftment. Blood 103:3313–3319

Bensinger W, Appelbaum F, Rowley S, et al (1995) Factors that influence collection and engraftment of autologous peripheral-blood stem cells. J Clin Oncol 13:2547–2555

Berry MF, Engler AJ, Woo YJ, et al (2006) Mesenchymal stem cell injection after myocardial infarction improves myocardial compliance. Am J Physiol Heart Circ Physiol 290:H2196–H2203

Bieback K, Kern S, Kluter H, et al (2004) Critical parameters for the isolation of mesenchymal stem cells from umbilical cord blood. Stem Cells 22:625–634

Bonsignore MR, Morici G, Santoro A, et al (2002) Circulating hematopoietic progenitor cells in runners. J Appl Physiol 93:1691–1697

Brown RA, Adkins D, Goodnough LT, et al (1997) Factors that influence the collection and engraftment of allogeneic peripheral-blood stem cells in patients with hematologic malignancies. J Clin Oncol 15:3067–3074

Broxmeyer HE, Orschell CM, Clapp DW, et al (2005) Rapid mobilization of murine and human hematopoietic stem and progenitor cells with AMD3100, a CXCR4 antagonist. J Exp Med 201:1307–1318

Calvi LM, Adams GB, Weibrecht KW, et al (2003) Osteoblastic cells regulate the haematopoietic stem cell niche. Nature 425:841–846

Campagnoli C, Roberts IA, Kumar S, et al (2001) Identification of mesenchymal stem/progenitor cells in human first-trimester fetal blood, liver, and bone marrow. Blood 98:2396–2402

Carlesso N, Aster JC, Sklar J, et al (1999) Notch1-induced delay of human hematopoietic progenitor cell differentiation is associated with altered cell cycle kinetics. Blood 93:838–848

Carter CRD, Whitmore KM, Thorpe R (2004) The significance of carbohydrates on G-CSF: differential sensitivity of G-CSFs to human neutrophil elastase degradation. J Leukoc Biol 75:515–522

Cashman J, Clark-Lewis I, Eaves A, et al (2002) Stromal-derived factor 1 inhibits the cycling of very primitive human hematopoietic cells in vitro and in NOD/SCID mice. Blood 99:792–799

Chan J, O'Donoghue K, de la Fuente J, et al (2005) Human fetal mesenchymal stem cells as vehicles for gene delivery. Stem Cells 23:93–102

Chen X, Xu HB, Wan C, et al (2006) Bioreactor expansion of human adult bone marrow-derived mesenchymal stem cells (MSCs). Stem Cells 24:2052–2059

Cline MJ, Golde DW (1977) Mobilization of hematopoietic stem cells (CFU-C) into the peripheral blood of man by endotoxin. Exp Hematol 5:186–190

de Haan G, Ausema A, Wilkens M, et al (2000) Efficient mobilization of haematopoietic progenitors after a single injection of pegylated recombinant human granulocyte colony-stimulating factor in mouse strains with distinct marrow-cell pool sizes. Br J Haematol 110:638–646

De Palma M, Venneri MA, Galli R, et al (2005) Tie2 identifies a hematopoietic lineage of proangiogenic monocytes required for tumor vessel formation and a mesenchymal population of pericyte progenitors. Cancer Cell 8:211–226

Demirer T, Bensinger WI (1995) Optimization of peripheral blood stem cell collection. Curr Opin Hematol 2:219–226

Deng YB, Liu XG, Liu ZG, et al (2006) Implantation of BM mesenchymal stem cells into injured spinal cord elicits de novo neurogenesis and functional recovery: evidence from a study in rhesus monkeys. Cytotherapy 8:210–214

Dennis JE, Charbord P (2002) Origin and differentiation of human and murine stroma. Stem Cells 20:205–214

Devine SM, Peter S, Martin BJ, et al (2001) Mesenchymal stem cells: stealth and suppression. Cancer J 7 [Suppl 2]:S76–S82

Devine SM, Flomenberg N, Vesole DH, et al (2004) Rapid mobilization of $CD34^+$ cells following administration of the CXCR4 antagonist AMD3100 to patients with multiple myeloma and non-Hodgkin's lymphoma. J Clin Oncol 22:1095–1102

El Ouriaghli F, Fujiwara H, Melenhorst JJ, et al (2003) Neutrophil elastase enzymatically antagonizes the in vitro action of G-CSF: implications for the regulation of granulopoiesis. Blood 101:1752–1758

Erices A, Conget P, Minguell JJ (2000) Mesenchymal progenitor cells in human umbilical cord blood. Br J Haematol 109:235–242

Eto T, Winkler I, Purton LE, et al (2005) Contrasting effects of P-selectin and E-selectin on the differentiation of murine hematopoietic progenitor cells. Exp Hematol 33:232–242

Fernandez M, Simon V, Herrera G, et al (1997) Detection of stromal cells in peripheral blood progenitor cell collections from breast cancer patients. Bone Marrow Transplant 20:265–271

Flomenberg N, Devine SM, DiPersio JF, et al (2005) The use of AMD3100 plus G-CSF for autologous hematopoietic progenitor cell mobilization is superior to G-CSF alone. Blood 106:1867–1874

Fuchs E, Tumbar T, Guasch G (2004) Socializing with the neighbors: stem cells and their niche. Cell 116:769–778

Glaspy JA, Shpall EJ, LeMaistre CF, et al (1997) Peripheral blood progenitor cell mobilization using stem cell factor in combination with filgrastim in breast cancer patients. Blood 90:2939–2951

Gotherstrom C, Ringden O, Tammik C, et al (2004) Immunologic properties of human fetal mesenchymal stem cells. Am J Obstet Gynecol 190:239–245

Gronthos S, Graves SE, Ohta S, et al (1994) The $STRO-1^+$ fraction of adult human bone marrow contains the osteogenic precursors. Blood 84:4164–4173

Grunewald M, Avraham I, Dor Y, et al (2006) VEGF-induced adult neovascularization: recruitment, retention, and role of accessory cells. Cell 124:175–189

Gutierrez-Rodriguez M, Reyes-Maldonado E, Mayani H (2000) Characterization of the adherent cells developed in Dexter-type long-term cultures from human umbilical cord blood. Stem Cells 18:46–52

Hattori K, Dias S, Heissig B, et al (2001a) Vascular endothelial growth factor and angiopoietin-1 stimulate postnatal hematopoiesis by recruitment of vasculogenic and hematopoietic stem cells. J Exp Med 193:1005–1014

Hattori K, Heissig B, Tashiro K, et al (2001b) Plasma elevation of stromal cell-derived factor-1 induces mobilization of mature and immature hematopoietic progenitor and stem cells. Blood 97:3354–3360

Haynesworth SE, Baber MA, Caplan AI (1992a) Cell surface antigens on human marrow-derived mesenchymal cells are detected by monoclonal antibodies. Bone 13:69–80

Haynesworth SE, Goshima J, Goldberg VM, et al (1992b) Characterization of cells with osteogenic potential from human marrow. Bone 13:81–88

Heeschen C, Aicher A, Lehmann R, et al (2003) Erythropoietin is a potent physiologic stimulus for endothelial progenitor cell mobilization. Blood 102:1340–1346

Heissig B, Hattori K, Dias S, et al (2002) Recruitment of stem and progenitor cells from the bone marrow niche requires MMP-9 mediated release of kit-ligand. Cell 109:625–637

Hendrix CW, Flexner C, MacFarland RT, et al (2000) Pharmacokinetics and safety of AMD-3100, a novel antagonist of the CXCR-4 chemokine receptor, in human volunteers. Antimicrob Agents Chemother 44:1667–1673

Hendrix CW, Collier AC, Lederman MM, et al (2004) Safety, pharmacokinetics, and antiviral activity of AMD3100, a selective CXCR4 receptor inhibitor, in HIV-1 infection. J Acquir Immune Defic Syndr 37:1253–1262

Horwitz EM, Le Blanc K, Dominici M, et al (2005) Clarification of the nomenclature for MSC: The International Society for Cellular Therapy position statement. Cytotherapy 7:393–395

Hosokawa K, Arai F, Suda T (2005) N-cadherin induces hematopoietic stem cells in a quiescent state in the bone marrow niche. Blood 106:141a (Abstr)

Ingram DA, Caplice NM, Yoder MC (2005) Unresolved questions, changing definitions, and novel paradigms for defining endothelial progenitor cells. Blood 106:1525–1531

Jacobsen K, Kravitz J, Kincade PW, et al (1996) Adhesion receptors on bone marrow stromal cells: in vivo expression of vascular cell adhesion molecule-1 by reticular cells and sinusoidal endothelium in normal and gamma-irradiated mice. Blood 87:73–82

Jin DK, Shido K, Kopp HG, et al (2006) Cytokine-mediated deployment of SDF-1 induces revascularization through recruitment of CXCR4$^+$ hemangiocytes. Nat Med 12:557–567

Johnstone B, Hering TM, Caplan AI, et al (1998) In vitro chondrogenesis of bone marrow-derived mesenchymal progenitor cells. Exp Cell Res 238:265–272

Kassis I, Zangi L, Rivkin R, et al (2006) Isolation of mesenchymal stem cells from G-CSF-mobilized human peripheral blood using fibrin microbeads. Bone Marrow Transplant 37:967–976

Katayama Y, Battista M, Kao WM, et al (2006) Signals from the sympathetic nervous system regulate hematopoietic stem cell egress from bone marrow. Cell 124:407–421

Kiel MJ, Yilmaz OH, Iwashita T, Yilmaz OH, Terhorst C, Morrison SJ (2005) SLAM family receptors distinguish hematopoietic stem and progenitor cells and reveal endothelial niches for stem cells. Cell 121:1109–1121

King AG, Horowitz D, Dillon SB, et al (2001) Rapid mobilization of murine hematopoietic stem cells with enhanced engraftment properties and evaluation of hematopoietic progenitor cell mobilization in rhesus monkeys by a single injection of SB-251353, a specific truncated form of the human CXC chemokine GRObeta. Blood 97:1534–1542

Kondo M, Weissman IL, Akashi K (1997) Identification of clonogenic common lymphoid progenitors in mouse bone marrow. Cell 91:661–672

Kondo M, Wagers AJ, Manz MG, et al (2003) Biology of hematopoietic stem cells and progenitors: implications for clinical application. Annu Rev Immunol 21:759–806

Kopp HG, Ramos CA, Rafii S (2006) Contribution of endothelial progenitors and proangiogenic hematopoietic cells to vascularization of tumor and ischemic tissue. Curr Opin Hematol 13:175–181

Korbling M, Anderlini P (2001) Peripheral blood stem cell versus bone marrow allotransplantation: does the source of hematopoietic stem cells matter? Blood 98:2900–2908

Krieger MS, Schiller G, Berenson JR, et al (1999) Collection of peripheral blood progenitor cells (PBPC) based on a rising WBC and platelet count significantly increases the number of CD34$^+$ cells. Bone Marrow Transplant 24:25–28

Kuznetsov SA, Mankani MH, Gronthos S, et al (2001) Circulating skeletal stem cells. J Cell Biol 153:1133–1140

Lataillade JJ, Clay D, Dupuy C, et al (2000) Chemokine SDF-1 enhances circulating CD34$^+$ cell proliferation in synergy with cytokines: possible role in progenitor survival. Blood 95:756–768

Lazarus HM, Haynesworth SE, Gerson SL, et al (1997) Human bone marrow-derived mesenchymal (stromal) progenitor cells (MPCs) cannot be recovered from peripheral blood progenitor cell collections. J Hematother 6:447–455

Lee OK, Kuo TK, Chen WM, et al (2004) Isolation of multipotent mesenchymal stem cells from umbilical cord blood. Blood 103:1669–1675

Lévesque JP, Zannettino AC, Pudney M, et al (1999) PSGL-1-mediated adhesion of human hematopoietic progenitors to P-selectin results in suppression of hematopoiesis. Immunity 11:369–378

Lévesque JP, Takamatsu Y, Nilsson SK, et al (2001) Vascular cell adhesion molecule-1 (CD106) is cleaved by neutrophil proteases in the bone marrow following hematopoietic progenitor cell mobilization by granulocyte colony-stimulating factor. Blood 98:1289–1297

Lévesque JP, Hendy J, Takamatsu Y, et al (2002) Mobilization by either cyclophosphamide or granulocyte colony-stimulating factor transforms the bone marrow into a highly proteolytic environment. Exp Hematol 30:430–439

Lévesque JP, Hendy J, Takamatsu Y, et al (2003a) Disruption of the CXCR4/CXCL12 chemotactic interaction during hematopoietic stem cell mobilization induced by GCSF or cyclophosphamide. J Clin Invest 111:187–196

Lévesque JP, Hendy J, Winkler IG, et al (2003b) Granulocyte colony-stimulating factor induces the release in the bone marrow of proteases that cleave c-KIT receptor (CD117) from the surface of hematopoietic progenitor cells. Exp Hematol 31:109–117

Lévesque JP, Liu F, Simmons PJ, et al (2004) Characterization of hematopoietic progenitor mobilization in protease-deficient mice. Blood 104:65–72

Lévesque JP, Winkler IG, Sims N, et al (2005) The inhibition of the osteoblast niche during hematopoietic stem cell mobilization is an indirect effect involving mature bone marrow leukocytes, IL6 and soluble IL6 receptor. Blood 106:557a [abstract]

Liles WC, Broxmeyer HE, Rodger E, et al (2003) Mobilization of hematopoietic progenitor cells in healthy volunteers by AMD3100, a CXCR4 antagonist. Blood 102:2728–2730

Lin Y, Weisdorf DJ, Solovey A, et al (2000) Origins of circulating endothelial cells and endothelial outgrowth from blood. J Clin Invest 105:71–77

Liu F, Poursine-Laurent J, Link DC (1997) The granulocyte colony-stimulating factor receptor is required for the mobilization of murine hematopoietic progenitors into peripheral blood by cyclophosphamide or interleukin-8 but not flt-3 ligand. Blood 90:2522–2528

Liu F, Poursine-Laurent J, Link DC (2000) Expression of the G-CSF receptor on hematopoietic progenitor cells is not required for their mobilization by G-CSF. Blood 95:3025–3031

Lord BI, Testa NG, Hendry JH (1975) The relative spatial distributions of CFUs and CFUc in the normal mouse femur. Blood 46:65–72

Minatoguchi S, Takemura G, Chen XH, et al (2004) Acceleration of the healing process and myocardial regeneration may be important as a mechanism of improvement of cardiac function and remodeling by postinfarction granulocyte colony-stimulating factor treatment. Circulation 109:2572–2580

Molineux G, Pojda Z, Dexter TM (1990) A comparison of hematopoiesis in normal and splenectomized mice treated with granulocyte colony-stimulating factor. Blood 75:563–569

Molineux G, Migdalska A, Szmitkowski M, et al (1991) The effects on hematopoiesis of recombinant stem cell factor (ligand for c-kit) administered in vivo to mice either alone or in combination with granulocyte colony-stimulating factor. Blood 78:961–966

Molineux G, McCrea C, Yan XQ, et al (1997) Flt-3 ligand synergizes with granulocyte colony-stimulating factor to increase neutrophil numbers and to mobilize peripheral blood stem cells with long-term repopulating potential. Blood 89:3998–4004

Morris ES, MacDonald KPA, Rowe V, et al (2004) Donor treatment with pegylated G-CSF augments the generation of IL-10-producing regulatory T cells and promotes transplantation tolerance. Blood 103:3573–3581

Morris ES, MacDonald KPA, Rowe V, et al (2005) NKT cell-dependent leukemia eradication following stem cell mobilization with potent G-CSF analogs. J Clin Invest 115:3093–3103

Morrison SJ, Weissman IL (1994) The long-term repopulating subset of hematopoietic stem cells is deterministic and isolatable by phenotype. Immunity 1:661–673

Nagasawa T (2006) Microenvironmental niches in the bone marrow required for B-cell development. Nat Rev Immunol 6:107–116

Nakamura Y, Tajima F, Ishiga K, et al (2004) Soluble c-kit receptor mobilizes hematopoietic stem cells to peripheral blood in mice. Exp Hematol 32:390–396

Nilsson SK, Johnston HM, Coverdale JA (2001) Spatial localization of transplanted hemopoietic stem cells: inferences for the localization of stem cell niches. Blood 97:2293–2299

Nilsson SK, Haylock DN, Johnston HM, et al (2003) Hyaluronan is synthesized by primitive hemopoietic cells, participates in their lodgment at the endosteum following transplantation, and is involved in the regulation of their proliferation and differentiation in vitro. Blood 101:856–862

Nilsson SK, Johnston HM, Whitty GA, et al (2005) Osteopontin, a key component of the hematopoietic stem cell niche and regulator of primitive hematopoietic progenitor cells. Blood 106:1232–1239

Orlic D, Kajstura J, Chimenti S, et al (2001) Mobilized bone marrow cells repair the infarcted heart, improving function and survival. Proc Natl Acad Sci USA 98:10344–10349

Papayannopoulou T, Nakamoto B (1993) Peripheralization of hemopoietic progenitors in primates treated with anti-VLA4 integrin. Proc Natl Acad Sci U S A 90:9374–9378

Papayannopoulou T, Craddock C, Nakamoto B, et al (1995) The VLA4/VCAM-1 adhesion pathway defines contrasting mechanisms of lodgement of transplanted murine hemopoietic progenitors between bone marrow and spleen. Proc Natl Acad Sci U S A 92:9647–9651

Papayannopoulou T, Priestley GV, Nakamoto B (1998) Anti-VLA4/VCAM-1-induced mobilization requires cooperative signaling through the kit/mkit ligand pathway. Blood 91:2231–2239

Passamonti F, Rumi E, Pietra D, et al (2006) Relation between JAK2 (V617F) mutation status, granulocyte activation, and constitutive mobilization of $CD34^+$ cells into peripheral blood in myeloproliferative disorders. Blood 107:3676–3682

Pelus LM, Bian H, King AG, et al (2004) Neutrophil-derived MMP-9 mediates synergistic mobilization of hematopoietic stem and progenitor cells by the combination of G-CSF and the chemokines GROβ/CXCL2 and GROβT/CXCL2Δ4. Blood 103:110–119

Pelus LM, Bian H, Fukuda S, et al (2005) The CXCR4 agonist peptide, CTCE-0021, rapidly mobilizes polymorphonuclear neutrophils and hematopoietic progenitor cells into peripheral blood and synergizes with granulocyte colony-stimulating factor. Exp Hematol 33:295–307

Perez LE, Alpdogan O, Shieh JH, et al (2004) Increased plasma levels of stromal-derived factor-1 (SDF-1/CXCL12) enhance human thrombopoiesis and mobilize human colony-forming cells (CFC) in NOD/SCID mice. Exp Hematol 32:300–307

Petit I, Szyper-Kravitz M, Nagler A, et al (2002) G-CSF induces stem cell mobilization by decreasing bone marrow SDF-1 and up-regulating CXCR4. Nat Immunol 3:687–694

Pittenger MF, Mackay AM, Beck SC, et al (1999) Multilineage potential of adult human mesenchymal stem cells. Science 284:143–147

Pochampally RR, Smith JR, Ylostalo J, et al (2004) Serum deprivation of human marrow stromal cells (hMSCs) selects for a subpopulation of early progenitor cells with enhanced expression of OCT-4 and other embryonic genes. Blood 103:1647–1652

Pruijt JF, Fibbe WE, Laterveer L, et al (1999) Prevention of interleukin-8-induced mobilization of hematopoietic progenitor cells in rhesus monkeys by inhibitory antibodies against the metalloproteinase gelatinase B (MMP-9). Proc Natl Acad Sci U S A 96:10863–10868

Pruijt JF, Verzaal P, van Os R, et al (2002) Neutrophils are indispensable for hematopoietic stem cell mobilization induced by interleukin-8 in mice. Proc Natl Acad Sci U S A 99:6228–6233

Purton LE, Mielcarek M, Torok-Storb B (1998) Monocytes are the likely candidate 'stromal' cell in G-CSF-mobilized peripheral blood. Bone Marrow Transplant 21:1075–1076

Qin G, Ii M, Silver M, et al (2006) Functional disruption of $\alpha 4$ integrin mobilizes bone marrow-derived endothelial progenitors and augments ischemic neovascularization. J Exp Med 203:153–163

Rafii S, Lyden D (2003) Therapeutic stem and progenitor cell transplantation for organ vascularization and regeneration. Nat Med 9:702–712

Roberts AW, DeLuca E, Begley CG, et al (1995) Broad inter-individual variations in circulating progenitor cell numbers induced by granulocyte colony-stimulating factor therapy. Stem Cells 13:512–516

Roberts AW, Foote S, Alexander WS, et al (1997) Genetic influences determining progenitor cell mobilization and leukocytosis induced by granulocyte colony-stimulating factor. Blood 89:2736–2744

Rochefort GY, Delorme B, Lopez A, et al (2006) Multipotential mesenchymal stem cells are mobilized into peripheral blood by hypoxia. Stem Cells 24:2202–2208

Sale GE, Storb R (1983) Bilateral diffuse pulmonary ectopic ossification after marrow allograft in a dog. Evidence for allotransplantation of hemopoietic and mesenchymal stem cells. Exp Hematol 11:961–966

Sato N, Sawada K, Takahashi TA, et al (1994) A time course study for optimal harvest of peripheral blood progenitor cells by granulocyte colony-stimulating factor in healthy volunteers. Exp Hematol 22:973–978

Schofield R (1970) A comparative study of the repopulating potential of grafts from various haemopoietic sources: CFU repopulation. Cell Tissue Kinet 3:119–130

Schofield R (1978) The relationship between the spleen colony-forming cell and the haemopoietic stem cell. Blood Cells 4:7–25

Schweitzer KM, Drager AM, van der Valk P, et al (1996) Constitutive expression of E-selectin and vascular cell adhesion molecule-1 on endothelial cells of hematopoietic tissues. Am J Pathol 148:165–175

Scott LM, Priestley GV, Papayannopoulou T (2003) Deletion of $\alpha 4$ integrins from adult hematopoietic cells reveals roles in homeostasis, regeneration, and homing. Mol Cell Biol 23:9349–9360

Semerad CL, Christopher MJ, Liu F, et al (2005) G-CSF potently inhibits osteoblast activity and CXCL12 mRNA expression in the bone marrow. Blood 106:3020–3027

Sheridan WP, Begley CG, To LB, et al (1994) Phase II study of autologous filgrastim (G-CSF)-mobilized peripheral blood progenitor cells to restore hemopoiesis after high-dose chemotherapy for lymphoid malignancies. Bone Marrow Transplant 14:105–111

Siena S, Schiavo R, Pedrazzoli P, et al (2000) Therapeutic relevance of CD34 cell dose in blood cell transplantation for cancer therapy. J Clin Oncol 18:1360–1377

Simmons PJ, Torok-Storb B (1991) Identification of stromal cell precursors in human bone marrow by a novel monoclonal antibody, STRO-1. Blood 78:55–62

Sipkins DA, Wei X, Wu JW, et al (2005) In vivo imaging of specialized bone marrow endothelial microdomains for tumour engraftment. Nature 435:969–973

Son BR, Marquez-Curtis LA, Kucia M, et al (2006) Migration of bone marrow and cord blood mesenchymal stem cells in vitro is regulated by stromal-derived factor-1-CXCR4 and hepatocyte growth factor-c-met axes and involves matrix metalloproteinases. Stem Cells 24:1254–1264

Stier S, Ko Y, Forkert R, et al (2005) Osteopontin is a hematopoietic stem cell niche component that negatively regulates stem cell pool size. J Exp Med 201:1781–1791

Stiff P, Gingrich R, Luger S, et al (2000) A randomized phase 2 study of PBPC mobilization by stem cell factor and filgrastim in heavily pretreated patients with Hodgkin's disease or non-Hodgkin's lymphoma. Bone Marrow Transplant 26:471–481

Takahashi T, Kalka C, Masuda H, et al (1999) Ischemia- and cytokine-induced mobilization of bone marrow-derived endothelial progenitor cells for neovascularization. Nat Med 5:434–438

Takamatsu Y, Simmons PJ, Moore RJ, et al (1998) Osteoclast-mediated bone resorption is stimulated during short-term administration of granulocyte colony-stimulating factor but is not responsible for hematopoietic progenitor cell mobilization. Blood 92:3465–3473

To LB, Haylock DN, Kimber RJ, et al (1984) High levels of circulating haemopoietic stem cells in very early remission from acute non-lymphoblastic leukaemia and their collection and cryopreservation. Br J Haematol 58:399–410

To LB, Dyson PG, Juttner CA (1986) Cell-dose effect in circulating stem-cell autografting. Lancet 2:404–405

To LB, Davy ML, Haylock DN, et al (1989) Autotransplantation using peripheral blood stem cells mobilized by cyclophosphamide. Bone Marrow Transplant 4:595–596

To LB, Haylock DN, Dowse T, et al (1994) A comparative study of the phenotype and proliferative capacity of peripheral blood (PB) $CD34^+$ cells mobilized by four different protocols and those of steady-phase PB and bone marrow $CD34^+$ cells. Blood 84:2930–2939

To LB, Haylock DN, Simmons PJ, et al (1997) The biology and clinical uses of blood stem cells. Blood 89:2233–2258

To LB, Bashford J, Durrant S, et al (2003) Successful mobilization of peripheral blood stem cells after addition of ancestim (stem cell factor) in patients who had failed a prior mobilization with filgrastim (granulocyte colony-stimulating factor) alone or with chemotherapy plus filgrastim. Bone Marrow Transplant 31:371–378

Tokoyoda K, Egawa T, Sugiyama T, et al (2004) Cellular niches controlling B lymphocyte behavior within bone marrow during development. Immunity 20:707–718

Torii Y, Nitta Y, Akahori H, et al (1998) Mobilization of primitive haemopoietic progenitor cells and stem cells with long-term repopulating ability into peripheral blood in mice by pegylated recombinant human megakaryocyte growth and development factor. Br J Haematol 103:1172–1180

Tumbar T, Guasch G, Greco V, et al (2004) Defining the epithelial stem cell niche in skin. Science 303:359–363

Ulyanova T, Scott LM, Priestley GV, et al (2005) VCAM-1 expression in adult hematopoietic and nonhematopoietic cells is controlled by tissue-inductive signals and reflects their developmental origin. Blood 106:86–94 van Der Auwera P, Platzer E, Xu ZX, et al (2001) Pharmacodynamics and pharmacokinetics of single doses of subcutaneous pegylated human G-CSF mutant (Ro 25–8315) in healthy volunteers: comparison with single and multiple daily doses of filgrastim. Am J Hematol 66:245–251

van der Ham AC, Benner R, Vos O (1977) Mobilization of B and T lymphocytes and haemopoietic stem cells by polymethacrylic acid and dextran sulphate. Cell Tissue Kinet 10:387–397

van Pel M, van Os R, Velders GA, et al (2006) Serpina1 is a potent inhibitor of IL-8-induced hematopoietic stem cell mobilization. Proc Natl Acad Sci USA 103:1469–1474

Vial T, Descotes J (1995) Clinical toxicity of cytokines used as haemopoietic growth factors. Drug Saf 13:371–406

Villalon L, Odriozola J, Larana JG, et al (2000) Autologous peripheral blood progenitor cell transplantation with <2×106 CD34$^+$/kg: an analysis of variables concerning mobilisation and engraftment. Hematol J 1:374–381

Visnjic D, Kalajzic I, Gronowicz G, et al (2001) Conditional ablation of the osteoblast lineage in Col2.3Δtk transgenic mice. J Bone Miner Res 16:2222–2231

Visnjic D, Kalajzic Z, Rowe DW, et al (2004) Hematopoiesis is severely altered in mice with an induced osteoblast deficiency. Blood 103:3258–3264

Wexler SA, Donaldson C, Denning-Kendall P, et al (2003) Adult bone marrow is a rich source of human mesenchymal 'stem' cells but umbilical cord and mobilized adult blood are not. Br J Haematol 121:368–374

Winkler IG, Snapp KR, Simmons PJ, et al (2004) Adhesion to E-selectin promotes growth inhibition and apoptosis of human and murine hematopoietic progenitor cells independent of PSGL-1. Blood 103:1685–1692

Winkler IG, Hendy J, Coughlin P, et al (2005) Serine protease inhibitors serpina1 and serpina3 are down-regulated in bone marrow during hematopoietic progenitor mobilization. J Exp Med 201:1077–1088

Wright DE, Wagers AJ, Gulati AP, et al (2001) Physiological migration of hematopoietic stem and progenitor cells. Science 294:1933–1936

Wynn RF, Hart CA, Corradi-Perini C, et al (2004) A small proportion of mesenchymal stem cells strongly expresses functionally active CXCR4 receptor capable of promoting migration to bone marrow. Blood 104:2643–2645

Xu M, Bruno E, Chao J, et al (2005) Constitutive mobilization of CD34$^+$ cells into the peripheral blood in idiopathic myelofibrosis may be due to the action of a number of proteases. Blood 105:4508–4515

Yin T, Li L (2006) The stem cell niches in bone. J Clin Invest 116:1195–1201

Yong KL, Watts M, Shaun Thomas N, et al (1998) Transmigration of CD34$^+$ cells across specialized and nonspecialized endothelium requires prior activation by growth factors and is mediated by PECAM-1 (CD31). Blood 91:1196–1205

Yu M, Xiao Z, Shen L, et al (2004) Mid-trimester fetal blood-derived adherent cells share characteristics similar to mesenchymal stem cells but full-term umbilical cord blood does not. Br J Haematol 124:666–675

Zhang J, Niu C, Ye L, et al (2003) Identification of the haematopoietic stem cell niche and control of the niche size. Nature 425:836–841

Zhu J, Emerson SG (2004) A new bone to pick: osteoblasts and the haematopoietic stem-cell niche. Bioessays 26:595–599

Zvaifler NJ, Marinova-Mutafchieva L, Adams G, et al (2000) Mesenchymal precursor cells in the blood of normal individuals. Arthritis Res 2:477–488

Role of Endothelial Nitric Oxide in Bone Marrow-Derived Progenitor Cell Mobilization

M. Monterio de Resende[1] · L.-Y. Huw[2] · H.-S. Qian[3] · K. Kauser[4] (✉)

[1] Medical College of Wisconsin, WI, USA
[2] Genentech, South San Francisco CA, USA
[3] Berlex Biosciences, Richmond CA, USA
[4] Boehringer Ingelheim Pharmaceuticals, Inc.,
175 Briar Ridge Road, Ridgefield CT, 06877, USA
kkauser@rdg.boehringer-ingelheim.com

1	Introduction	37
2	EDNO Dependent Bone Marrow Cell Mobilization	38
3	EDNO-Independent Bone Marrow Cell Mobilization	40
4	Summary	41
	References	41

Abstract Mobilization and recruitment of bone marrow-derived progenitor cells (BMDPCs) play an important role in postischemic tissue repair. Patients with coronary artery disease (CAD) or peripheral vascular disease (PVD) exhibit endothelial dysfunction, and as a result are likely to have a reduced number of progenitor cells mobilized in their peripheral circulation following ischemic injury. Identification of eNOS independent pathways for BMDPC mobilization may have important therapeutic value in this patient population. To identify such mechanisms we investigated the effect of granulocyte-colony stimulating factor (G-CSF) and stem cell factor (SCF) in eNOS-KO mice with and without surgical hind-limb ischemia. Our results suggest that BMDPC mobilization can be achieved via activation of NO-independent pathways.

Keywords Bone marrow-derived progenitor cells · VEGF · G-CSF · eNOS · CLI

1
Introduction

Mobilization and recruitment of bone marrow-derived progenitor cells (BMD-PCs) play an important role in postischemic tissue repair. There is accumulating evidence supporting the feasibility of using adult, autologous BMDPCs to induce therapeutic revascularization and tissue repair in patients with severe peripheral arterial disease as well as acute myocardial infarction (Tateishi-Yuyama et al. 2002; Barbash et al. 2003; Kawamoto et al. 2003; Losordo and Dimmeler 2004; Lenk et al. 2005). However, it has also been demonstrated that patients with coronary artery disease (CAD) exhibit reduced levels and functional impairment of endothelial progenitor cells (EPCs), one of the most frequently used autologous bone marrow-derived cell types for therapeutic angiogenesis (Vasa et al. 2001; Heeschen et al. 2004; Hill et al. 2003; Werner and Nickenig 2006). Most patients, likely to become candidates for biological revascularization therapy (cell or gene delivery), exhibit severe endothelial dysfunction, and reduced bioavailability of endothelium-derived nitric oxide (EDNO), due to the presence of cardiovascular risk factors, advanced age, or both (Vita et al. 1990; Zeiher et al. 1993; Boger et al. 1997; Schachinger et al. 2000).

EDNO is produced by the endothelial isoform of nitric oxide synthase enzyme (eNOS), which is a cytochrome p450-like heme-containing enzyme. It requires several cofactors [NADPH (the reduced form of nicotinamide adenine dinucleotide phosphate), FAD (flavin adenine dinucleotide), calmodulin, and tetrahydrobiopterin] for proper function. The endothelial cell-specific form of nitric oxide synthases (eNOS) catalyzes the oxidation of L-arginine to L-citrulline and NO. EDNO is an important, short-lived molecule that plays a key role in the maintenance of endothelial/vascular integrity (for review see Kauser and Rubanyi 2002). Besides contributing to endothelial cell survival, EDNO is a key mediator of angiogenesis as a downstream effector of angiogenic growth factors (Murohara et al. 1998; Dimmeler et al. 1999; Qian et al. 2001). Genetic deficiency of endothelial nitric oxide synthase (eNOS-KO) results in developmental limb defects (Gregg et al. 1998), and severe tissue necrosis in response to ischemia (Murohara et al. 1998; Qian et al. 2001). The ischemic damage due to surgical femoral artery ligation in eNOS-KO mice portrays a similar phenotype to that seen in patients with Fontaine stage III and IV peripheral arterial disease (Murohara et al. 1998; Qian et al. 2001; Aicher et al. 2003). eNOS has been shown to be essential for BMDPC mobilization in response to VEGF as well as for BMDPC function at the site of the ischemic injury (Aicher et al. 2003, 2004). Cardiovascular risk factors have shown a strong inverse correlation with EPC function (Vasa et al. 2001; Heeschen et al. 2004; Werner et al. 2005), but data also support the possibility that increased levels of EPC in the peripheral circulation is able to counterbalance endothelial dysfunction associated with the increased risk factors (Hill et al. 2003).

2
EDNO Dependent Bone Marrow Cell Mobilization

Vascular endothelial growth factor (VEGF) and stromal cell-derived factor-1 (SDF-1) seem to play an important role in endogenous, postischemic mobilization of BMDPCs. EDNO, in addition to its importance in mediating the local angiogenic response following ischemic injury, has been shown to mediate a systemic effect of VEGF on EPC mobilization and function (Murohara et al. 1998; Aicher et al. 2003). eNOS-KO mice develop severe necrosis following ischemic injury, and respond with attenuated mobilization of BMDPCs to VEGF (Aicher et al. 2003). This implies that patients with CAD or peripheral vascular disease (PVD) and endothelial dysfunction are likely to have a reduced number of progenitor cells mobilized in their peripheral circulation following ischemic injury.

Most studies focus on EPCs as the key progenitor cell type to "drive" the neovascularization process in the setting of ischemic damage (Losordo and Dimmeler 2004; Takahashi et al. 1999). EDNO is an essential contributor to EPC function (Aicher et al. 2003, 2004; Werner et al. 2005). EPCs isolated from eNOS-KO mice are impaired in function, and their functional deficit has been shown to contribute to the inadequate angiogenic response in the eNOS-KO mouse hind-limb ischemia model that was investigated following bone marrow transplantation of a selected $CD45^+$ bone marrow population into lethally irradiated mice (Aicher et al. 2003). There are reports, however, that confirm the role of other progenitor cell types from the bone marrow in the revascularization process (Kinnaird et al. 2004; Al-Khaldi et al. 2003). These cells are not as likely to become structural elements of the developing or remodeled vasculature, but play an essential role by providing a permissive milieu for revascularization, serving as a source of cytokines and growth factors at an optimal site and time needed for the regenerative process (Kinnaird et al. 2004; Jo et al. 2003).

Impaired BMDPC mobilization has also been described in matrix metalloproteinase-9 knockout (MMP-9-KO) mice (Heissig et al. 2002). Both in eNOS-KO and MMP-9-KO mice postischemic injury, VEGF-induced BMDPC mobilization was impaired. eNOS-KO mice showed a decreased level of preMMP-9 in the bone marrow, suggesting an important role for MMP-9 as a downstream effector of this pathway (Aicher et al. 2003).

eNOS likely plays a critical role in the bone marrow compartment. Bone marrow osteoblastic cells as well as stromal cells express eNOS, which may also contribute to the proper maturation and differentiation of different lineage bone marrow cells (Ozuyaman et al. 2005).

In addition to ischemic cytokine mediators, pharmacological substances such as 3-hydroxy-3-methyl-glutaryl-CoA reductase (HMG-CoA) reductase inhibitors, or statins, have been shown to induce mobilization of EPCs from bone marrow and enhance left ventricular function following myocardial infarction (Llevadot et al. 2001; Dimmeler et al. 2001). A similar benefit of statins has been shown in patients with stable coronary artery disease (Urbich and Dimmeler 2005). The increase of circulating EPC numbers by statin has been indicated to require eNOS function (Landmesser et al. 2004). Statins increase eNOS expression by stabilizing messenger RNA (mRNA) expression (Gonzalez-Fernandez et al. 2001), which may be the mechanism that is responsible for the benefit to EPC function.

Besides pharmacological stimulation, physical exercise also has been shown to result in increased EPC levels (Laufs et al. 2004). Interestingly, similar physiological stimulation of EPC mobilization could not be demonstrated in eNOS-KO mice, suggesting the involvement of EDNO in this phenomenon as well (Laufs et al. 2004).

EDNO deficiency in the severe CAD patient population is likely to limit the effectiveness of therapeutic revascularization approaches using autologous BMDPCs either exogenously delivered or endogenously mobilized. Identification of eNOS-independent pathways for BMDPC mobilization could have important therapeutic value in this patient population.

3
EDNO-Independent Bone Marrow Cell Mobilization

Pharmacological agents that increase the level of circulating BMDPCs have been demonstrated to enhance cardiac performance in different animal models of cardiovascular injury (Kinnaird et al. 2004; Orlic et al. 2001; Takahashi et al. 1999). Systemic cytokine treatment with granulocyte-colony stimulating factor (G-CSF) and stem cell factor (SCF) or granulocyte macrophage-colony stimulating factor (GM-CSF) have been shown to promote cardiac repair and enhance vascularization of severely ischemic tissues (Orlic et al. 2001; Takahashi et al. 1999). We have shown that treatment with G-CSF and SCF can result in increased BMDPC mobilization in the absence of eNOS (L.-Y. Hu et al., submitted). There was no difference between CFU-GM and EPC counts between C57Bl/J6 and eNOS-KO mice in response to G-CSF and SCF, suggesting that BMDPC mobilization, irrespective of the progenitor cell type, is independent of eNOS expression and subsequent EDNO availability. SDF-1 and its receptor, CXCR4, are critical for the retention of progenitor cells within the bone marrow, and represent the downstream pathway for G-CSF-induced BMDPC mobilization (Rabbany et al. 2003; Semerad et al. 2005; Lévesque et al. 2004). We found an equal quantity of CXCR4 receptors on bone marrow cells harvested from C57Bl/J6 and eNOS-KO mice (L.-Y. Hu et al., submitted). The CXCR4 antagonist, AMD-3100, led to a similarly efficient mobilization of BMDPCs compared to G-CSF in eNOS-KO mice (L.-Y. Hu et al., submitted), suggesting that this pathway is independent of the presence of eNOS, unlike VEGF-induced EPC mobilization (Aicher et al. 2003).

We could demonstrate efficient BMDPC mobilization by G-CSF and SCF in eNOS-KO mice concurrent with a significant therapeutic benefit by this treatment following surgical femoral artery ligation (L.-Y. Hu et al., submitted). The ischemic tissue damage in this strain of mice is refractory to growth factor treatment (Murohara et al. 1998). In fact this model has not yet shown response to any treatment besides substitution of eNOS by gene delivery, indicating the essential role of EDNO in mediating revascularization and tissue repair process by many different factors (Yu et al. 2005; Qian et al. 2006). In addition, exogenously delivered autologous bone marrow cells were able to mimic the effect on blood flow recovery, indicating that availability of an increased number of autologous BMDPCs in the peripheral circulation was able to contribute to postischemic revascularization irrespective of genetic deficiency in eNOS.

4
Summary

Severe peripheral arterial diseases and heart failure represent the increasing medical need for efficient novel regenerative therapies. Biological revascularization strategies (i.e., cell and gene therapy) aim to achieve restoration of function by triggering natural repair processes using growth factors and adult bone BMDPCs. The restorative value of autologous cells is dependent on the overall health of the patient, the presence of risk factors, and the availability of EDNO (Vasa et al. 2001; Heeschen et al. 2004; Hill et al. 2003; Werner and Nickenig 2006; Vita et al. 1990; Zeiher et al. 1993; Boger et al. 1997). It is therefore important to identify molecular pathways that are able to bypass the need for EDNO to improve the therapeutic outcome of these treatment approaches. Investigation of eNOS-independent pathways for BMDPC mobilization could have important therapeutic value in patients with severe endothelial dysfunction.

BMDPC mobilization by G-CSF and SCF may represent an eNOS-independent mechanism. Combined G-CSF and SCF treatment in eNOS-KO mice, subjected to surgical ischemia, resulted in limb salvage and significant blood flow recovery. This response was similar to the benefit achieved by intravenous transfer of autologous BMDPCs. CXCR4 antagonists, which mimic downstream G-SCF signaling, induced a similar degree of BMDPC release compared to G-CSF to the peripheral blood in eNOS-KO mice, offering a potential molecular target for BMDPC mobilization in CAD or PVD patients with severe endothelial dysfunction.

References

Aicher A, Heeschen C, Mildner-Rihm C, Urbich C, Ihling C, Technau-Ihling K, Zeiher AM, Dimmeler S (2003) Essential role of endothelial nitric oxide synthase for mobilization of stem and progenitor cells. Nat Med 9:1370–1376

Aicher A, Heeschen C, Dimmeler S (2004) The role of NOS3 in stem cell mobilization. Trends Mol Med 10:421–425

Al-Khaldi A, Al-Sabti H, Galipeau J, Lachapelle K (2003) Therapeutic angiogenesis using autologous bone marrow stromal cells: improved blood flow in a chronic limb ischemia model. Ann Thorac Surg 75:204–209

Barbash IM, Chouraqui P, Baron J, Feinberg MS, Etzion S, Tessone A, Miller L, Guetta E, Zipori D, Kedes LH, et al (2003) Systemic delivery of bone marrow-derived mesenchymal stem cells to the infracted myocardium: feasibility, cell migration, and body distribution. Circulation 108:863–868

Boger RH, Bode-Boger SM, Thiele W, Junker W, Alexander K, Frolich JC (1997) Biochemical evidence for impaired nitric oxide synthesis in patients with peripheral arterial occlusive disease. Circulation 95:2068–2074

Dimmeler S, Hermann C, Galle J, Zeiher AM (1999) Upregulation of superoxide dismutase and nitric oxide synthase mediates the apoptosis-suppressive effects of shear stress on endothelial cells. Arterioscler Thromb Vasc Biol 19:656–664

Dimmeler S, Aicher A, Vasa M, Mildner-Rihm C, Adler K, Tiemann M, Rutten H, Fichtlscherer S, Martin H, Zeiher AM (2001) HMG-CoA reductase inhibitors (statins) increase endothelial progenitor cells via the PI 3-kinase/Akt pathway. J Clin Invest 108:391–397

Gonzalez-Fernandez F, Jimenez A, Lopez-Blaya A, Velasco S, Arriero MM, Celdran A, Rico L, Farre J, Casado S, Lopez-Farre A (2001) Cerivastatin prevents tumor necrosis factor-alpha-induced downregulation of endothelial nitric oxide synthase: role of endothelial cytosolic proteins. Atherosclerosis 155:61–70

Gregg AR, Schauer A, Shi O, Liu Z, Lee CG, O'Brien WE (1998) Limb reduction defects in endothelial nitric oxide synthase-deficient mice. Am J Physiol 275:H2319–H2324

Heeschen C, Lehmann R, Honold J, Assmus B, Aicher A, Walter DH, Martin H, Zeiher AM, Dimmeler S (2004) Profoundly reduced neovascularization capacity of bone marrow mononuclear cells derived from patients with chronic ischemic heart disease. Circulation 109:1615–1622

Heissig B, Hattori K, Dias S, Friedrich M, Ferris B, Hackett NR, Crystal RG, Besmer P, Lyden D, Moore MA, et al (2002) Recruitment of stem and progenitor cells from the bone marrow niche requires MMP-9 mediated release of kit-ligand. Cell 109:625–637

Hill JM, Zalos G, Halcox JP, Schenke WH, Waclawiw MA, Quyyumi AA, Finkel T (2003) Circulating endothelial progenitor cells, vascular function, and cardiovascular risk. N Engl J Med 348:593–600

Jo DY, Hwang JH, Kim JM, Yun HJ, Kim S (2003) Human bone marrow endothelial cells elaborate non-stromal-cell-derived factor-1 (SDF-1)-dependent chemoattraction and SDF-1-dependent transmigration of haematopoietic progenitors. Br J Haematol 121:649–652

Kauser K, Rubanyi GM (2002) "Nitric oxide deficiency" in cardiovascular diseases. Cardiovascular protection by restoration of endothelial nitric oxide production. In: Rubanyi GM (ed) Mechanisms of vasculoprotection. Springer-Verlag, Berlin Heidelberg New York, pp 1–31

Kawamoto A, Tkebuchava T, Yamaguchi J, Nishimura H, Yoon YS, Milliken C, Uchida S, Masuo O, Iwaguro H, Ma H, et al (2003) Intramyocardial transplantation of autologous endothelial progenitor cells for therapeutic neovascularization of myocardial ischemia. Circulation 107:461–468

Kinnaird T, Stabile E, Burnett MS, Shou M, Lee CW, Barr S, Fuchs S, Epstein SE (2004) Local delivery of marrow-derived stromal cells augments collateral perfusion through paracrine mechanisms. Circulation 109:1543–1549

Landmesser U, Engberding N, Bahlmann FH, Schaefer A, Wiencke A, Heineke A, Spiekermann S, Hilfiker-Kleiner D, Templin C, Kotlarz D, Mueller M, Fuchs M, Hornig B, Haller H, Drexler H (2004) Statin-induced improvement of endothelial progenitor cell mobilization, myocardial neovascularization, left ventricular function, and survival after experimental myocardial infarction requires endothelial nitric oxide synthase. Circulation 110:1933–1939

Laufs U, Werner N, Link A, Endres M, Wassmann S, Jurgens K, Miche E, Bohm M, Nickenig G (2004) Physical training increases endothelial progenitor cells, inhibits neointima formation, and enhances angiogenesis. Circulation 109:220–226

Lenk K, Adams V, Lurz P, Erbs S, Linke A, Gielen S, Schmidt A, Scheinert D, Biamino G, Emmrich F, et al (2005) Therapeutical potential of blood-derived progenitor cells in patients with peripheral arterial occlusive disease and critical limb ischaemia. Eur Heart J 26:1903–1909

Lévesque JP, Liu F, Simmons PJ, Betsuyaku T, Senior RM, Pham C, Link DC (2004) Characterization of hematopoietic progenitor mobilization in protease-deficient mice. Blood 104:65–72

Llevadot J, Murasawa S, Kureishi Y, Uchida S, Masuda H, Kawamoto A, Walsh K, Isner JM, Asahara T (2001) HMG-CoA reductase inhibitor mobilizes bone marrow—derived endothelial progenitor cells. J Clin Invest 108:399–405

Losordo DW, Dimmeler S (2004) Therapeutic angiogenesis and vasculogenesis for ischemic disease. Part II: cell-based therapies. Circulation 109:2692–2697

Murohara T, Asahara T, Silver M, Bauters C, Masuda H, Kalka C, Kearney M, Chen D, Symes JF, Fishman MC, et al (1998) Nitric oxide synthase modulates angiogenesis in response to tissue ischemia. J Clin Invest 101:2567–2578

Orlic D, Kajstura J, Chimenti S, Limana F, Jakoniuk I, Quaini F, Nadal-Ginard B, Bodine DM, Leri A, Anversa P (2001) Mobilized bone marrow cells repair the infarcted heart, improving function and survival. Proc Natl Acad Sci U S A 98:10344–10349

Ozuyaman B, Ebner P, Niesler U, Ziemann J, Kleinbongard P, Jax T, Godecke A, Kelm M, Kalka C (2005) Nitric oxide differentially regulates proliferation and mobilization of endothelial progenitor cells but not of hematopoietic stem cells. Thromb Haemost 94:770–772

Qian HS, Liu P, Kauser K, et al (2001) Nitric oxide deficiency leads to impaired angiogenesis and severe dysfunction of microcirculation in a mouse hind limb ischemia model. Proceedings of the 7th World Congress of Microcirculation Sydney, Australia. Monduzzi Editore, Sydney,pp 525–529

Qian HS, Liu P, Huw LY, Orme A, Halks-Miller M, Hill SM, Jin F, Kretschmer P, Blasko E, Cashion L, Szymanski P, Vergona R, Harkins R, Yu J, Sessa WC, Dole WP, Rubanyi GM, Kauser K (2006) Effective treatment of vascular endothelial growth factor refractory hindlimb ischemia by a mutant endothelial nitric oxide synthase gene. Gene Ther 13:1342–1350

Rabbany SY, Heissig B, Hattori K, Rafii S (2003) Molecular pathways regulating mobilization of marrow-derived stem cells for tissue revascularization. Trends Mol Med 9:109–117

Schachinger V, Britten MB, Zeiher AM (2000) Prognostic impact of coronary vasodilator dysfunction on adverse long-term outcome of coronary heart disease. Circulation 101:1899–1906

Semerad CL, Christopher MJ, Liu F, Short B, Simmons PJ, Winkler I, Lévesque JP, Chappel J, Ross FP, Link DC (2005) G-CSF potently inhibits osteoblast activity and CXCL12 mRNA expression in the bone marrow. Blood 106:3020–3027

Takahashi T, Kalka C, Masuda H, Chen D, Silver M, Kearney M, Magner M, Isner JM, Asahara T (1999) Ischemia- and cytokine-induced mobilization of bone marrow-derived endothelial progenitor cells for neovascularization. Nat Med 5:434–438

Tateishi-Yuyama E, Matsubara H, Murohara T, Ikeda U, Shintani S, Masaki H, Amano K, Kishimoto Y, Yoshimoto K, Akashi H, et al (2002) Therapeutic angiogenesis for patients with limb ischaemia by autologous transplantation of bone-marrow cells: a pilot study and a randomised controlled trial. Lancet 360:427–435

Urbich C, Dimmeler S (2005) Risk factors for coronary artery disease, circulating endothelial progenitor cells, and the role of HMG-CoA reductase inhibitors. Kidney Int 67:1672–1676

Vasa M, Fichtlscherer S, Aicher A, Adler K, Urbich C, Martin H, Zeiher AM, Dimmeler S (2001) Number and migratory activity of circulating endothelial progenitor cells inversely correlate with risk factors for coronary artery disease. Circ Res 89:E1–7

Vita JA, Treasure CB, Nabel EG, McLenachan JM, Fish RD, Yeung AC, Vekshtein VI, Selwyn AP, Ganz P (1990) Coronary vasomotor response to acetylcholine relates to risk factors for coronary artery disease. Circulation 81:491–497

Werner N, Nickenig G (2006) Influence of cardiovascular risk factors on endothelial progenitor cells: limitations for therapy? Arterioscler Thromb Vasc Biol 26:257–266

Werner N, Kosiol S, Schiegl T, Ahlers P, Walenta K, Link A, Bohm M, Nickenig G (2005) Circulating endothelial progenitor cells and cardiovascular outcomes. N Engl J Med 353:999–1007

Yu J, deMuinck ED, Zhuang Z, Drinane M, Kauser K, Rubanyi GM, Qian HS, Murata T, Escalante B, Sessa WC (2005) Endothelial nitric oxide synthase is critical for ischemic remodeling, mural cell recruitment, and blood flow reserve. Proc Natl Acad Sci U S A 102:10999–11004

Zeiher AM, Drexler H, Saurbier B, Just H (1993) Endothelium-mediated coronary blood flow modulation in humans. Effects of age, atherosclerosis, hypercholesterolemia, and hypertension. J Clin Invest 92:652–662

Immune Plasticity of Bone Marrow-Derived Mesenchymal Stromal Cells

J. Stagg · J. Galipeau (✉)

Sir Mortimer B. Davis Jewish General Hospital Lady Davis Research Institute, McGill University, 3755 Cote Ste-Catherine Road, Montreal QC, H3T 1E2, Canada
jacques.galipeau@mcgill.ca

1	Mesenchymal Stromal Cells: An Overview	46
	1.1 Definition	46
	1.2 Identification	47
	1.3 Maintenance and Differentiation of MSCs	47
	1.4 MSCs and Hematopoiesis	48
	1.5 MSCs in Tissue Repair and Cancer	49
2	Immune-Modulating Properties of MSCs	50
	2.1 Immunosuppression by Soluble Factors	50
	2.2 MSCs Induce T Cell Anergy	52
	2.3 MSCs Inhibit B Cells	52
	2.4 MSCs Inhibit Resting NK Cells but Are Killed by Activated NK Cells	53
	2.5 MSCs Inhibit Professional Antigen-Presenting Cells	53
	2.6 MSCs Suppress Immune Responses In Vivo	54
	2.7 Regulating MSC-Mediated Immunomodulation	55
	2.8 MSCs Act as APCs	56
	2.9 MSCs Induce Protective Immunity In Vivo	57
	2.10 Nonhematopoietic APCs	58
3	Clinical Perspective: MSC-Based Cell Therapy	59
4	Spontaneous Transformation of Human MSCs	60
5	Conclusion	61
References		62

Abstract Isolated from simple bone marrow aspirates, mesenchymal stromal cells (MSCs) can be easily expanded ex vivo and differentiated into various cell lineages. Because they are present in humans of all ages, are harvested in the absence of prior mobilization and preserve their plasticity following gene modification, MSCs are particularly attractive for cell-based medicine. One of the most fascinating properties of ex vivo expanded MSCs is their ability to suppress ongoing immune responses, both in vitro and in vivo. Although not fully understood, the immunosuppressive properties of MSCs have been reported to affect the function of a broad range of immune cells, including T cells, antigen-presenting cells, natural killer cells and B cells. Whereas successful harnessing of these immunosuppressive properties might one day open the door to the development of new cell-based strategies for the control of graft-versus-host and other autoimmune diseases, recent studies suggest that the immune-modulating properties of MSCs are far more complex than first thought. Reminiscent of the dichotomy of function of dendritic cells (DCs), which can act as potent activators or potent suppressors of immune responses, new studies including our own work has shown that MSCs in fact possess the dual ability to suppress or activate immune responses. In this review, we summarize the different biological properties of MSCs and discuss the current literature on the complex mechanism of immune modulation mediated by ex vivo expanded MSCs.

Keywords Mesenchymal stem cells · Adult stem cells · Antigen presentation · Immune suppression · Graft-vs-host disease

1
Mesenchymal Stromal Cells: An Overview

1.1
Definition

Bone marrow stroma consists of a heterogeneous population of cells that include adipocytes, reticular cells, macrophages, vascular endothelial cells, smooth muscle cells, and mesenchymal stromal cells (MSCs) (Clark and Keating 1995). The first report of a population of stem cells in the bone marrow stroma is attributed to Friedenstein et al. (1976). He observed that a small fraction of adherent cells, which he called colony forming unit fibroblasts (CFU-F), multiplied rapidly after 2–4 days of in vitro culture and had the ability to differentiate into bone- and cartilage-like colonies. These initial observations have been confirmed by several studies that demonstrated that cells isolated following Friedenstein's technique were able to differentiate into various cell types, including osteoblasts, chondroblasts, adipocytes, myocytes, astrocytes, neurons, endothelial cells, and lung epithelial cells (Pittenger et al. 1999; Reyes et al. 2001; Al-Khaldi et al. 2003). Because of their mesenchymal plasticity, adherent cells from MSC cultures are often referred to as mesenchymal stem cells. However, since most unfractionated adherent cells from bone marrow aspirates do not meet the criteria of self-renewal and clonal plasticity of a stem cell, we refer to these colony-forming adherent cells as mesenchymal stromal cells in accordance with the consensus recommendation of the International Society for Cellular Therapy (Horwitz et al. 2005).

When plated at low cell density (1–10 cells/cm^2), adherent MSCs can be cloned as single-cell-derived colonies, each individual colony displaying a variable degree of plasticity (Reyes et al. 2001; Colter et al. 2001; Pittenger et al. 1999). MSC cultures are thus heterogeneous, containing subpopulations of early and more committed progenitors. Darwin Prockop and colleagues have reported that two morphologically distinctive cell types can be found within MSC cultures: small rapidly self-renewing multipotent cells (RS-MSCs) and more mature slowly replicating larger cells (SR-MSCs) (Colter et al. 2000, 2001; Digirolamo et al. 1999). When maintained at low cell-density, MSC cultures remain rich in multipotent RS-MSCs, most likely as a result of the Wnt signaling inhibitor Dickkopf-1 (Dkk-1) and its receptor LRP6, the expression of which has been shown to inversely correlate with cell-density of MSC cultures (Sekiya et al. 2002). Added as an exogenous factor, recombinant Dkk-1-derived peptides can stimulate the proliferation of undifferentiated MSCs (Gregory et al. 2005). The effect of Dkk-1 on the intrinsic properties of MSCs, however, remains to be determined.

1.2
Identification

No single marker has yet been described that identifies and localizes MSCs in vivo. MSCs are therefore identified in vitro based on their ability to differentiate into mesenchymal lineage cells (adipocytic, osteogenic, and chondrogenic) and to express specific membrane-bound surface antigens (Deans and Moseley 2000). Both human and mouse MSCs express CD105 (endoglin), CD73 (membrane-bound ecto-5'-nucleotidase), and CD44 (hyaluronate receptor). In order to rule out any contamination of MSC preparations by hematopoietic or endothelial cells, MSCs are routinely tested for the absence of expression of CD45 (common leukocyte antigen) and CD31 [platelet/endothelial cell adhesion molecule (PECAM)-1; highly expressed on endothelial cells]. The expression of the hematopoietic stem cell (HSC) marker CD34 is also commonly used to assess the presence of contaminating HSC in MSC preparations. While human and BALB/c MSCs are consistently negative for CD34, we (and others) have observed that C57BL/6 mouse MSCs express CD34 in a heterologous fashion (10%–20% positive)(Peister et al. 2004). Hence, MSCs can be defined as being positive for the expression of CD105, CD73, and CD44, negative for the expression of CD45 and CD31, and negative for the expression CD34 with the exception of C57BL/6 mice. MSCs further express low levels of major histocompatibility complex (MHC) class I molecules while, as a general rule, they do not constitutively express MHC class II molecules (Tse et al. 2003; Di Nicola et al. 2002). However, constitutive MHC class II expression on some MSC populations has been observed (Potian et al. 2003; Stagg et al. 2006). Both MHC class I and class II molecules generally get upregulated following interferon (IFN)-γ treatment, with a more heterogeneous expression between individual cells for MHC class II molecules (Gotherstrom et al. 2004; Krampera et al. 2003; Meisel et al. 2004). Costimulatory molecules such as CD80, CD86, CD40, and CD40L are not known to be expressed or induced on human MSCs, while our own group and others have observed that mouse MSCs can constitutively express CD80 (Krampera et al. 2003; Stagg et al. 2006).

1.3
Maintenance and Differentiation of MSCs

Among the different factors governing the development and differentiation of stem cells, transforming growth factor (TGF)-β-family proteins have been revealed to be crucial players (Ruscetti et al. 2005). The TGF-β family comprises at least 30 related growth factors and differentiation factors, including TGF-βs, activins, and bone morphogenic proteins (BMPs) (Massague et al. 2000; Mishra et al. 2005). While TGF-β itself promotes the expansion of undifferentiated MSCs, other TGF-β-family members promote the differentiation of MSCs. Transcriptional modulators cooperate with TGF-β-family proteins in governing cell type-specific differentiation of MSCs. Osteoblasts and adipocytes, for instance, originate through alternative activation of transcription factors (Nakashima and de Crombrugghe 2003; Rosen et al. 2000). Indeed, while the transcription factor Runx2 drives MSCs to differentiate into osteoblasts, the differentiation of MSCs into adipocytes relies on expression of the transcription factor peroxisome proliferator-activated receptor (PPAR)γ. Importantly, this transcriptionally controlled differentiation is mutually exclusive. Recently, Hong et al. (2005) identified the transcriptional coactivator with PDZ-binding motif (TAZ) as a key "molecular rheostat" that modulates this mutually exclusive differentiation of MSCs. In primary MSC cultures, it was demonstrated that TAZ coactivates Runx2-dependent transcription while repressing PPARγ-dependent transcription. The study described an absolute requirement for TAZ in osteoblastic differentiation. It is our contention that the identification and thorough analysis of other "stemness" and differentiation signaling pathways will be of great value for developing new methods to maintain and propagate MSCs in vitro—and perhaps in vivo—as well as methods to guide or regulate their differentiation for cell-based therapy.

1.4
MSCs and Hematopoiesis

It is widely accepted that one of the main biological functions of marrow MSCs is to provide the microenvironment necessary for hematopoiesis, including cell-to-cell interactions and the secretion of cytokines and growth factors such as macrophage-CSF (M-CSF), Flt-3L, stem-cell factor (SCF), interleukin (IL)-6, IL-7, IL-8, IL-11, IL-12, IL-14, and IL-15. Upon IL-1α stimulation, MSCs can further produce IL-1α, leukemia inhibitory factor (LIF), G-CSF, and GM-CSF (Deans and Moseley 2000). MSCs are also implicated in lymphopoiesis (Torlakovic et al. 2005). Maturation of B cells, for instance, occurs in the bone marrow and involves interactions with MSCs, which provides them with SCF and IL-7 (Milne et al. 2004). Through T cell–MSC interactions, MSCs might also be involved in extrathymic T cell lymphopoiesis (Barda-Saad et al. 1996). Recent studies have demonstrated that in addition to supporting hematopoiesis, bone marrow plays an important role in the induction of endogenous immune responses. Indeed, bone marrow has been shown to be the preferred site of induction of adaptive immunity in response to blood-born antigens (Mazo et al. 2005). It was also demonstrated that central memory $CD8^+$ T cells are preferentially recruited to the bone marrow. Notably, homing of memory $CD8^+$ T cell was mediated by CXCL12 (a.k.a. SDF-1), a chemokine abundantly produced by MSCs (Mazo et al. 2005). Bone marrow is thus being revealed as a unique lymphoid organ able to activate naïve T cells and to recruit memory T cells. In this regard, MSCs may constitute a previously unrecognized player of endogenous immune responses.

1.5
MSCs in Tissue Repair and Cancer

Another important property of MSCs is their ability to home to injured tissues and actively participate in tissue repair. It is believed that MSCs can replace dying cells of a given organ in order to maintain homeostasis. Because signaling through the TGF-β pathway is activated in response to tissue injury, and since TGF-β is known to promote the expansion of undifferentiated MSCs, TGF-β may play an important role in the ability of MSCs to perform wound repair. MSCs can repair tissues by three mechanisms: (1) by differentiating into the phenotype of the damaged cells; (2) by secreting growth factors and cytokines that enhances repair of endogenous cells; and (3) by undergoing cell fusion (Prockop et al. 2003; Terada et al. 2002; Spees et al. 2003).

Because the microenvironment of a solid tumor is similar to the environment of injured/stressed tissue, it has been proposed that the intrinsic wound-healing properties of MSCs would trigger them to migrate to sites of solid tumors. Michael Andreff's group has shown that intravenously delivered human MSCs are capable of integrating into xenografted human tumors in immunocompromised mice (Studeny et al. 2002). Recently, Nakamizo et al. (2005) demonstrated that human MSCs, following intraarterial or intracranial delivery, can localize to human gliomas. In both studies, immunocompromised mice were used as hosts. Whether naïve MSCs can migrate to the site of a solid tumor in immunocompetent hosts and whether they can modulate tumor growth by themselves is still undefined.

Some observations suggest that MSCs can promote tumor growth through the release of tumor and vascular growth factors and by releasing immunosuppressive factors (Djouad et al. 2003). However, at least one study has demonstrated that intravenously injected MSCs can induce potent antitumor effects by a contact-dependent inhibition of Akt activity in tumor cells (Khakoo et al. 2006). In that study, intravenously injected human MSCs were shown to home to human Kaposi's sarcoma tumors implanted subcutaneously in immunodeficient mice and to significantly delay tumor growth. Notably, the inhibitory effect of MSCs on tumor cell growth was dependent on E-cadherin, as a neutralizing antibody against E-cadherin abrogated the inhibition of Akt activation by MSCs in vitro. MSCs, however, were unable to inhibit Akt activation in two other tumor cell lines (i.e., PC-3 prostate cancer cells and MCF-7 breast cancer cells). Interestingly, cell–cell interactions through E-cadherin have been reported to activate Akt in a mechanism dependent upon phosphatidylinositol 3 kinase and epidermal growth factor receptor (Reddy et al. 2005). One of the effects of MSCs might thus be to sequester E-cadherin away from cell–cell interactions, thereby preventing proper survival signaling in specific cancer cells.

The molecules and receptors governing MSC migration—in response to injury or tumors—are still not fully characterized. One of the difficulties in studying the migratory properties of MSCs comes from the fact that ex vivo cultured MSCs often loose expression of chemokine receptors and responsiveness to chemokines. Indeed, early passage MSCs, but not long-term cultured MSCs, have been described to express a broad range of chemokine receptors, including CCR1, CCR7, CCR9, CX3CL1, CXCR4, CXCR5, and CXCR6, and to migrate in response to CXCL12, CXCL13, CXCL16, CCL19, CCL5, and CCL25 (Honczarenko et al. 2006).

2
Immune-Modulating Properties of MSCs

The first studies of MSC-mediated immunosuppression reported that ex vivo expanded third-party human MSCs could potently inhibit the in vitro proliferation of allogeneic lymphocytes when incorporated into mixed lymphocytes cultures. Several independent groups have since demonstrated that ex vivo-expanded MSCs can suppress T cell proliferation induced by allogeneic peripheral blood mononuclear cells (PBMCs), allogeneic splenocytes, mitogens, concanavalin A, and anti-CD3/anti-CD28 antibodies (Le Blanc et al. 2003; Tse et al. 2003; Di Nicola et al. 2002; Krampera et al. 2003; Djouad et al. 2003). In addition to their suppressive effect of T cells, MSCs have been shown to negatively modulate the function of B cells (Corcione et al. 2006), natural killer (NK) cells (Spaggiari et al. 2006), and dendritic cells (Jiang et al. 2005) (Fig. 1). In a landmark case study, Le Blanc and colleagues (2004b) exploited the immunosuppressive properties of MSCs in an attempt to treat a young patient from severe grade IV acute graft-vs-host disease (GVHD). Remarkably, repeated administration of purified haploidentical human MSCs (from the patient's mother) following allogeneic stem cell transplantation completely reversed the GVHD. The patient was still free of GVHD 1 year following treatment, and had no minimal residual disease of his leukemia in blood and bone marrow. Of interest, the patient failed to develop immune responses against the injected allogeneic MSCs, suggesting that MSCs may be immunoprivileged in addition to being immunosuppressive. Notwithstanding this observation, our group and others have shown that mouse MSCs are potently rejected when transplanted into allogeneic hosts (Eliopoulos et al. 2005; Nauta et al. 2006).

2.1
Immunosuppression by Soluble Factors

The exact mechanism responsible for MSC-mediated immunosuppression remains, at best, imprecise. Nonetheless, it is generally accepted that: (1) soluble factors secreted by MSCs play a major role in their immunosuppressive effect, and (2) MSCs need to be "activated" (or stimulated) in order to become immunosuppressive. For instance, when incorporated into mixed lymphocyte cultures, MSCs need to be in contact with the allogeneic cells in order to produce

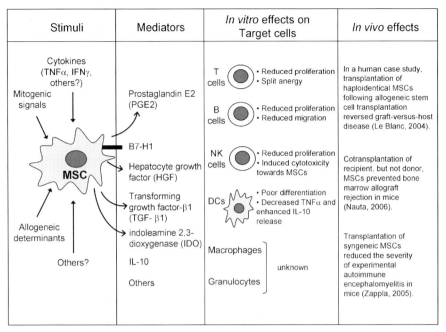

Fig. 1 Immunosuppression mediated by ex vivo-expanded MSCs. Primary mouse and human MSCs have been shown in vitro to inhibit the proliferation and function of several immune effector cells, including T cells, B cells, NK cells, and dendritic cells (DCs). MSC-induced immunosuppression depends on a broad range of factors, including prostaglandin E2 (PGE2), B7-H1, hepatocyte growth factor (HGF), TGF-β1, IDO, IL-10, and other unidentified factors. In addition, several exogenous stimuli, such as tumor necrosis factor (TNF)-α, IFN-γ, and mitogenic and allogeneic determinants, modulate the nature of the immunosuppression induced by MSCs. In vivo, transplanted MSCs have been shown to reverse graft-vs-host disease (GVHD), prevent allograft rejection, and reduce the severity of experimental autoimmune encephalomyelitis, a model of multiple sclerosis

immunosuppressive factors that inhibit lymphocyte proliferation (Le Blanc et al. 2003; Di Nicola et al. 2002). While key soluble factors have been identified, others remain anonymous (Tse et al. 2003; Djouad et al. 2003; Le Blanc et al. 2004a). Using neutralizing monoclonal antibodies, independent studies have demonstrated that hepatocyte growth factor (HGF), prostaglandin E2 (PGE2), TGF-β1, indoleamine 2,3-dioxygenase (IDO), and IL-10 are all key players in MSC-mediated immunosuppression in vitro (Aggarwal and Pittenger 2005; Beyth et al. 2005). Few studies have suggested that contact-dependent mechanisms are implicated, including the B7-H1/PD-1 pathway (Krampera et al. 2003; Augello et al. 2005). It is our hypothesis that multiple immunosuppressive factors can be released or expressed by ex vivo expanded MSCs and that the nature of these factors is dependent on the type of stimuli received by MSCs (e.g., allogeneic determinants, membrane-bound proteins, mitogens, cytokines, etc.). In support of this hypothesis, Rasmusson et al. (2005) recently reported that MSCs use different mechanisms to inhibit lymphocyte proliferation induced by mitogens or alloantigens, with the former, but not the latter, relying on the release of prostaglandins.

2.2
MSCs Induce T Cell Anergy

In addition to their ability to inhibit lymphocyte proliferation, ex vivo expanded MSCs have been shown to induce split T cell anergy. Anergic T cells are characterized by an absence of proliferation and cytokine production in response to antigenic stimulation, generally as a result of insufficient costimulation. In contrast to a classical form of anergy, which can be overcome by the addition of exogenous IL-2, Glennie et al. (2005) demonstrated that allogeneic MSCs can induce T cells to enter an anergic state only partly reversed by exogenous IL-2. They showed that removal of allogeneic MSCs combined to the addition of exogenous IL-2 restored IFN-γ production but not proliferation of the lymphocytes in mixed lymphocytes cultures.

2.3
MSCs Inhibit B Cells

Recently, Corcione et al. (2006) investigated the ability of ex vivo expanded human MSCs to modulate the proliferation and function of human B cells. In their first set of experiments, in vitro expanded human MSCs were cocultured with purified CD19$^+$ human B cells in the presence of a cocktail of stimuli (i.e., CpG oligonucleotides, recombinant CD40L, anti-human Ig goat antibodies, IL-2, IL-4, and IL-10). After 3 days of coculture, the proliferation of B cells was significantly inhibited by the addition of allogeneic human MSCs. Intriguingly, maximum inhibition was observed at the B cell/MSC ratio of 1:1 and disappeared at ratios of 1:5 and 1:10. This is in marked contrast with the inhibition of T cell proliferation by MSCs, which has been shown in different studies to be directly dose-dependent (i.e., the more MSCs added to mixed lymphocytes cultures, the stronger the suppression on T cell proliferation; Djouad et al. 2003). Corcione et al. (2006) showed that MSCs inhibit B cell proliferation through a mechanism that blocked the cells in the G_0/G_1 phase of the cell cycle independent of apoptosis. Akin to the effect on T cells, MSCs cocultured with B cells in transwell plates (thus physically separated) also inhibited B cell proliferation. The authors reported that allogeneic human MSCs further inhibited the production of immunoglobulins by B cells, although this could be the result of decreased B cell proliferation. Finally, the study showed that B cells cocultured with MSCs had a decreased expression of CXCR4, CXCR5, and CXCR7 chemokine receptors and exhibited a significant inhibition to migrate in response to the given chemokines.

2.4
MSCs Inhibit Resting NK Cells but Are Killed by Activated NK Cells

Spaggiari et al. (2006) investigated the modulatory effect of ex vivo-expanded human MSCs on human NK cells. NK cells represent one of our first lines of immunological defense against pathogens and tumors and are emerging as crucial players in regulating adaptive immune responses via their interaction with dendritic cells (DCs). NK cells mediate their protective effect by recognizing and lysing target cells that express stress-induced ligands or down-regulate MHC class I molecules, and through the provision of cytokines such as IFN-γ. NK cells are the major source of IFN-γ during an immune response. Coculture experiments demonstrated that allogeneic MSCs completely inhibited cytokine-induced proliferation of resting NK cells in response to IL-2 (100 U/ml) or IL-15 (100 ng/ml). In contrast, proliferation of preactivated NK cells with IL-2 was only partly inhibited by MSCs. Since the expression levels of IL-2 and IL-15 receptors on MSCs have never been determined, we cannot rule out the possibility that recombinant IL-2 and IL-15 can be sequestered by MSCs, thus indirectly inhibiting proliferation of resting NK cells known to require high levels of cytokine in order to proliferate.

Interestingly, the study demonstrated that human MSCs are highly susceptible to NK cell-mediated lysis due to the expression of a broad range of surface ligands involved in NK cell activation. In particular, human MSCs express multiple NKG2D ligands as well as DNAX accessory molecule (DNAM)-1 ligands. However, the nature of the ligands expressed by MSCs varied considerably between individual preparations. One exception to this was the NKG2D ligand ULBP3 that was always found to be expressed on MSCs preparations. The study demonstrated that IL-2 activated NK cells, but not resting NK cells, could efficiently kill autologous and allogeneic MSCs and produce IFN-γ when cocultured together. As the authors reported, "this finding is reminiscent of previous data regarding the susceptibility of immature DC to lysis by autologous NK cells." NK cell-mediated lysis of MSCs was dependent upon the engagement of a combination of NK cell receptors, given that only a cocktail of neutralizing antibodies (and not single antibodies) to NKG2D, NKp30, and DNAM could completely inhibit MSCs lysis. Remarkably, MSCs preexposed to IFN-γ, which had upregulated MHC class I expression, were significantly less susceptible to NK cell-mediated lysis and failed to induce IFN-γ release by NK cells. These observations suggest that IFN-γ is a potent modulator of MSC/NK cell crosstalk.

2.5
MSCs Inhibit Professional Antigen-Presenting Cells

DCs are considered the most potent antigen-presenting cells (APCs), as they are able to uptake, process, and present antigens and to potently activate naïve T cells. Jiang et al. (2005) investigated the effect of allogeneic human MSCs on the generation and function of monocyte-derived DCs. DCs can be generated from blood monocytes cultured in the presence of granulocyte-macrophage colony-stimulating factor (GM-CSF) and IL-4. In a first set of experiments, they demonstrated that MSCs cocultured with blood monocytes inhibited the generation of differentiated DCs in a contact-independent manner. This suppressive effect was maintained even when "strengthen maturation stimuli" were used, i.e., doubled concentration of GM-CSF, IL-4, and lipopolysaccharide (LPS). Interestingly, fibroblasts have also been shown to skew the differentiation of monocytes to macrophages in a process that could be reversed by adding tumor necrosis factor (TNF)-α. However, adding TNF-α to MSC/monocyte cocultures did not induce the generation of DCs. In another set of experiments, the authors investigated whether allogeneic human MSCs could inhibit the function of LPS-matured DCs. Coculture of MSCs with mature DCs caused a significant decreased expression of MHC class II and costimulatory molecules CD80 and CD86, implying that MSCs could reverse mature DCs into an immature phenotype. Importantly, these observations are in agreement with another study (Zhang et al. 2004). Also in agreement with these studies, Aggarwal and Pittenger (2005) demonstrated that allogeneic human MSCs/DC cocultures produced significantly less TNF-α in response to LPS than cultures with DCs alone (50% inhibition). Using type 2 plasmacytoid DCs, Aggarwal et al. (2005) also demonstrated that MSCs/type 2 DC cocultures produced significantly greater levels of IL-10 than cultures with DCs alone (140% increase). Taken together, these studies suggest that allogeneic MSCs can induce an immunologic tolerance state by inhibiting the generation and function of DCs.

2.6
MSCs Suppress Immune Responses In Vivo

Despite extensive descriptions of the in vitro immunosuppressive effects of MSCs, only a few studies have demonstrated MSCs' ability to suppress clinically relevant immune responses following in vivo transplantation. As exemplified by the case study reported by Le Blanc et al. (2004b), MSCs may possess the intrinsic ability to reverse self-destructing immune responses such as GVHD. In order to investigate the in vivo effects of MSCs, Nauta et al. (2006) recently assessed whether ex vivo MSCs could enhance the engraftment of allogeneic bone marrow transplantation in sublethally irradiated mice. They observed that coadministration of syngeneic MSCs—with respect to the recipient—significantly enhanced long-term engraftment of allogeneic bone marrow cells. The effect of syngeneic MSCs was significantly more pronounced in preventing rejection through minor histocompatibility antigens compared to rejection through major MHC antigens. Syngeneic MSCs appeared to have induced tolerance against host and donor antigens, as shown by the absence of proliferation of recipient splenocytes during mixed lymphocytes cultures. Surprisingly, however, the study demonstrated that MSCs need to be syngeneic to the host, i.e., MHC-matched to the recipient, in order to exert their graft-facilitating effect. Indeed, cotransplantation of allogeneic MSCs with allogeneic bone marrow cells not only failed to prevent graft rejection, but increased it. In support of our previous findings (Eliopoulos et al. 2005), they demonstrated that a single injection of allogeneic MSCs is sufficient to induce a memory T cell response able to eliminate subsequently injected allogeneic MSCs. Taken together, these observations strongly suggest that, at least in the mouse system, cotransplantation of recipient MSCs can prevent allogeneic graft rejection and that MSCs are not intrinsically immunoprivileged.

In another study, Zappia et al. (2005) evaluated the capacity of MSCs to prevent another severe autoimmune disease, namely experimental autoimmune encephalomyelitis (EAE). EAE is an autoimmune disease of the central nervous system mediated by T cells and macrophages that constitutes a model for multiple sclerosis. EAE is induced in mice by the administration of myelin oligodendrocyte glycoprotein (MOG) 35–55. In that study, ex vivo expanded MSCs (10^6 cells) were injected intravenously into syngeneic immunocompetent mice at day 3 and 8 after immunization with MOG 35–55. Remarkably, administration of syngeneic MSCs significantly reduced disease severity without affecting disease onset. Furthermore, the injection of MSCs was associated with a reduction of demyelination in the brain and in the spinal cord, and was associated with a decrease in T cell and macrophage infiltration. When splenocytes were analyzed for proliferative response, T cells from mice injected with the syngeneic MSCs were significantly unresponsive to anti-CD3/anti-CD28 antibodies or concanavalin A, suggesting induction of T cell anergy in agreement with previous in vitro findings.

2.7
Regulating MSC-Mediated Immunomodulation

Ex vivo expanded MSCs, in response to environmental cues, can thus inhibit multiple aspects of an immune response and make use of different pathways in order to achieve this. Recent studies suggest that the nature of the stimulus that activates MSC-mediated immune suppression influences the type of suppressive pathway that is used by MSCs. It was shown that MSCs rely mainly on the release of prostaglandins to inhibit lymphocyte proliferation induced by mitogens but not to inhibit lymphocyte proliferation induced by alloantigens (Rasmusson et al. 2005). The nature of the stimulus that induces MSC-mediated immune suppression thus influences the type of suppressive pathway that is activated by MSCs. As previously mentioned, an important biological property of MSCs is their ability to secrete a broad range of growth factors and cytokines as a consequence of their supportive role during hematopoiesis. Most importantly, MSCs can respond to their microenvironment by altering the nature of the growth factors they release. It is therefore not surprising that cytokines play a crucial role in regulating MSC-mediated immunosuppression. Aggarwal and Pittenger (2005) recently showed that TNF-α and IFN-γ can enhance by as much as 100-fold the production of immunosuppressive prostaglandins by MSCs. In apparent contradiction, Djouad et al. (2005) reported that addition of TNF-α is sufficient to reverse the immunosuppressive effect of MSCs on T cell proliferation. This apparent paradox perhaps reflects the sensitivity of MSCs for varying doses of TNF-α and/or the divergent effects of TNF-α on MSC-mediated immunosuppression induced by different stimuli.

An important cytokine that regulates MSC-mediated immunomodulation is IFN-γ. Compelling studies have now shown that IFN-γ plays an active role in the immunosuppression mediated by MSCs. As reported by Aggarwal and Pittenger (2005), IFN-γ directly induces MSCs to release prostaglandins. In agreement with this, Krampera et al. (2006) demonstrated that IFN-γ, secreted by NK cells and T cells when in the presence of MSCs, is required for the antiproliferative effect of MSCs. Although counterintuitive due the immune-activating nature of IFN-γ, these observations are supported by independent studies demonstrating that treatment with high doses of IFN-γ renders human MSCs more immunosuppressive against allogeneic PBMCs (Chan et al. 2006). The suppressive effect of IFN-γ is partly dependent on its ability to stimulate IDO production from MSCs. IDO causes depletion of tryptophan, an essential factor for lymphocyte proliferation. In addition, high levels of IFN-γ (100 U/ml) was shown to significantly decrease the expression of MHC class II molecules on the surface of human MSCs (Chan et al. 2006). At low doses (10 U/ml), however, IFN-γ upregulates MHC class I and II expression on both mouse and human MSCs (Chan et al. 2006; Stagg et al. 2006).

2.8
MSCs Act as APCs

Several studies have shown that irradiated human MSCs are unable to induce proliferation of allogeneic PBMCs in mixed lymphocytes cultures, suggesting that allogeneic MSCs are poor APCs. One study, however, suggested that MSCs may behave differently in autologous conditions. Beyth et al. (2005) reported that human MSCs cocultured with autologous purified $CD4^+$ T cells in the presence of a superantigen could induce the proliferation of the $CD4^+$ T cells. Despite the fact that antigen processing and presentation was not required in these experiments in order to induce T cell proliferation, it suggested that human MSCs are able to provide sufficient T cell activation signals.

In order to assess the possibility that ex vivo expanded MSCs might behave as APCs in autologous conditions, we recently investigated the effect of purified MSCs on the syngeneic activation, in vitro and in vivo, of antigen-specific immune responses (Stagg et al. 2006). In a first set of experiments, we tested whether IFN-γ could modulate the syngeneic immune properties of MSCs. We demonstrated that IFN-γ induced mouse MSCs to: (1) process a soluble exogenous protein, (2) present antigenic peptides on MHC class II molecules,

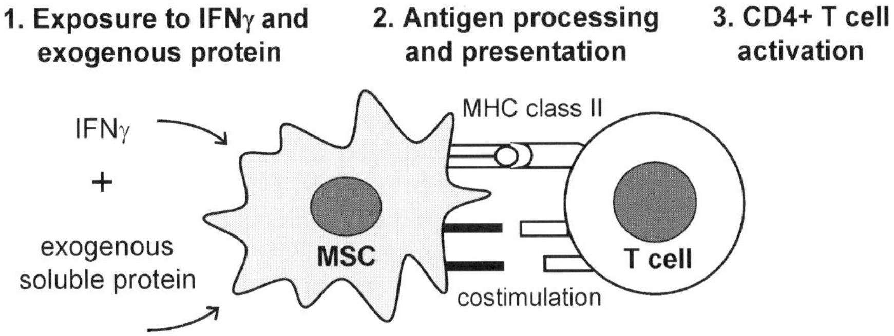

1. Exposure to IFNγ and exogenous protein **2. Antigen processing and presentation** **3. CD4+ T cell activation**

Fig. 2 MSCs act as antigen-presenting cells in vitro. Primary mouse and human MSCs, when stimulated in vitro with IFN-γ, can process soluble proteins, present antigenic peptides on MHC class II molecules and activate antigen-specific T cells. (Stagg et al. 2006)

and (3) activate in vitro antigen-specific primary naïve T cells (Fig. 2). Notably, MSCs that were not treated with IFN-γ failed to induce T cell activation. MSCs isolated from different preparations, as well as distinct clonal MSCs, were equally effective at behaving as APCs when exposed to IFN-γ. Our data showed that the antigen-presenting property of MSCs was endocytosis-dependent. However, MSCs may also use the phagocytic pathway to present exogenous antigens, as studies revealed that MSCs can phagocytose large particles (Chan et al. 2006; Hill et al. 2003).

2.9
MSCs Induce Protective Immunity In Vivo

In our study (Stagg et al. 2006), we also investigated whether antigen-pulsed IFN-γ-treated MSCs could induce antigen-specific immune responses in vivo. MSCs or control fibroblasts derived from C57BL/6 mice were stimulated with recombinant IFN-γ in the presence of soluble ovalbumin and injected intraperitoneally into syngeneic mice. Two weeks later, mice were injected a second time with the corresponding cells, and one week after that the mice were challenged with an injection of tumor cells that expressed ovalbumin (i.e., EG.7 cells). Remarkably, in vivo injection of MSCs induced a complete protection against ovalbumin-expressing tumors. Using restimulated splenocytes, we demonstrated that antigen-pulsed IFN-γ-treated MSCs efficiently induced the generation of antigen-specific $CD8^+$ cytotoxic T cells (CTL) in vivo. The fact that MSCs can induce a strong $CD8^+$ CTL response despite being unable to perform antigen cross-presentation efficiently in vitro suggests a role for host APCs in the generation of CTL. Indeed, host APCs are known to internalize and present exogenous antigens acquired from other cell types (a phenomenon known as cross-priming). Effective cross-priming of CTL and subsequent secondary expansion of CTL upon antigen reencounter are dependent upon proper activation of CD4 helper T cells. We thus believe that $CD4^+$ T cell activation by MSCs can enhance host-derived CTL cross-priming, resulting in the generation of a strong antigen-specific protective immunity.

We finally assessed whether human MSCs also acquire antigen-presenting functions following IFN-γ stimulation. Human MSCs were isolated from a healthy HLA class II DR1$^+$ donor and used in antigen presentation assays. When DR1$^+$ IFN-γ-treated human MSCs were cocultured for 24 h with antigen-specific DR1-restricted T cells in the presence of the antigen, significant levels of activation was detected. Our results thus indicated that IFN-γ-treated human MSCs can efficiently process exogenous antigens and present antigen-derived peptides to MHC class II-restricted T cells. Taken together, our results strongly suggest that in syngeneic conditions, ex vivo-expanded MSCs behave as conditional antigen-presenting cells able to activate antigen-specific immune responses. Our data that MSCs can behave as APCs have recently been supported by Chan et al. (2006) who demonstrated that human MSCs pulsed with two different recall antigens, i.e., *Candida Albicans* and *Tetanus toxoid*, efficiently induced secondary proliferation of purified antigen-specific autologous CD4$^+$ T cells.

2.10
Nonhematopoietic APCs

It is our contention that MSCs constitute a novel subset of nonhematological APCs. Few other cell types have been described to possess similar functions in the presence of pro-inflammatory stimuli. The best described are vascular endothelial cells, which have been shown to activate in vivo CD8$^+$ T cells in a CD80-dependent fashion upon IFN-γ stimulation (Kreisel et al. 2002; Pober et al. 2001). Interestingly, IFN-γ-treated endothelial cells can inhibit T cell activation through B7-H1 expression (Rodig et al. 2003). Similarly, we have observed significant upregulation of B7-H1 expression on the surface of mouse and human MSCs upon IFN-γ stimulation. B7-H1 (PD-L1) belongs to the B7 family members and is a ligand for programmed cell death-1 (PD-1) receptor expressed on activated T, B, and myeloid cells. *B7-H1$^{-/-}$* mice suggest an essential role for B7-H1 in negatively regulating T cell activation (Latchman et al. 2004). Augello et al. (2005) reported that blocking the PD-1/B7-H1 pathway could restore 50% of T cell proliferation in response to phytohaemagglutinin (PHA) when cocultured with irradiated MSCs. These observations suggest a negatively regulating role for B7-H1 expression on MSCs.

Our studies have shown that costimulatory molecules distinct of CD86 and CD80 might play a role in MSC-mediated antigen presentation (Stagg et al. 2006). Laouar et al. (2005) identified a unique population of APCs of nonhematopoietic origin that are found in the lamina propria of the gut and depend upon CD70 for antigen presentation. CD70 was thus identified as a new costimulatory molecule essential for antigen-presentation for these APCs. Whether CD70 represents an alternative costimulatory molecule on

MSCs is currently unknown. Of particular interest is the fact that the lamina propria nonhematopoietic APCs described by Laouar et al. (2005) share many common surface antigens with MSCs. An important aspect of the biology of MSCs that will need further investigation is whether MSC-mediated antigen presentation actually occurs in bone marrow and, if so, whether it plays a significant role during endogenous immune responses. As bone marrow is being revealed as a unique lymphoid organ able to activate naïve T cells and to induce systemic immunity (Mazo et al. 2005), MSCs may represent a previously unrecognized player of physiological immune responses.

3
Clinical Perspective: MSC-Based Cell Therapy

The plasticity, immunomodulatory properties, and migratory properties of MSCs make them ideal candidates for regenerative medicine and tissue engineering. Clinical studies based on the transplantation of MSCs have already generated promising results for the treatment of several illnesses, including bone, cardiovascular, and brain diseases.

One of the first diseases for which transplantation of MSCs has shown therapeutic benefits is osteogenesis imperfecta (OI). OI is a genetic disorder in which osteoblasts produce defective type I collagen, thereby inducing numerous fractures. Horwitz et al. (1999) were the first to demonstrate that allogeneic bone marrow transplantation can significantly improve the condition of OI patients via engraftment of mesenchymal progenitor cells. Horwitz et al. (2002) further demonstrated that infusion of purified allogeneic MSCs enhanced the clinical benefits of allogeneic marrow transplantation for the treatment of OI. Remarkably, the study provided evidence that transplanted allogeneic MSCs could engraft in the bone and differentiate into osteoblasts without the requirement for preparative chemotherapy.

The plasticity of MSCs has also been exploited in the context of tendon regeneration. Currently, therapeutic options to repair torn ligaments consist of autografts, allografts, and synthetic prostheses, none of which provide long-term therapeutic benefits. A recent study revealed the important role for Smad8 in the differentiation of MSCs into tendon (Hoffmann et al. 2006) and demonstrated that gene-modifying MSCs coexpressing active Smad8 and BMP2 can be exploited in order to promote tendon repair.

Another field of interest in regenerative medicine is the development of MSC-based therapy for the treatment of cardiovascular diseases. Since adult cardiomyocytes have limited regenerative capacity, implantation of progenitor cells with cardiac plasticity has been suggested for the regeneration of damaged cardiac cells after myocardial infarction (Anversa and Nadal-Ginard 2002). Studies have suggested that MSCs have the ability to home to the infarcted myocardium (Saito et al. 2002; Forrester et al. 2003). The factors responsible for inducing MSC migration, however, have yet to be identified. It has been proposed that trophic factors secreted by MSCs are mainly responsible for the therapeutic effects of MSCs after cardiac infarction. In a recent issue of *Nature Medicine*, Victor Dzau's group provided evidence in support of this hypothesis (Gnecchi et al. 2005). They demonstrated that supernatant from hypoxia-exposed MSCs (and to a greater extent Akt gene-modified MSCs) can significantly prevent apoptosis of cardiomyocytes after myocardial infarction in rats.

Since ex vivo expanded MSCs have been reported to cross the blood–brain barrier and to migrate preferentially to an ischemic cortex, their therapeutic potential in the treatment of stroke has also been investigated. Once in the ischemic cortex, MSCs have been shown to differentiate into microglia and astroglia (Eglitis et al. 1999). Li et al. (2002) reported that intravenous injection of human MSCs 1 day after stroke can improve functional outcome in rats. They observed, however, that less than 2% of injected MSCs express neuronal differentiation markers, leading them to hypothesize that trophic factors secreted by MSCs, and not differentiation of MSCs, may be responsible for the observed therapeutic benefits. Another explanation is that the interaction of MSCs with the host brain may lead MSCs or parenchymal cells to produce abundant growth factors such as brain-derived neurotrophic factor (BDNF) and nerve growth factor (NGF), both of which have been detected in MSC-injected regions.

4
Spontaneous Transformation of Human MSCs

It is important to point out that a study published by Rubio et al. (2005) reported that human adult MSCs derived from adipose tissue can spontaneously transform after long-term in vitro culture. This study was the first report of spontaneous transformation of adult human stem cells. The authors demonstrated that virtually all ex vivo expanded human MSCs have the ability to bypass cellular senescence. Surprisingly, 30% of human MSCs preparations that bypass senescence display trisomy of chromosome 8. Normally, when cells bypass senescence, they continue to grow until telomeres become too short and then enter a crisis phase, characterized by chromosome instability followed by cell death. Remarkably, it was observed that 50% of human MSC preparations can bypass this crisis phase if maintained 4–5 months in culture. When these long-term expanded MSCs were injected into immunodeficient mice, all mice developed tumors, suggesting that the MSCs had spontaneously transformed. Karyotype analysis of the transformed MSCs revealed nonrandom chromosome rearrangements. Notably, telomerase activity was detected in all transformed MSCs samples. This study highlights the importance of better defining the biology of MSCs in order to establish safe criteria for their use in clinical settings.

5
Conclusion

As we have discussed, ex vivo-expanded MSCs isolated from the bone marrow constitute an important source of adult stem cells that display multifaceted immunomodulatory properties. On one hand, MSCs inhibit the proliferation of T cells, induce T cell anergy, inhibit the proliferation and migration of B cells, inhibit the proliferation of naïve NK cells, and negatively modulate the differentiation and function of antigen-presenting DCs. On the other hand, MSCs can acquire antigen-presenting functions when exposed to IFN-γ and thereby induce antigen-specific protective immunity when transplanted into syngeneic hosts. Whereas antigen-presenting functions of MSCs crucially depends on IFN-γ stimulation, their immunosuppressive properties rely on a broad range of mediators, including secreted soluble factors such as PGE2, HGF, TGF-β1, IDO, and IL-10, as well as contact-dependent mechanisms such as B7-H1/PD-1 signaling. In response to specific stimulation, MSCs can thus inhibit or induce immune responses and make use of different pathways in order to achieve this. In this regard, cytokines such as TNFα and IFN-γ have been described to significantly influence the nature of the immunomodulation induced by MSCs. The duality of action of MSCs is reminiscent of the duality of professional APCs, which may selectively adopt a tolerogenic or stimulatory phenotype, depending upon microenvironmental context (Geijtenbeek et al. 2004). It will be of great interest to further investigate the potential role of other membrane-bound and secreted proteins such as Toll-like receptors, costimulatory molecules, cytokines, and chemokines in the immunomodulation mediated by MSCs. While there has been extensive description of the in vitro immunosuppressive effects of MSCs, only a few studies have investigated the ability of MSCs to suppress immune responses in vivo. One study demonstrated that syngeneic MSCs (to the host) enhanced the engraftment of allogeneic bone marrow, while another study demonstrated that syngeneic MSCs significantly reduced the severity of EAE. More in-depth preclinical investigation of the in vivo effect of transplanted MSCs is thus required. While doing so, we must keep in mind that numerous disparities exist between mouse and human MSCs. For instance, mouse and human MSCs possess phenotypically distinct features. As an example, CD80 can be found constitutively expressed on mouse MSCs whereas it is never expressed on human MSCs. Following in vivo transplantation, mouse and human MSCs also appear to induce distinctive effects: whereas transplanted human MSCs have been reported to suppress allogeneic immune responses (Le Blanc et al. 2004a), mouse MSCs have been reported not only to fail to suppress allogeneic immune responses, but to trigger them (Eliopoulos et al. 2005; Nauta et al. 2006). In conclusion, the immunomodulating properties of ex vivo-expanded MSCs are far more complex than first thought and merit greater investigation into their role in physiological immune responses in health and disease.

References

Aggarwal S, Pittenger MF (2005) Human mesenchymal stem cells modulate allogeneic immune cell responses. Blood 105:1815–1822

Al-Khaldi A, Eliopoulos N, Martineau D, et al (2003) Postnatal bone marrow stromal cells elicit a potent VEGF-dependent neo-angiogenic response in vivo. Gene Ther 10:621–629

Anversa P, Nadal-Ginard B (2002) Myocyte renewal and ventricular remodeling. Nature 415:240–243

Augello A, Tasso R, Negrini SM, Amateis A, Indiveri F, Cancedda R, Pennesi G (2005) Bone marrow mesenchymal progenitor cells inhibit lymphocyte proliferation by activation of the programmed death 1 pathway. Eur J Immunol 35:1482–1490

Barda-Saad M, Rozenszajn LA, Globerson A, Zhang AS, Zipori D (1996) Selective adhesion of immature thymocytes to bone marrow stromal cells: relevance to T cell lymphopoiesis. Exp Hematol 24:386–391

Beyth S, Borovsky Z, Mevorach D, Liebergall M, Gazit Z, Aslan H, Galun E, Rachmilewitz J (2005) Human mesenchymal stem cells alter antigen-presenting cell maturation and induce T-cell unresponsiveness. Blood 105:2214–2219

Chan JL, Tang KC, Patel AP, Bonilla LM, Pierobon N, Ponzio NM, Rameshwar P (2006) Antigen-presenting property of mesenchymal stem cells occurs during a narrow window at low levels of interferon-gamma. Blood 107:4817–4824

Clark BR, Keating A (1995) Biology of bone marrow stroma. Ann N Y Acad Sci 770:70–78

Colter DC, Class R, DiGirolamo CM, Prockop DJ (2000) Rapid expansion of recycling stem cells in cultures of plastic-adherent cells from human bone marrow. Proc Natl Acad Sci U S A 97:3213–3218

Colter DC, Sekiya I, Prockop DJ (2001) Identification of a subpopulation of rapidly self-renewing and multipotential adult stem cells in colonies of human marrow stromal cells. Proc Natl Acad Sci U S A 98:7841–7845

Corcione A, Benvenuto F, Ferretti E, Giunti D, Cappiello V, Cazzanti F, Risso M, Gualandi F, Mancardi GL, Pistoia V, Uccelli A (2006) Human mesenchymal stem cells modulate B-cell functions. Blood 107:367–372

Deans RJ, Moseley AB (2000) Mesenchymal stem cells: biology and potential clinical uses. Exp Hematol 28:875–884

Di Nicola M, Carlo-Stella C, Magni M, Milanesi M, Longoni PD, Matteucci P, Grisanti S, Gianni AM (2002) Human bone marrow stromal cells suppress T-lymphocyte proliferation induced by cellular or nonspecific mitogenic stimuli. Blood 99:3838–3843

Digirolamo CM, Stokes D, Colter D, Phinney DG, Class R, Prockop DJ (1999) Propagation and senescence of human marrow stromal cells in culture: a simple colony-forming assay identifies samples with the greatest potential to propagate and differentiate. Br J Haematol 107:275–281

Djouad F, Plence P, Bony C, Tropel P, Apparailly F, Sany J, Noel D, Jorgensen C (2003) Immunosuppressive effect of mesenchymal stem cells favors tumor growth in allogeneic animals. Blood 102:3837–3844

Djouad F, Fritz V, Apparailly F, Louis-Plence P, Bony C, Sany J, Jorgensen C, Noel D (2005) Reversal of the immunosuppressive properties of mesenchymal stem cells by tumor necrosis factor alpha in collagen-induced arthritis. Arthritis Rheum 52:1595–1603

Eglitis MA, Dawson D, Park KW, Mouradian MM (1999) Targeting of marrow-derived astrocytes to the ischemic brain. Neuroreport 10:1289–1292

Eliopoulos N, Stagg J, Lejeune L, Pommey S, Galipeau J (2005) Allogeneic marrow stromal cells are immune rejected by MHC class I- and class II-mismatched recipient mice. Blood 106:4057–4065

Forrester JS, Price MJ, Makkar RR (2003) Stem cell repair of infarcted myocardium: an overview for clinicians. Circulation 108:1139–1145

Friedenstein AJ, Gorskaja JF, Kulagina NN (1976) Fibroblast precursors in normal and irradiated mouse hematopoietic organs. Exp Hematol 4:267–274

Geijtenbeek TB, van Vliet SJ, Engering A, 't Hart BA, van Kooyk Y (2004) Self- and nonself-recognition by C-type lectins on dendritic cells. Annu Rev Immunol 22:33–54

Glennie S, Soeiro I, Dyson PJ, Lam EW, Dazzi F (2005) Bone marrow mesenchymal stem cells induce division arrest anergy of activated T cells. Blood 105:2821–2827

Gnecchi M, He H, Liang OD, Melo LG, Morello F, Mu H, Noiseux N, Zhang L, Pratt RE, Ingwall JS, Dzau VJ (2005) Paracrine action accounts for marked protection of ischemic heart by Akt-modified mesenchymal stem cells. Nat Med 11:367–368

Gotherstrom C, Ringden O, Tammik C, Zetterberg E, Westgren M, Le Blanc K (2004) Immunologic properties of human fetal mesenchymal stem cells. Am J Obstet Gynecol 190:239–245

Gregory CA, Perry AS, Reyes E, Conley A, Gunn WG, Prockop DJ (2005) Dkk-1-derived synthetic peptides and lithium chloride for the control and recovery of adult stem cells from bone marrow. J Biol Chem 280:2309–2323

Hill JM, Dick AJ, Raman VK, et al (2003) Serial cardiac magnetic resonance imaging of injected mesenchymal stem cells. Circulation 108:1009–1014

Hoffmann A, Pelled G, Turgeman G, Eberle P, Zilberman Y, Shinar H, Keinan-Adamsky K, Winkel A, Shahab S, Navon G, Gross G, Gazit D (2006) Neotendon formation induced by manipulation of the Smad8 signalling pathway in mesenchymal stem cells. J Clin Invest 116:940–952

Honczarenko M, Le Y, Swierkowski M, Ghiran I, Glodek AM, Silberstein LE (2006) Human bone marrow stromal cells express a distinct set of biologically functional chemokine receptors. Stem Cells 24:1030–1041

Hong JH, Hwang ES, McManus MT, Amsterdam A, Tian Y, Kalmukova R, Mueller E, Benjamin T, Spiegelman BM, Sharp PA, Hopkins N, Yaffe MB (2005) TAZ, a transcriptional modulator of mesenchymal stem cell differentiation. Science 309:1074–1078

Horwitz EM, Prockop DJ, Fitzpatrick LA, Koo WW, Gordon PL, Neel M, Sussman M, Orchard P, Marx JC, Pyeritz RE, Brenner MK (1999) Transplantability and therapeutic effects of bone marrow-derived mesenchymal cells in children with osteogenesis imperfecta. Nat Med 5:309–313

Horwitz EM, Gordon PL, Koo WK, Marx JC, Neel MD, McNall RY, Muul L, Hofmann T (2002) Isolated allogeneic bone marrow-derived mesenchymal cells engraft and stimulate growth in children with osteogenesis imperfecta: Implications for cell therapy of bone. Proc Natl Acad Sci U S A 99:8932–8937

Horwitz EM, Le Blanc K, Dominici M, Mueller I, Slaper-Cortenbach I, Marini FC, Deans RJ, Krause DS, et al (2005) Clarification of the nomenclature for MSC: The International Society for Cellular Therapy position statement. Cytotherapy 7:393–395

Jiang XX, Zhang Y, Liu B, Zhang SX, Wu Y, Yu XD, Mao N (2005) Human mesenchymal stem cells inhibit differentiation and function of monocyte-derived dendritic cells. Blood 105:4120–4126

Khakoo AY, Pati S, Anderson SA, Reid W, Elshal MF, Rovira II, Nguyen AT, Malide D, Combs CA, Hall G, Zhang J, Raffeld M, Rogers TB, Stetler-Stevenson W, Frank JA, Reitz M, Finkel T (2006) Human mesenchymal stem cells exert potent antitumorigenic effects in a model of Kaposi's sarcoma. J Exp Med 203:1235–1247

Krampera M, Glennie S, Dyson J, Scott D, Laylor R, Simpson E, Dazzi F (2003) Bone marrow mesenchymal stem cells inhibit the response of naive and memory antigen-specific T cells to their cognate peptide. Blood 101:3722–3729

Krampera M, Cosmi L, Angeli R, Pasini A, Liotta F, Andreini A, Santarlasci V, Mazzinghi B, Pizzolo G, Vinante F, Romagnani P, Maggi E, Romagnani S, Annunziato F (2006) Role for interferon-gamma in the immunomodulatory activity of human bone marrow mesenchymal stem cells. Stem Cells 24:386–398

Kreisel D, Krupnick AS, Balsara KR, Riha M, Gelman AE, Popma SH, Szeto WY, Turka LA, Rosengard BR (2002) Mouse vascular endothelium activates $CD8^+$ T lymphocytes in a B7-dependent fashion. J Immunol 169:6154–6161

Laouar A, Harias V, Vargas D, Zhinan X, Chaplin D, van Lier RA, Manjunath N (2005) $CD70^+$ antigen-presenting cells control the proliferation and differentiation of T cells in the intestinal mucosa. Nat Immunol 6:698–706

Latchman YE, Liang SC, Wu Y, Chernova T, Sobel RA, Klemm M, Kuchroo VK, Freeman GJ, Sharpe AH (2004) PD-L1-deficient mice show that PD-L1 on T cells, antigen-presenting cells, and host tissues negatively regulates T cells. Proc Natl Acad Sci U S A 101:10691–10696

Le Blanc K, Tammik C, Rosendahl K, Zetterberg E, Ringden O (2003) HLA expression and immunologic properties of differentiated and undifferentiated mesenchymal stem cells. Exp Hematol 10:890–896

Le Blanc K, Rasmusson I, Gotherstrom C, Seidel C, Sundberg B, Sundin M, Rosendahl K, Tammik C, Ringden O (2004a) Mesenchymal stem cells inhibit the expression of CD25 (interleukin-2 receptor) and CD38 on phytohaemagglutinin-activated lymphocytes. Scand J Immunol 60:307–315

Le Blanc K, Rasmusson I, Sundberg B, Gotherstrom C, Hassan M, Uzunel M, Ringden O (2004b) Treatment of severe acute graft-versus-host disease with third party haploidentical mesenchymal stem cells. Lancet 363:1439–1441

Li Y, Chen J, Chen XG, Wang L, Gautam SC, Xu YX, Katakowski M, Zhang LJ, Lu M, Janakiraman N, Chopp M (2002) Human marrow stromal cell therapy for stroke in rat: neurotrophins and functional recovery. Neurology 59:514–523

Massague J, Blain SW, Lo RS (2000) TGFbeta signaling in growth control, cancer, and heritable disorders. Cell 103:295–309

Mazo IB, Honczarenko M, Leung H, Cavanagh LL, Bonasio R, Weninger W, Engelke K, Xia L, McEver RP, Koni PA, Silberstein LE, von Andrian UH (2005) Bone marrow is a major reservoir and site of recruitment for central memory $CD8^+$ T cells. Immunity 22:259–270

Meisel R, Zibert A, Laryea M, Gobel U, Daubener W, Dilloo D (2004) Human bone marrow stromal cells inhibit allogeneic T-cell responses by indoleamine 2,3-dioxygenase-mediated tryptophan degradation. Blood 103:4619–4621

Milne CD, Fleming HE, Zhang Y, Paige CJ (2004) Mechanisms of selection mediated by interleukin-7, the preBCR, and hemokinin-1 during B-cell development. Immunol Rev 197:75–88

Mishra L, Derynck R, Mishra B (2005) Transforming growth factor-beta signaling in stem cells and cancer. Science 310:68–71

Nakamizo A, Marini F, Amano T, Khan A, Studeny M, Gumin J, Chen J, Hentschel S, Vecil G, Dembinski J, Andreeff M, Lang FF (2005) Human bone marrow-derived mesenchymal stem cells in the treatment of gliomas. Cancer Res 65:3307–3318. Erratum in: Cancer Res 2006 Jun 1 66:5975

Nakashima K, de Crombrugghe B (2003) Transcriptional mechanisms in osteoblast differentiation and bone formation. Trends Genet 19:458–466

Nauta AJ, Westerhuis G, Kruisselbrink AB, Lurvink EG, Willemze R, Fibbe WE (2006) Donor-derived mesenchymal stem cells are immunogenic in an allogeneic host and stimulate donor graft rejection in a non-myeloablative setting. Blood 108:2114–2120

Peister A, Mellad JA, Larson BL, Hall BM, Gibson LF, Prockop DJ (2004) Adult stem cells from bone marrow (MSCs) isolated from different strains of inbred mice vary in surface epitopes, rates of proliferation, and differentiation potential. Blood 103:1662–1668

Pittenger MF, Mackay AM, Beck SC, Jaiswal RK, Douglas R, Mosca JD, Moorman MA, Simonetti DW, Craig S, Marshak DR (1999) Multilineage potential of adult human mesenchymal stem cells. Science 284:143–147

Pober JS, Kluger MS, Schechner JS (2001) Human endothelial cell presentation of antigen and the homing of memory/effector T cells to skin. Ann N Y Acad Sci 941:12–25

Potian JA, Aviv H, Ponzio NM, Harrison JS, Rameshwar P (2003) Veto-like activity of mesenchymal stem cells: functional discrimination between cellular responses to alloantigens and recall antigens. J Immunol 171:3426–3434

Prockop DJ, Gregory CA, Spees JL (2003) One strategy for cell and gene therapy: harnessing the power of adult stem cells to repair tissues. Proc Natl Acad Sci U S A 100 [Suppl 1]:11917–11923

Rasmusson I, Ringden O, Sundberg B, Le Blanc K (2005) Mesenchymal stem cells inhibit lymphocyte proliferation by mitogens and alloantigens by different mechanisms. Exp Cell Res 305:33–41

Reddy P, Liu L, Ren C, Lindgren P, Boman K, Shen Y, Lundin E, Ottander U, Rytinki M, Liu K (2005) Formation of E-cadherin-mediated cell-cell adhesion activates AKT and mitogen activated protein kinase via phosphatidylinositol 3 kinase and ligand-independent activation of epidermal growth factor receptor in ovarian cancer cells. Mol Endocrinol 19:2564–2578

Reyes M, Lund T, Lenvik T, Aguiar D, Koodie L, Verfaillie CM (2001) Purification and ex vivo expansion of postnatal human marrow mesodermal progenitor cells. Blood 98:2615–2625

Rodig N, Ryan T, Allen JA, Pang H, Grabie N, Chernova T, Greenfield EA, Liang SC, Sharpe AH, Lichtman AH, Freeman GJ (2003) Endothelial expression of PD-L1 and PD-L2 down-regulates $CD8^+$ T cell activation and cytolysis. Eur J Immunol 33:3117–3126

Rosen ED, Walkey CJ, Puigserver P, Spiegelman BM (2000) Transcriptional regulation of adipogenesis. Genes Dev 14:1293–1307

Rubio D, Garcia-Castro J, Martin MC, de la Fuente R, Cigudosa JC, Lloyd AC, Bernad A (2005) Spontaneous human adult stem cell transformation. Cancer Res 65:3035–3039

Ruscetti FW, Akel S, Bartelmez SH (2005) Autocrine transforming growth factor-beta regulation of hematopoiesis: many outcomes that depend on the context. Oncogene 24:5751–5763

Saito T, Kuang JQ, Bittira B, Al-Khaldi A, Chiu RC (2002) Xenotransplant cardiac chimera: immune tolerance of adult stem cells. Ann Thorac Surg 74:19–24

Sekiya I, Larson BL, Smith JR, Pochampally R, Cui JG, Prockop DJ (2002) Expansion of human adult stem cells from bone marrow stroma: conditions that maximize the yields of early progenitors and evaluate their quality. Stem Cells 20:530–541

Spaggiari GM, Capobianco A, Becchetti S, Mingari MC, Moretta L (2006) Mesenchymal stem cell-natural killer cell interactions: evidence that activated NK cells are capable of killing MSCs, whereas MSCs can inhibit IL-2-induced NK-cell proliferation. Blood 107:1484–1490

Spees JL, Olson SD, Ylostalo J, Lynch PJ, Smith J, Perry A, Peister A, Wang MY, Prockop DJ (2003) Differentiation, cell fusion, and nuclear fusion during ex vivo repair of epithelium by human adult stem cells from bone marrow stroma. Proc Natl Acad Sci U S A 100:2397–2402

Stagg J, Pommey S, Eliopoulos N, Galipeau J (2006) Interferon-gamma-stimulated marrow stromal cells: a new type of nonhematopoietic antigen-presenting cell. Blood 107:2570–2577

Studeny M, Marini FC, Champlin RE, et al (2002) Bone marrow-derived mesenchymal stem cells as vehicles for interferon-beta delivery into tumors. Cancer Res. Stem Cells 62:3603–3608

Terada N, Hamazaki T, Oka M, Hoki M, Mastalerz DM, Nakano Y, Meyer EM, Morel L, Petersen BE, Scott EW (2002) Bone marrow cells adopt the phenotype of other cells by spontaneous cell fusion. Nature 416:542–545

Torlakovic E, Tenstad E, Funderud S, Rian E (2005) $CD10^+$ stromal cells form B-lymphocyte maturation niches in the human bone marrow. J Pathol 205:311–317

Tse WT, Pendleton JD, Beyer WM, Egalka MC, Guinan EC (2003) Suppression of allogeneic T-cell proliferation by human marrow stromal cells: implications in transplantation. Transplantation 75:389–397

Zappia E, Casazza S, Pedemonte E, Benvenuto F, Bonanni I, Gerdoni E, Giunti D, Ceravolo A, Cazzanti F, Frassoni F, Mancardi G, Uccelli A (2005) Mesenchymal stem cells ameliorate experimental autoimmune encephalomyelitis inducing T-cell anergy. Blood 106:1755–1761

Zhang W, Ge W, Li C, You S, Liao L, Han Q, Deng W, Zhao RC (2004) Effects of mesenchymal stem cells on differentiation, maturation, and function of human monocyte-derived dendritic cells. Stem Cells Dev 13:263–271

Bone Marrow-Derived Cells: The Influence of Aging and Cellular Senescence

C. Beauséjour

Department of Pharmacology, Université de Montréal & Centre Hospitalier Universitaire Sainte-Justine, 3175 Cote Ste-Catherine Road, Montreal QC, H3T 1C5, Canada
c.beausejour@umontreal.ca

1	Introduction to Cellular Senescence, Aging and Bone Marrow-Derived Cells	68
2	The Role of Telomeres and Telomerase in Cell Proliferation	69
3	Genetic and the Hematopoietic System	71
4	Influence of Aging on Bone Marrow Homeostasis	73
4.1	The Effects of Aging on Hematopoietic Repopulation Capacity	73
4.2	The Effects of Aging on Bone Marrow Stem and Progenitor Cells Numbers	74
4.3	Mobilization and Aging	75
5	Bone Marrow Cells and Oxidative Stress	76
5.1	Stress Imposed by Culture Conditions	76
5.2	Effects of Antioxidants on Bone Marrow Cell Proliferation	77
5.3	Cyclin-Dependent Kinase Inhibitors and Senescence Induction	78
5.4	Effects of Radiation/Chemotherapy on Bone Marrow Cell Senescence	80
6	Consequences of Aging on Bone Marrow-Derived Cellular Therapies	81
7	Conclusion	82
	References	83

Abstract During the course of an entire lifespan, tissue repair and regeneration is made possible by the presence of adult stem cells. Stem cell expansion, maintenance, and differentiation must be tightly controlled to assure longevity. Hematopoietic stem cells (HSC) are greatly solicited given the daily high blood cell turnover. Moreover, several bone marrow-derived cells including HSC, mesenchymal stromal cells (MSC), and endothelial progenitor cells (EPC) also significantly contribute to peripheral tissue repair and regeneration, including tumor formation. Therefore, factors influencing bone marrow-derived cell proliferation and functions are likely to have a broad impact. Aging has been identified as one of these factors. One hypothesis is that aging directly affects stem cells as a consequence of exhaustive proliferation. Alternatively, it is also possible that aging indirectly affects stem cells by acting on their microenvironment. Cellular senescence is believed to have evolved as a tumor suppressor mechanism capable of arresting growth to reduce risk of malignancy. In opposition to apoptosis, senescent cells accumulate in tissues. Recent evidence suggests their accumulation contributes to the phenotype of aging. Senescence can be activated by both telomere-dependent and telomere-independent pathways. Genetic alteration, genome-wide DNA damage, and oxidative stress are inducers of senescence and have recently been identified as occurring in bone marrow-derived cells. Below is a review of the link between cellular senescence, aging, and bone marrow-derived cells, and the possible consequences aging may have on bone marrow transplantation procedures and emerging marrow-derived cell-based therapies.

Keywords Senescence · Aging · Bone marrow-derived cells · Oxidative stress · Cyclin-dependent kinase inhibitor

1
Introduction to Cellular Senescence, Aging and Bone Marrow-Derived Cells

Stem cells can proliferate into specific lineages of differentiated cells while preserving self-renewing property. In opposition to embryonic pluripotent stem cells, the adult bone marrow represents an attractive and more ethical source for the purification of multipotent potential stem cells. Along with the marrow, the peripheral and cord blood contains a high number of stem and progenitor cells that can be obtained easily. In the bone marrow, there are hematopoietic stem cells (HSC) that can form all blood cell lineages and nonhematopoietic mesenchymal stromal cells (MSC) capable of differentiating in various connective tissues such as fat, bone, and cartilage (Pittenger et al. 1999). Both HSC and MSC comprise a very small fraction of marrow-nucleated cells, with an estimated frequency in mice of only one in $1-3\times10^4$ cells. Therefore, development of ex vivo expansion protocols is likely to be required to foster therapeutic usage of bone marrow-derived cells. Consequently, understanding the key molecular events that control bone marrow cell proliferation and differentiation is essential. Of all the factors thus far identified to have a role in stem cell proliferation, cellular aging is certainly an important one.

Cellular aging, also referred as cellular senescence, is arguably best defined as a cell's diminished replicative capacity and altered functionality. Tissue regeneration and repair involve intense cellular proliferation with a consequence of possibly developing hyperproliferative diseases such as cancer. Over time, organisms with renewable tissues evolved, and cancer became a threat to their gain in longevity. This danger was counteracted by the evolution of tumor-suppressor mechanisms (Campisi 2003, 2005). These mechanisms either eliminate potential cancer cells by apoptosis or cause them to withdraw from the cell cycle (cellular senescence). Cellular senescence was first properly described in 1961 by Hayflick as the process that limits the replicative lifespan of normal human cells in culture (Hayflick 1965). Senescent cells typically assume an enlarged (and flattened) morphology and stain positive for the senescence-associated β-galactosidase (SA-β-gal) (Dimri et al. 1995). Currently we know senescence can be induced by various stimuli and involves essentially the permanent arrest of cell proliferation into a phenotype that cannot be reversed by physiological mitogens (Beausejour et al. 2003). Early studies on cellular senescence suggested this process might contribute to organismal aging (Campisi 1997; Hayflick 1965; Smith and Pereira-Smith 1996). Hence, senescent growth arrest would, over time, limit the ability of renewable tissues to replace damaged or dysfunctional cells and reduce their capacity for repair (Smith and Pereira-Smith 1996; Stanulis-Praeger 1987).

Studies reporting the presence of bone marrow-derived cells at the site of regenerated tissues (Aghi and Chiocca 2005; Peters et al. 2005) generated great enthusiasm, although the exact phenotype of the involved bone marrow-derived cells remains largely debated and is likely to differ from one experimental setting to another. In fact, contrary to what was initially thought, bone marrow-derived cells are unlikely to differentiate into various epithelial cell types but rather play a role in the formation of the tissue vasculature and stromal support. Of great interest is the correlation between delayed tissue repair and accumulation of senescent cells with age. In opposition to apoptosis, where cells are rapidly eliminated, recent evidence suggests metabolically active senescent cells accumulate over time (Herbig et al. 2006). Thus, not only does cellular senescence directly limit tissue renewal, but accumulation of senescent cells could contribute to the process of aging through their senescent secretory phenotype (J.P. Coppé, C. Patil, F. Rodier, and J. Campisi, manuscript in preparation—Fig. 1). Indeed, senescent cells secrete biologically active molecules that can disrupt normal tissue microenvironments and affect the behavior of neighboring cells (Krtolica et al. 2001). This secretion phenotype, together with the accumulation of senescent cells in vivo, suggests that senescence may influence tissue homeostasis. Thus, cellular senescence could limit the replication of HSC and, through its secretion profile, affect the recruitment of bone marrow-derived cells involved in physiological processes. Overall, the following chapter will review our current understanding of the causes and consequences that the senescent and aging phenotype may have on bone marrow cells' expansion and contribution to tissue regeneration.

2
The Role of Telomeres and Telomerase in Cell Proliferation

With each cell division, telomeres (i.e., stretches of DNA that cap the end of chromosomes) get shorter due to incomplete DNA replication. After several divisions, replicative senescence occurs when cells acquire one or a few short dysfunctional telomeres (Karlseder et al. 2002; Martens et al. 2000; Wright and Shay 2001). Cells can elongate their telomeres by expressing an enzyme called telomerase. Telomerase is undetectable or not expressed in most human adult somatic tissues, its expression being restricted to germ cells and tissue stem cells (Broccoli et al. 1995; Wright et al. 1996). Usually, this particularity forces tumor cells to overcome the telomere replication limit by expressing telomerase. Mouse cells are more promiscuous than human cells in expressing telomerase (Prowse and Greider 1995; Weng and Hodes 2000). Telomeres of inbred mice can exceed a length of 60 kb in comparison to 5–15 kb in humans (Sharpless and DePinho 2004). Nonetheless, mouse and telomerase-expressing human cells can undergo senescence using telomere-dependent and telomere-independent mechanisms (Pearson et al. 2000; Wei and Sedivy 1999).

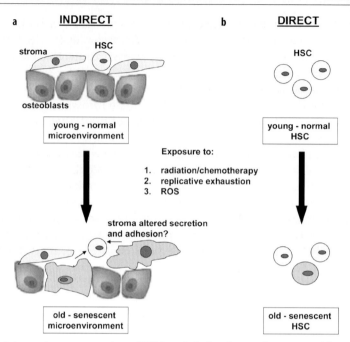

Fig. 1a, b Schematic representation of HSC and their microenvironment with age. **a** Impaired hematopoiesis following senescence induction of stromal and osteoblast cells in the hematopoietic stem cell niche. Hypothetically, senescent cells through their altered secretory phenotype may indirectly affect HSC proliferation and adhesion to the niche. **b** HSC senescence and impaired hematopoiesis following direct exposure to senescent stimuli

Telomerase is upregulated in response to cytokine-induced proliferation and cell cycle activation in human $CD34^+$ cells cultured in vitro. In these cells, an average telomere loss of 1 to 2 kb is observed after 4 weeks of culture (Engelhardt et al. 1997). Also, $CD34^+$ cells isolated from young tissues such as cord blood have significantly longer telomeres compared to peripheral or marrow $CD34^+$ cells isolated from older individuals. Hence, telomerase activity at best partially abrogates telomeres' shortening upon robust proliferation or normal aging.

While mouse cells express telomerase, human MSC seem to be devoid of its expression (Zimmermann et al. 2003). Consequently, human MSC can divide for up to 50 population doublings, but often much less, whereas murine MSC seem to be able to proliferate indefinitely (Meirelles Lda and Nardi 2003; Stenderup et al. 2003). In fact, human MSC loose the equivalent of half their total replicative lifespan after only a short in vitro expansion period, an event offset by the ectopic expression of telomerase (Baxter et al. 2004). Human HSC express telomerase (Morrison et al. 1996), and yet their telomere length was shown to decrease during serial transplantation (Notaro et al. 1997). Interestingly, the telomere length of peripheral blood granulocytes derived from engrafted bone marrow was significantly reduced and inversely correlated with the number of nucleated cells infused. This suggests telomerase activity is insufficient to maintain telomere length of human HSC and MSC subjected to intense proliferation.

Not surprisingly, HSC from telomerase-deficient mice showed reduced long-term repopulating capacity, concomitant with an increase in genetic instability in serial and competitive transplantation assays (Samper et al. 2002). In fact, telomerase-deficient HSC can be serially transplanted for two rounds, whereas wild-type HSC can support at least four rounds of serial transplantation (Allsopp et al. 2003a). As previously demonstrated, telomere shortening in wild-type HSC also occurs following serial transplantation despite robust telomerase expression in these cells (Allsopp et al. 2001). The use of transgenic mice overexpressing telomerase best clarifies the role of telomere length in hematopoiesis (Allsopp et al. 2003b). In these mice, telomere length from HSC remains stable during serial transplantation. However, maintenance of telomere length did not improve transplantation potential of transgenic cells. Thus, telomere-independent factors must have caused premature senescence of HSC. Taken together, these results suggest that both mouse and human bone marrow-derived cells have sufficient intrinsic renewable capacities to sustain a normal lifespan, but these can be rapidly saturated under forced proliferation.

3
Genetic and the Hematopoietic System

Compelling evidence suggests that HSC have a measurable and progressive replicative impairment with age. The effects of aging on HSC functions have been studied using a variety of assays, often resulting in contradictory conclusions. This controversy was alleviated by the observation that hematopoietic deficiencies are in part due to genetic differences. Indeed, BALB/c and DBA/2 mouse strains have an age-associated reduction in both repopulating ability and self-renewal of HSC, a phenotype not observed in C57BL/6 mice (Chen et al. 2000). Furthermore, HSC exhaustion is more severe for BALB/c than for C57BL/6 mice during development (Yuan et al. 2005).

Using a genetic approach, Geiger and colleagues identified a genetic locus on chromosome 2 with significant linkage to the difference in the HSC of inbred mice strains (Geiger et al. 2001). They recently confirmed the role of this locus and argued that stem cell aging is caused by the loss of DNA integrity or the accumulation of DNA damage (Geiger et al. 2005). C57BL/6 mice congenic for the distal part of chromosome 2 from DBA/2 mice (C57BL/6.D2) were much more sensitive to low-dose irradiation (1–2 Gy) compared to control C57BL/6 mice. The authors reasoned that stem cell aging could result from the loss of DNA integrity and accumulation of DNA damage. This hypothesis is of interest as it links impaired HSC function to what is possibly the best phenotypic marker of cellular senescence: permanent DNA damage foci (Narita et al. 2003). A more detailed analysis of the DNA repair capacity of cells derived from congenic C57BL/6.D2 needs to be completed to validate this hypothesis. Nonetheless, these findings are very interesting and help explain the observed discrepancy between the telomere/telomerase theory and mouse tissue renewable capabilities.

Supporting the idea that DNA damage directly affects HSC function, it was shown that marrow cells isolated from sublethally irradiated donor mice performed poorly compared to cells obtained from nonirradiated mice in a competitive repopulating assay. This observation was somewhat surprising given that the total stem cell number—in sublethally irradiated donor mice—recovered prior to the marrow purification (Harrison and Astle 1982). Thus, these results support the hypothesis that permanent DNA damage, possibly through activation of cyclin-dependent kinase inhibitors (CDKI—see Sect. 5.3), impairs stem cell-repopulating abilities by inducing senescence. Finally, some DNA-damaging chemotherapeutic drugs, such as cyclophosphamide, bis-chloronitrosourea, and vincristine, also cause long-term proliferative defects in HSC (Gardner et al. 1993; Ross et al. 1982).

In the early 1990s, Van Zant and colleagues generated chimeric mice from C57BL/6 and DBA/2 embryos and observed that within the course of a lifetime, DBA/2 HSC function declined to the point where all blood formation was C57BL/6 derived (Van Zant et al. 1990, 1992). Interestingly, bone marrow collected from old chimeras, in which no DBA/2 blood cells had been detected for months, could still form DBA/2-derived blood lineage cells after being transplanted into an irradiated host. Thus, in this model, the mechanism by which aging affects HSC is by the induction of quiescence rather than senescence. In contrast, the extent of bone marrow stromal chimerism was maintained continuously, suggesting the decline in DBA/2 blood cells was not dependent on similar changes in the stroma. It is possible DBA/2 cells have intrinsic differences making them more reactive than C57BL/6 cells to the aging stroma, independently of its chimerism.

There is currently not enough information to determine whether human HSC will act like HSC derived from inbred mouse strains upon aging. Given the genetic diversity of the human population, one could argue that aging is likely to generate a high degree of variation in the phenotype of human HSC. Still, when human HSC from fetal liver, umbilical cord blood, and adult marrow were compared in vitro, their differentiation capacity was inversely correlated with age (Lansdorp et al. 1993). Of interest, dietary restriction was found to somewhat alleviate the aging phenotype of BALB/c mice (Chen et al. 2003). In this particular experiment, 2 years of dietary restriction (up to 25% of the calorie intake) improved BALB/c-derived HSC functional abilities and total cell number to a level comparable to what is found in young mice. Reduced calorie intake is the only known effective method that regularly extends lifespan and counteracts the effects of aging (Bordone and Guarente 2005). Given that the dietary restriction must take place for almost the entire lifespan of the organism before a significant effect is observed, it is doubtful that a similar approach will ever be tested or useful in humans.

Finally, increased incidence of myeloid leukemia and the overall tendency of progenitor cells to differentiate along the myeloid lineage are significant consequences of aging on the hematopoietic system. Recently, microarray studies identified genetic cell-intrinsic alterations which potentially explain such age-dependent effects. Gene expression profiling of long-term HSC between young and old mice revealed a downregulation of genes mediating lymphoid differentiation (Rossi et al. 2005). Conversely, upregulation of genes mediating myeloid differentiation and leukemic transformation was also identified. It was also confirmed that age-associated myeloid differentiation of HSC is influenced by the aged microenvironment (Liang et al. 2005).

4
Influence of Aging on Bone Marrow Homeostasis

4.1
The Effects of Aging on Hematopoietic Repopulation Capacity

Marrow stem cells can be serially transplanted up to five times before their proliferative capacity is severely reduced (Siminovitch et al. 1964). This observation is surprising knowing that mice have long telomeres (albeit a few strains) and express telomerase (Sharpless and DePinho 2004). Furthermore, mice deficient in telomerase activity can be crossed up to five generations before a significant phenotype appears, suggesting that the replicative potential, at least in mice, should not be an issue during the course of a normal lifespan (Blasco et al. 1997). Nonetheless, hematopoietic proliferative capacity is substantially impaired after a single marrow transplant (Harrison et al. 1978). In fact, using a competitive repopulating assay, HSC isolated from transplanted mice were only 10%–20% as effective compared to HSC purified from non-transplanted animals (Harrison and Astle 1982). It is unclear why marrow stem cells, after a single transplant, lose their ability to compete with untransplanted competitors regardless of the age of the original donor (Harrison and Astle 1982). Parabiotic studies ruled out the possibility that artificial handling procedures impair stem cell function. Thus, the most probable explanations for these findings are: (1) an excessive differentiation of stem cells shortly after transplantation causes stem cell exhaustion; (2) lethal irradiation of the recipient causes permanent bone marrow microenvironment damage preventing complete stem cell regeneration and/or function. Indeed, the first hypothesis has been confirmed by the observation that previously transplanted marrow cells lead to hematopoietic reconstitution with reduced stem cell number and repopulating abilities (Harrison et al. 1990). The stem cell microenvironment integrity hypothesis is technically more difficult to evaluate. Still, it was shown that stem cells from old mice were slightly defective in producing phytohemagglutinin (PHA)-responsive cells in vitro, a defect attributed to the aged microenvironment (Harrison et al. 1978). Indeed, the deficiency in PHA response was not intrinsic to the old stem cells since these cells were fully capable of producing PHA when transplanted and then isolated from a young recipient (Harrison et al. 1978). Similar observations were made using mouse spermatogonial and muscular stem cells in which a decline in self-renewal was associated with age and reversed when stem cells were maintained in a young microenvironment (Conboy et al. 2005; Reyes et al. 2001). Finally, the seeding efficiency of young HSC in the bone marrow of old mice is only 33%–50% of that measured in young mice (Liang et al. 2005). This observation is the only quantitative demonstration of the influence of an old microenvironment on hematopoiesis.

4.2
The Effects of Aging on Bone Marrow Stem and Progenitor Cells Numbers

The effects of aging on HSC functions vary greatly depending on the mouse strain (see Chap. 3). Furthermore, many conflicting data also exist when looking at the effect of aging within the same strain. The repopulating activity of C57BL/6-derived bone marrow cells from aged mice is superior to that of cells from young mice (Harrison 1983; Harrison et al. 1989; Sudo et al. 2000). Conversely, HSC from old C57BL/6 mice were not as efficient in engrafting the bone marrow of irradiated mice (Morrison et al. 1996; Yilmaz et al. 2006). One reason that could explain these apparent contradictory capacities of young and old bone marrow cells may derive from the marrow cellular fraction used to perform the limiting dilution reconstitution assay. In fact, when using highly purified Lin^- $Sca-1^+$ $c-kit^+$ (hereafter refer as LSK) $Thy-1^{lo}$ cells from C57BL/6 mice, fourfold less LSK cells collected from young compared to old donors were necessary to show comparable engraftment and marrow repopulation levels (Morrison et al. 1996; Xing et al. 2006). Alternatively, the same assay performed using nonpurified whole bone marrow cells gave a higher level of marrow reconstitution using cells collected from an old donor (Morrison et al. 1996). One logical explanation for such discrepancies came from the observation that the number of $CD34^-$ LSK cells within the bone marrow is increased 17-fold after 18 months of life (Sudo et al. 2000). Thus, if one performs a repopulation assay using a defined number of nonpurified bone marrow cells, cells collected from an old donor are likely to perform better given the high number of LSK cells within the cell population. Conversely, when the same assay was performed using a defined number of LSK cells, young donor cells perform better, suggesting an age-associated impairment. In other words, the increased LSK cell fraction with age counteracts the age-associated decline observed at the single cell level.

Interestingly, the 17-fold increase in LSK cells with age does not match the increased repopulating potential (at most twofold higher) of bone marrow cells collected from old animals. This observation leads Sudo and colleagues to argue that this difference in the number of multilineage repopulating cells reflects the accumulation of myeloid lineage-committed precursor cells within the LSK cell fraction. Secondary transplant experiments confirmed that these putative myeloid-committed progenitor cells maintained their self-renewal potential while reducing their lymphoid lineage potential. These results correlate nicely with the previously observed overall tendency of progenitor cells to differentiate along the myeloid lineage over time (Morrison et al. 1996).

4.3
Mobilization and Aging

The propensity of young and old hematopoietic stem/progenitor cells to mobilize following granulocyte colony stimulating factor (G-CSF) injection was analyzed. Using mice hematopoietically reconstituted with an equal number of young and old bone marrow cells, Xing and colleagues found that progenitor cells that were derived from old donors mobilized fivefold better than cells from young donors (Xing et al. 2006). This experiment strongly suggests against an important role for the microenvironment in the mobilization process. However, we cannot rule out the possibility that the increased mobilization of old cells is the consequence of their previous interaction with an aged microenvironment. The authors also provided in vitro data suggesting a correlation between mobilization of aged HSC and their reduced adhesion (two-fold) to stromal cells. HSC from old mice are only one-quarter as efficient as HSC from young mice at homing and engrafting the bone marrow of irradiated recipients, again supporting the adhesion defect theory (Morrison et al. 1996). Collectively, the observations that old HSC have reduced adhesion and increased mobilization agree well with the original observations that HSC are at least five times as frequent in old mice (Harrison 1983; Harrison et al. 1989; Morrison et al. 1996).

Finally, a new family of receptors (signaling lymphocyte activation molecule or SLAM) able to distinguish hematopoietic stem and progenitor cells was recently identified (Kiel et al. 2005). The CD48 and CD150 SLAM receptors were differentially expressed between stem and progenitor cells. Selection for LSK CD48$^-$ CD150$^+$ significantly increased the purity of HSC as demonstrated by repopulation studies. Notably, the increased purity procured by the SLAM receptor selection was also observed using bone marrow-derived cells from old or reconstituted mice (Yilmaz et al. 2006). The total number of LSK cells increases with age while their repopulating abilities decrease. This suggests that pure HSC within this subpopulation get diluted in old mice. An estimated modest threefold increase in HSC frequency was also observed in the SLAM-purified cell subpopulation of aged mice, suggesting a commensurate reduction in their engraftment ability.

5
Bone Marrow Cells and Oxidative Stress

5.1
Stress Imposed by Culture Conditions

Cellular-based therapy is likely to require that cells be cultured ex vivo for a short period of time, either to expand their total number or to allow their genetic modification. In its present form, genetic engineering of bone marrow progenitor or stem cells is best done using replication-incompetent viruses requiring a minimum of 1 or 2 days of ex vivo culture (Piacibello et al. 2002). Serum-free culture conditions have been optimized through the use of purified recombinant proteins. Cells cultured in these conditions have so far been carried in humidified incubators and exposed to ambient oxygen concentrations of nearly 20%. On average, cells are confronted with a much lower oxygen tension, likely to be 3%–6% depending on their microenvironment (Atkuri et al. 2005).

HSC reside in the most hypoxic areas of the bone, possibly keeping them in the quiescent state (Cipolleschi et al. 1993). Given the relationship between high oxygen level and oxidative stress, researchers have explored the consequences of culturing hematopoietic progenitor cells at a low or even hypoxic oxygen concentration. Murine HSC better preserved their ability to repopulate the bone marrow of lethally irradiated mice when cultured at low oxygen (1.5%) compared to normoxic conditions (20%) (Cipolleschi et al. 1993). The effect of low oxygen tension on human hematopoietic cells was also analyzed and reported to increase the clonogenic potential of progenitor cells (Cipolleschi et al. 1997; Ivanovic et al. 2000). It is only recently that the benefit of low oxygen culture on human hematopoietic cells was confirmed at the true stem cell level, using the severe combined immunodeficiency (SCID) repopulating cells (SRC) assay (Danet et al. 2003). It was found that the proportion of SRC from human $CD34^+CD38^-$ cells cultured for 4 days in serum-free conditions under a 1.5% oxygen level increased more than fivefold compared to cells cultured at 20% oxygen (Danet et al. 2003). Interestingly, the number of SRC following hypoxic culture was still fourfold higher compared to freshly isolated $CD34^+CD38^-$ cells. Because SRC numbers are determined based on limiting dilutions and statistical analysis, we cannot rule out that the benefit from culturing cells at low oxygen actually favors stem cell homing rather than stem cell expansion. Both SDF-1 and CXCR4, the best-defined cytokine/receptor combination known to influence homing of hematopoietic cells, have been shown to be upregulated in hypoxia (Ceradini et al. 2004; Staller et al. 2003). Still, one should remember that low oxygen exposure is no magic bullet and that following 4 days of ex vivo culture, no SRCs could be maintained independently of the oxygen level used (Danet et al. 2003).

Another potential benefit from hypoxic exposure was demonstrated using MSC. These cells are normally isolated from the bone marrow, although they also have been shown to be accessible from peripheral or umbilical cord blood (Bieback et al. 2004; Erices et al. 2000). Using a rat model, the circulating MSC pool size was consistently and markedly increased (by almost 15-fold) when animals were exposed to chronic hypoxia (Rochefort et al. 2006). Importantly, the immunophenotype and the differentiation potential of circulating mobilized MSC were similar to those of bone marrow-derived MSC. Notably, total circulating hematopoietic progenitor cells were not significantly increased, thus suggesting that hypoxia-induced mobilization is specific to MSC. Overall, the increased colony-forming unit and repopulating ability of hematopoietic cells under low to hypoxic conditions suggest that oxidative stress, through possibly reactive oxygen species (ROS), strongly impairs these cells during ex vivo manipulation.

5.2
Effects of Antioxidants on Bone Marrow Cell Proliferation

In the context of ROS being damageable to cells, the addition of antioxidants to culture media, or the overexpression of ROS scavenger enzymes by hematopoietic cells, should help reduce oxidative damage imposed by in vitro procedures. Hydrogen peroxide formed from the combination of superoxide radicals and water is neutralized by enzymes such as glutathione peroxidases (GPxs), thioredoxin reductases (TrxRs), and catalase. Most of the GPx and TrxR enzymes have the particularity to be selenium-dependent (Ebert et al. 2006). Hence, it was recently shown that selenium supplementation of human MSC culture media could reduce intracellular ROS production and the stress-associated phenotype (Ebert et al. 2006). No data were provided as to whether selenite supplementation could also further prolong the stem cell lifespan. Still, cell culture protocols could easily be modified for the addition of selenite, which should help ex vivo manipulation of stem cells.

Concomitantly, the addition of recombinant purified catalase to mouse bone marrow cells resulted in a greater than 200-fold LSK cell number increase compared to controls (Gupta et al. 2006). Interestingly, the authors also showed that a prolonged catalase treatment (4–5 weeks) results in intense hematopoiesis early on followed by entry into a reversible quiescent state. They hypothesized that catalase may promote progenitor and stem cell survival by inhibiting both differentiation and apoptosis. Notably, the benefit of catalase addition was observed in the presence of a supportive stromal layer, which may have indirectly affected catalase treatment as it was not reproduced in a stromal layer-free culture system. The addition of superoxide dismutase, an enzyme known to produce hydrogen peroxide from the conversion of superoxide anion and water, did not have any effect on cell proliferation. This suggests that hydrogen peroxide rather than superoxide anions elimination is the limiting step in protecting cells from ROS. If this is the case, it becomes very important to measure the protective effect of catalase in stromal-free cultures, as excess hydrogen peroxide may originate from the irradiated stromal layer. In summary, these results make a solid argument for the presence of a tight peroxide-sensitive regulatory mechanism playing an important role in controlling hematopoiesis in vitro.

5.3
Cyclin-Dependent Kinase Inhibitors and Senescence Induction

Oxidative stress is a known senescence inducer of both mouse and human primary cell lines (Chen et al. 1995; Parrinello et al. 2003). One pathway by which excessive oxidative stress limits hematopoietic growth could be through telomere-independent premature induction of cellular senescence (Campisi 2005). Recently, Zhang and colleagues demonstrated that in vitro transient exposure to hypoxia (4 h) followed by reoxygenation (48 h) was sufficient to induce premature senescence in hematopoietic cells (Zhang et al. 2005b). Bone marrow failure is a frequent symptom of Fanconi anemia patients. Stirred by this observation, LSK cells derived from $Fancc^{-/-}$ mice were prone, much more than controlled wild-type cells, to premature senescence following reoxygenation. The senescence phenotype was revealed by the expression of SA-β-gal and the accumulation of DNA damage in $Fancc^{-/-}$ bone marrow cells. However, a classical DNA damage marker such as phosphorylated H2AX histone was not verified (Peterson and Cote 2004). Senescence induction correlated with strong accumulation of p53 and p21^{WAF1}, to a lower extent p16^{INK4a}, while p19^{ARF} did not change (Zhang et al. 2005b). Interestingly, the use N-acetyl-cysteine, a ROS scavenger, dramatically diminished reoxygenation-associated DNA damage (Zhang et al. 2005a). Inhibition of apoptosis through the use of a pan-caspase inhibitor (Z-VAD-FMK) allowed the identification of a higher proportion of senescent cells, as presumably both senescence and apoptosis pathways overlap. Inversely, inhibition of the ataxia telangiectasia mutated gene (Atm) by short interfering RNA (siRNA) reduced SA-β-gal staining and increased apoptosis of reoxygenated progenitor cells. To my knowledge, these results are the first to demonstrate that hematopoietic progenitor cells undergo premature senescence in culture.

Added proof for the role of oxidative stress and ROS formation on hematopoiesis came from the work of Ito and colleagues using Atm-mutated mice (Ito et al. 2004). The Atm gene is thought to regulate cell cycle checkpoint following DNA damage or genomic instability (Shiloh 2003). Deficiency in ATM also leads to an increased ROS level, although the mechanism implicated remains undetermined. Atm-deficient mice have progressive bone marrow failure, a defect alleviated by the administration of N-acetyl-cysteine (Ito et al. 2004). Bone marrow failure was attributed to elevated level of p16^{INK4a} and p19^{ARF} protein in LSK cells starting at 6 months of age, or as soon as 2 days following in vitro stimulation with cytokines. Antioxidant treatments reduced p16^{INK4a} and p19^{ARF} protein levels, suggesting that bone marrow failure is caused by CDKI activation. Inhibition of the retinoblastoma (Rb) protein through expression of the human papilloma virus type E16 E7 protein restored the repopulating capacity of *Atm*-deficient cells. Similarly, direct normalization of p16^{INK4a} and p19^{ARF} protein by overexpression of *Bmi-1*, a transcriptional repressor of the p16^{INK4a} and p19^{ARF} genes, also restored the repopulating capacity in these cells (Ito et al. 2004; Jacobs et al. 1999). In opposition to what Zhang and colleagues observed, p21^{WAF1} upregulation was not implicated. This suggests that the effects of ROS accumulation in *Atm*-deficient cells are different from those of ROS accumulation following cell reoxygenation, even if the effects of both accumulations were inhibited by antioxidant treatments. One explanation for this apparent discrepancy may be that Zhang's group used either wild-type or *Fancc*-deficient cells in which they showed ATM was required to trigger p53 and p21^{WAF1} accumulation (Zhang et al. 2005a). Here, Ito's group used *Atm*-deficient mice, suggesting that in the absence of ATM, ROS-induced oxidative damage is detected by p16^{INK4a}/p19^{ARF} rather than p21^{WAF1}. Both p53/p21 and p16/Rb pathways have been reported to overlap during senescence induction of human fibroblasts (Stein et al. 1999).

Ito and colleagues further defined the mechanism by which oxidative stress may limit the lifespan of HSC. Remarkably, they demonstrated that increased ROS levels lead to activation of p38 mitogen-activated protein kinase (MAPK) and subsequent expression of the tumor suppressor genes p16^{INK4a} and p19^{ARF} (Ito et al. 2006). These results were obtained either in wild-type cells depleted of intracellular glutathione or *Atm*-deficient cells. p38 MAPK activation was limited to LSK-enriched stem cell populations. p38 MAPK activation also triggered the accumulation of other *Ink4a* family members, including p15^{INK4b} and p18^{INK4c} proteins. Notably, activation of a strong mitogenic signal, such as p38 MAPK and oncogenic Ras, can cause upregulation of p16^{INK4a} (Bulavin et al. 2004; Serrano et al. 1997). Unexpectedly, wild-type cells from old mice (over 24 months) were also capable of upregulating p38 MAPK without the need to deplete intracellular glutathione. Treatment with antioxidants or an inhibitor of p38 MAPK abrogated the *Ink4a* member's accumulation and extended the lifespan or repopulation capacity of wild-type cells in serial transplantation. Transplantation of wild-type bone marrow cells is known to be limited to 4–5 serial transplantations, after which bone marrow is believed to be exhausted (Siminovitch et al. 1964). ROS levels were found to gradually increase in each round of transplantation conjointly with associated decreased repopulation ability. p38 MAPK activation and p16^{INK4a} and p19^{ARF} protein accumulation were noticed after the third round of transplantation and could be prevented, in this case again, by the administration of an antioxidant. In normal conditions, true stem cells are known to undergo limited cell division in vivo, making it possible that ROS accumulation in these cells originates intrinsically and/or extrinsically from presumably excessive ROS released from the surrounding aging microenvironment.

Ito and colleagues reasoned that p38 MAPK activation pushes HSC to replicative exhaustion prior to entering a $p16^{INK4a}$- and $p19^{ARF}$-dependent senescence. Although possible, this hypothesis is unlikely given the average telomere length in mice and the fact that forced telomerase activity does not prevent senescence (Allsopp et al. 2003b). Because direct induction of senescence through SA-β-gal staining was not measured, it is more likely that gradual accumulation of ROS ultimately triggered $p16^{INK4a}$ and $p19^{ARF}$ protein accumulation, thus causing premature senescence without stem cell exhaustion. Also, to explain the decreased quiescence observed in young Atm-deficient cells compared to wild-type cells, it is possible that low (possibly undetectable) p38 MAPK activation forces cells to cycle until the mitogenic signal gets too strong, activating instead tumor suppressor pathways. Human fibroblasts grow faster following weak overexpression of oncogenic Ras, and they senesce following its strong overexpression (C. Beauséjour, unpublished data). In summary, these results further strengthen the hypothesis that hematopoietic cell senescence is regulated primarily by increased oxidative stress rather than by telomere shortening.

5.4
Effects of Radiation/Chemotherapy on Bone Marrow Cell Senescence

Most if not all cancer patients will at one point undergo chemo- or radiation therapy, or need bone marrow transplantation. These treatments hopefully eliminate cancer cells, but are likely to cause significant collateral damage, as neither of them is specific to cancer cells. In fact, ionizing radiation and alkylating drugs, such as busulfan, induce premature senescence in murine bone marrow hematopoietic cells (Meng et al. 2003). Indeed, 5 weeks following the exposure of LSK cells to 4 Gy sublethal ionizing radiation or 30 μm busulfan, surviving nonapoptotic cells showed increased SA-β-gal staining, and a large proportion (30%–40%) of them had undergone cellular senescence. Elevated expression levels of $p16^{INK4a}$, $p19^{ARF}$ and $p21^{WAF1}$ were thought to have induced senescence, although the latter was only expressed following radiation. These results were confirmed in vivo using LSK cells isolated directly from sublethally (6.5 Gy) irradiated C57BL/6 mice (Wang et al. 2006). Importantly, this phenotype was restricted to LSK cells (Wang et al. 2006). Interestingly, 28 days following irradiation, mice bone marrow mononuclear cell counts were back to normal while LSK cells were still depleted 56 days post irradiation. Competitive repopulating assays using a fixed number of LSK cells, either nonirradiated or irradiated 28 days prior to purification, revealed severe deficiency in the ability of the irradiated cells to repopulate recipient mice. To my knowledge, this observation is the first to demonstrate that a population of enriched wild-type stem cells is fully capable of undergoing premature senescence in vivo without being manipulated ex vivo. These results may explain why many patients exposed to moderate total body irradiation, or undergoing chemotherapy, exhibit long-term bone marrow damage (Testa et al. 1985).

Overall, these results are somewhat unexpected since bone marrow cells obtained from $p16^{INK4a}/p19^{ARF}$-deficient mice did not show any significant advantages compared to wild-type cells in repopulation studies (Stepanova and Sorrentino 2005). However, when forced to replicate under serial transplantation, the $p16^{INK4a}/p19^{ARF}$-deficient cells had a modestly extended lifespan. Such a small difference could be explained by premature exhaustion of the stem cell pool in these mice, a phenomenon already observed in $p21^{WAF1}$-deficient mice (Cheng et al. 2000). Since expression of $p16^{INK4a}$ is a biomarker of aging, it would be interesting to assay the repopulating activity of $p16^{INK4a}/p19^{ARF}$-deficient cells in aged animals (Krishnamurthy et al. 2004). Janzen and colleagues observe increased $p16^{INK4a}$ expression in stem cells from old mice and following bone marrow transplantation (Janzen et al. 2006). The benefit from $p16^{INK4a}$ inactivation was not observed in young mice, suggesting $p16^{INK4a}$ expression is turned on only following exhaustive repopulation with age or accumulated stress. Therefore, propensity of stem cells to activate $p16^{INK4a}$ following transplantation may be a good indicator of the graft clinical outcome.

6
Consequences of Aging on Bone Marrow-Derived Cellular Therapies

Cell-based therapy for the repair of vascular injury certainly represents a promising new clinical approach. The ideal therapy would involve direct injection of a limited number of cells, i.e., bone marrow-derived cells, capable of adequate homing and function to the injured tissue. Although several groups have demonstrated the benefit of cell therapy for heart and vascular diseases, little information is available with regards to the effects of aging on such therapy. In rabbits, aging was shown to affect angiogenesis and reendothelialization after arterial injury (Gennaro et al. 2003; Rivard et al. 1999). These age-associated defects were mostly associated with reduced vascular endothelial growth factor (VEGF) secretion by the endothelium with no mention of the potential contribution of bone marrow-derived cells.

Using a model of accelerated atherosclerosis, namely ApoE-deficient mice being fed a high fat diet, the effects of aging on cell therapy were demonstrated (Rauscher et al. 2003). In this model, mice develop atherosclerosis after only 3–4 months of age, a phenotype that can be attenuated by the intravenous injection of bone marrow-derived cells (every 2 weeks starting at 3 weeks of age, total 6 injections). While injection of wild-type donor cells was clearly more efficient than injection of $ApoE^{-/-}$ cells in reducing atherosclerosis formation, old compared to young $ApoE^{-/-}$ donor cells were significantly impaired. Hence, aged bone marrow-derived cells are deficient in an aspect other than bone marrow repopulation. Remarkably, the effect was observed using 6-months-old ApoE$^{-/-}$ donor cells, an age at which wild-type cells usually have no apparent defect, including homing and engraftment to the bone marrow. A reduced number of CD31$^+$/CD45$^-$ progenitor cells in 6-month-old compared to 1-month-old ApoE$^{-/-}$ cells were reported as possibly causing the phenotype. Interestingly, injected ApoE$^{-/-}$ donor cells were localized almost exclusively at the atherosclerosis-prone region, with minimal engrafted cells found in atherosclerosis-free regions of the aorta. Unfortunately, no measure of the homing potential of young versus old donor cells at the atherosclerotic site was reported. Other than poor homing, enhanced sensitivity of aged bone marrow-derived cells toward apoptotic stimuli may help explain the observed difference. Indeed, aged human umbilical vein endothelial cells were shown to have reduced endothelial NO synthase protein expression, concomitantly making them more sensitive to apoptosis (Hoffmann et al. 2001).

Similarly, the secretion of cardiac myocyte platelet-derived growth factor (PDGF) by endothelial cells is decreased in aged hearts and associated with reduced cardiac angiogenic function (Edelberg et al. 2002). Convincing evidence demonstrated a role for circulating endothelial progenitor cells (EPC) in the development and maintenance of the heart vasculature (Asahara et al. 1997; Kalka et al. 2000). Indeed, direct injection of bone marrow-derived EPC can reverse the aging-associated decline in myocardial angiogenic activity (Edelberg et al. 2002). Surprisingly, in opposition to 3-month-old derived EPCs, injection of 18-month-old EPC did not restore PDGF expression, nor did it improve cardiac angiogenesis in aged mice. Therefore, age-associated impairment PDGF secretion may represent another mechanism by which bone marrow-derived cells from old donors are deficient. Finally, using the *klotho* mouse model of aging, it was shown that incorporation of transplanted homologous bone marrow cells into capillaries of ischemic tissues is impaired in *klotho* compare to wild-type mice (Shimada et al. 2004). All together, these experiments suggest that autologous cell therapy in elderly patients may have limited potential and emphasize the need to support fundamental research looking at the cause of this age-associated decline in cell function.

7
Conclusion

Increasing evidence demonstrates that senescent cells accumulate in vivo, either in premalignant or aged tissues. Telomere-dependent replicative exhaustion and stress-induced premature senescence are likely involved. The consequences of aging and associated accumulation of senescent cells may have serious deleterious effects such as age-associated diseases. Given the technically challenging task of detecting senescent cells in vivo, we only begin the identification of various cellular compartments in which a significant proportion of cells are senescent. Bone marrow-derived stem cells are the best-characterized adult stem cells. The possible consequences of their intense replication and manipulation are currently being identified. Interestingly, these cells are fully capable of undergoing cellular senescence—both in vitro and in vivo—by upregulating the expression of $p16^{INK4a}$, most likely in response to oxidative stress. Since expression of $p16^{INK4a}$ is a biomarker of aging, control of its expression by pharmacological approaches may help attenuate age-related phenotypes. Indeed, inhibition of p38 MAPK and subsequent $p16^{INK4a}$ expression alleviated age-dependent anemia in A*tm*-deficient mice (Ito et al. 2006). Genetic studies suggest that the propensity of marrow-derived cells to undergo senescence will likely be heterogeneous and depend on genetic background. A better understanding of the consequences of aging and senescence on bone marrow-derived cells will certainly benefit our understanding of numerous diseases, both those that are age-associated and those that will benefit from cellular therapy.

References

Aghi M, Chiocca EA (2005) Contribution of bone marrow-derived cells to blood vessels in ischemic tissues and tumors. Mol Ther 12:994–1005

Allsopp RC, Cheshier S, Weissman IL (2001) Telomere shortening accompanies increased cell cycle activity during serial transplantation of hematopoietic stem cells. J Exp Med 193:917–924

Allsopp RC, Morin GB, DePinho R, Harley CB, Weissman IL (2003a) Telomerase is required to slow telomere shortening and extend replicative lifespan of HSCs during serial transplantation. Blood 102:517–520

Allsopp RC, Morin GB, Horner JW, DePinho R, Harley CB, Weissman IL (2003b) Effect of TERT over-expression on the long-term transplantation capacity of hematopoietic stem cells. Nat Med 9:369–371

Asahara T, Murohara T, Sullivan A, Silver M, van der Zee R, Li T, Witzenbichler B, Schatteman G, Isner JM (1997) Isolation of putative progenitor endothelial cells for angiogenesis. Science 275:964–967

Atkuri KR, Herzenberg LA, Herzenberg LA (2005) Culturing at atmospheric oxygen levels impacts lymphocyte function. Proc Natl Acad Sci U S A 102:3756–3759

Baxter MA, Wynn RF, Jowitt SN, Wraith JE, Fairbairn LJ, Bellantuono I (2004) Study of telomere length reveals rapid aging of human marrow stromal cells following in vitro expansion. Stem Cells 22:675–682

Beausejour CM, Krtolica A, Galimi F, Narita M, Lowe SW, Yaswen P, Campisi J (2003) Reversal of human cellular senescence: roles of the p53 and p16 pathways. EMBO J 22:4212–4222

Bieback K, Kern S, Kluter H, Eichler H (2004) Critical parameters for the isolation of mesenchymal stem cells from umbilical cord blood. Stem Cells 22:625–634

Blasco MA, Lee HW, Hande MP, Samper E, Lansdorp PM, DePinho RA, Greider CW (1997) Telomere shortening and tumor formation by mouse cells lacking telomerase RNA. Cell 91:25–34

Bordone L, Guarente L (2005) Calorie restriction, SIRT1 and metabolism: understanding longevity. Nat Rev Mol Cell Biol 6:298–305

Broccoli D, Young JW, de Lange T (1995) Telomerase activity in normal and malignant hematopoietic cells. Proc Natl Acad Sci U S A 92:9082–9086

Bulavin DV, Phillips C, Nannenga B, Timofeev O, Donehower LA, Anderson CW, Appella E, Fornace AJ Jr (2004) Inactivation of the Wip1 phosphatase inhibits mammary tumorigenesis through p38 MAPK-mediated activation of the p16(Ink4a)-p19(Arf) pathway. Nat Genet 36:343–350

Campisi J (1997) The biology of replicative senescence. Eur J Cancer 33:703–709

Campisi J (2003) Cancer and ageing: rival demons? Nat Rev Cancer 3:339–349

Campisi J (2005) Senescent cells, tumor suppression, and organismal aging: good citizens, bad neighbors. Cell 120:513–522

Ceradini DJ, Kulkarni AR, Callaghan MJ, Tepper OM, Bastidas N, Kleinman ME, Capla JM, Galiano RD, Levine JP, Gurtner GC (2004) Progenitor cell trafficking is regulated by hypoxic gradients through HIF-1 induction of SDF-1. Nat Med 10:858–864

Chen J, Astle CM, Harrison DE (2000) Genetic regulation of primitive hematopoietic stem cell senescence. Exp Hematol 28:442–450

Chen J, Astle CM, Harrison DE (2003) Hematopoietic senescence is postponed and hematopoietic stem cell function is enhanced by dietary restriction. Exp Hematol 31:1097–1103

Chen Q, Fischer A, Reagan JD, Yan LJ, Ames BN (1995) Oxidative DNA damage and senescence of human diploid fibroblast cells. Proc Natl Acad Sci U S A 92:4337–4341

Cheng T, Rodrigues N, Shen H, Yang Y, Dombkowski D, Sykes M, Scadden DT (2000) Hematopoietic stem cell quiescence maintained by p21cip1/waf1. Science 287:1804–1808

Cipolleschi MG, Dello Sbarba P, Olivotto M (1993) The role of hypoxia in the maintenance of hematopoietic stem cells. Blood 82:2031–2037

Cipolleschi MG, D'Ippolito G, Bernabei PA, Caporale R, Nannini R, Mariani M, Fabbiani M, Rossi-Ferrini P, Olivotto M, Dello Sbarba P (1997) Severe hypoxia enhances the formation of erythroid bursts from human cord blood cells and the maintenance of BFU-E in vitro. Exp Hematol 25:1187–1194

Conboy IM, Conboy MJ, Wagers AJ, Girma ER, Weissman IL, Rando TA (2005) Rejuvenation of aged progenitor cells by exposure to a young systemic environment. Nature 433:760–764

Danet GH, Pan Y, Luongo JL, Bonnet DA, Simon MC (2003) Expansion of human SCID-repopulating cells under hypoxic conditions. J Clin Invest 112:126–135

Dimri GP, Lee X, Basile G, Acosta M, Scott G, Roskelley C, Medrano EE, Linskens M, Rubelj I, Pereira-Smith OM, Peacocke M, Campisi J (1995) A novel biomarker identifies senescent human cells in culture and in aging skin in vivo. Proc Natl Acad Sci USA 92:9363–9367

Ebert R, Ulmer M, Zeck S, Meissner-Weigl J, Schneider D, Stopper H, Schupp N, Kassem M, Jakob F (2006) Selenium supplementation restores the antioxidative capacity and prevents cell damage in bone marrow stromal cells in vitro. Stem Cells 24:1226–1235

Edelberg JM, Lee SH, Kaur M, Tang L, Feirt NM, McCabe S, Bramwell O, Wong SC, Hong MK (2002) Platelet-derived growth factor-AB limits the extent of myocardial infarction in a rat model: feasibility of restoring impaired angiogenic capacity in the aging heart. Circulation 105:608–613

Engelhardt M, Kumar R, Albanell J, Pettengell R, Han W, Moore MA (1997) Telomerase regulation, cell cycle, and telomere stability in primitive hematopoietic cells. Blood 90:182–193

Erices A, Conget P, Minguell JJ (2000) Mesenchymal progenitor cells in human umbilical cord blood. Br J Haematol 109:235–242

Gardner RV, Lerner C, Astle CM, Harrison DE (1993) Assessing permanent damage to primitive hematopoietic stem cells after chemotherapy using the competitive repopulation assay. Cancer Chemother Pharmacol 32:450–454

Geiger H, True JM, de Haan G, Van Zant G (2001) Age- and stage-specific regulation patterns in the hematopoietic stem cell hierarchy. Blood 98:2966–2972

Geiger H, Rennebeck G, Van Zant G (2005) Regulation of hematopoietic stem cell aging in vivo by a distinct genetic element. Proc Natl Acad Sci U S A 102:5102–5107

Gennaro G, Menard C, Michaud SE, Rivard A (2003) Age-dependent impairment of reendothelialization after arterial injury: role of vascular endothelial growth factor. Circulation 107:230–233

Gupta R, Karpatkin S, Basch RS (2006) Hematopoiesis and stem cell renewal in long-term bone marrow cultures containing catalase. Blood 107:1837–1846

Harrison DE (1983) Long-term erythropoietic repopulating ability of old, young, and fetal stem cells. J Exp Med 157:1496–1504

Harrison DE, Astle CM (1982) Loss of stem cell repopulating ability upon transplantation. Effects of donor age, cell number, and transplantation procedure. J Exp Med 156:1767–1779

Harrison DE, Astle CM, Delaittre JA (1978) Loss of proliferative capacity in immunohemopoietic stem cells caused by serial transplantation rather than aging. J Exp Med 147:1526–1531

Harrison DE, Astle CM, Stone M (1989) Numbers and functions of transplantable primitive immunohematopoietic stem cells. Effects of age. J Immunol 142:3833–3840

Harrison DE, Stone M, Astle CM (1990) Effects of transplantation on the primitive immunohematopoietic stem cell. J Exp Med 172:431–437

Hayflick L (1965) The limited in vitro lifetime of human diploid cell strains. Exp Cell Res 37:614–636

Herbig U, Ferreira M, Condel L, Carey D, Sedivy JM (2006) Cellular senescence in aging primates. Science 311:1257

Hoffmann J, Haendeler J, Aicher A, Rossig L, Vasa M, Zeiher AM, Dimmeler S (2001) Aging enhances the sensitivity of endothelial cells toward apoptotic stimuli: important role of nitric oxide. Circ Res 89:709–715

Ito K, Hirao A, Arai F, Matsuoka S, Takubo K, Hamaguchi I, Nomiyama K, Hosokawa K, Sakurada K, Nakagata N, Ikeda Y, Mak TW, Suda T (2004) Regulation of oxidative stress by ATM is required for self-renewal of haematopoietic stem cells. Nature 431:997–1002

Ito K, Hirao A, Arai F, Takubo K, Matsuoka S, Miyamoto K, Ohmura M, Naka K, Hosokawa K, Ikeda Y, Suda T (2006) Reactive oxygen species act through p38 MAPK to limit the lifespan of hematopoietic stem cells. Nat Med 12:446–451

Ivanovic Z, Dello Sbarba P, Trimoreau F, Faucher JL, Praloran V (2000) Primitive human HPCs are better maintained and expanded in vitro at 1 percent oxygen than at 20 percent. Transfusion 40:1482–1488

Jacobs JJ, Kieboom K, Marino S, DePinho RA, van Lohuizen M (1999) The oncogene and Polycomb-group gene bmi-1 regulates cell proliferation and senescence through the ink4a locus. Nature 397:164–168

Janzen V, Forkert R, Fleming HE, Saito Y, Waring MT, Dombkowski DM, Cheng T, DePinho RA, Sharpless NE, Scadden DT (2006) Stem-cell ageing modified by the cyclin-dependent kinase inhibitor p16INK4a. Nature 443:404–405

Kalka C, Masuda H, Takahashi T, Kalka-Moll WM, Silver M, Kearney M, Li T, Isner JM, Asahara T (2000) Transplantation of ex vivo expanded endothelial progenitor cells for therapeutic neovascularization. Proc Natl Acad Sci U S A 97:3422–3427

Karlseder J, Smogorzewska A, de Lange T (2002) Senescence induced by altered telomere state, not telomere loss. Science 295:2446–2449

Kiel MJ, Yilmaz OH, Iwashita T, Yilmaz OH, Terhorst C, Morrison SJ (2005) SLAM family receptors distinguish hematopoietic stem and progenitor cells and reveal endothelial niches for stem cells. Cell 121:1109–1121

Krishnamurthy J, Torrice C, Ramsey MR, Kovalev GI, Al-Regaiey K, Su L, Sharpless NE (2004) Ink4a/Arf expression is a biomarker of aging. J Clin Invest 114:1299–1307

Krtolica A, Parrinello S, Lockett S, Desprez PY, Campisi J (2001) Senescent fibroblasts promote epithelial cell growth and tumorigenesis: a link between cancer and aging. Proc Natl Acad Sci U S A 98:12072–12077

Lansdorp PM, Dragowska W, Mayani H (1993) Ontogeny-related changes in proliferative potential of human hematopoietic cells. J Exp Med 178:787–791

Liang Y, Van Zant G, Szilvassy SJ (2005) Effects of aging on the homing and engraftment of murine hematopoietic stem and progenitor cells. Blood 106:1479–1487

Martens UM, Chavez EA, Poon SS, Schmoor C, Lansdorp PM (2000) Accumulation of short telomeres in human fibroblasts prior to replicative senescence. Exp Cell Res 256:291–299

Meirelles Lda S, Nardi NB (2003) Murine marrow-derived mesenchymal stem cell: isolation, in vitro expansion, and characterization. Br J Haematol 123:702–711

Meng A, Wang Y, Van Zant G, Zhou D (2003) Ionizing radiation and busulfan induce premature senescence in murine bone marrow hematopoietic cells. Cancer Res 63:5414–5419

Morrison SJ, Prowse KR, Ho P, Weissman IL (1996) Telomerase activity in hematopoietic cells is associated with self-renewal potential. Immunity 5:207–216

Narita M, Nunez S, Heard E, Narita M, Lin AW, Hearn SA, Spector DL, Hannon GJ, Lowe SW (2003) Rb-mediated heterochromatin formation and silencing of E2F target genes during cellular senescence. Cell 113:703–716

Notaro R, Cimmino A, Tabarini D, Rotoli B, Luzzatto L (1997) In vivo telomere dynamics of human hematopoietic stem cells. Proc Natl Acad Sci U S A 94:13782–13785

Parrinello S, Samper E, Krtolica A, Goldstein J, Melov S, Campisi J (2003) Oxygen sensitivity severely limits the replicative lifespan of murine fibroblasts. Nat Cell Biol 5:741–747

Pearson M, Carbone R, Sebastiani C, Cioce M, Fagioli M, Saito S, Higashimoto Y, Appella E, Minucci S, Pandolfi PP, Pelicci PG (2000) PML regulates p53 acetylation and premature senescence induced by oncogenic RAS. Nature 406:207–210

Peters BA, Diaz LA, Polyak K, Meszler L, Romans K, Guinan EC, Antin JH, Myerson D, Hamilton SR, Vogelstein B, Kinzler KW, Lengauer C (2005) Contribution of bone marrow-derived endothelial cells to human tumor vasculature. Nat Med 11:261–262

Peterson CL, Cote J (2004) Cellular machineries for chromosomal DNA repair. Genes Dev 18:602–616

Piacibello W, Bruno S, Sanavio F, Droetto S, Gunetti M, Ailles L, Santoni de Sio F, Viale A, Gammaitoni L, Lombardo A, Naldini L, Aglietta M (2002) Lentiviral gene transfer and ex vivo expansion of human primitive stem cells capable of primary, secondary, and tertiary multilineage repopulation in NOD/SCID mice. Nonobese diabetic/severe combined immunodeficient. Blood 100:4391–4400

Pittenger MF, Mackay AM, Beck SC, Jaiswal RK, Douglas R, Mosca JD, Moorman MA, Simonetti DW, Craig S, Marshak DR (1999) Multilineage potential of adult human mesenchymal stem cells. Science 284:143–147

Prowse KR, Greider CW (1995) Developmental and tissue-specific regulation of mouse telomerase and telomere length. Proc Natl Acad Sci USA 92:4818–4822

Rauscher FM, Goldschmidt-Clermont PJ, Davis BH, Wang T, Gregg D, Ramaswami P, Pippen AM, Annex BH, Dong C, Taylor DA (2003) Aging, progenitor cell exhaustion, and atherosclerosis. Circulation 108:457–463

Reyes M, Lund T, Lenvik T, Aguiar D, Koodie L, Verfaillie CM (2001) Purification and ex vivo expansion of postnatal human marrow mesodermal progenitor cells. Blood 98:2615–2625

Rivard A, Fabre JE, Silver M, Chen D, Murohara T, Kearney M, Magner M, Asahara T, Isner JM (1999) Age-dependent impairment of angiogenesis. Circulation 99:111–120

Rochefort GY, Delorme B, Lopez A, Herault O, Bonnet P, Charbord P, Eder V, Domenech J (2006) Multipotential mesenchymal stem cells are mobilized into peripheral blood by hypoxia. Stem Cells 24:2202–2208

Ross EA, Anderson N, Micklem HS (1982) Serial depletion and regeneration of the murine hematopoietic system. Implications for hematopoietic organization and the study of cellular aging. J Exp Med 155:432–444

Rossi DJ, Bryder D, Zahn JM, Ahlenius H, Sonu R, Wagers AJ, Weissman IL (2005) Cell intrinsic alterations underlie hematopoietic stem cell aging. Proc Natl Acad Sci U S A 102:9194–9199

Samper E, Fernandez P, Eguia R, Martin-Rivera L, Bernad A, Blasco MA, Aracil M (2002) Long-term repopulating ability of telomerase-deficient murine hematopoietic stem cells. Blood 99:2767–2775

Serrano M, Lin AW, McCurrach ME, Beach D, Lowe SW (1997) Oncogenic ras provokes premature cell senescence associated with accumulation of p53 and p16INK4a. Cell 88:593–602

Sharpless NE, DePinho RA (2004) Telomeres, stem cells, senescence, and cancer. J Clin Invest 113:160–168

Shiloh Y (2003) ATM and related protein kinases: safeguarding genome integrity. Nat Rev Cancer 3:155–168

Shimada T, Takeshita Y, Murohara T, Sasaki K, Egami K, Shintani S, Katsuda Y, Ikeda H, Nabeshima Y, Imaizumi T (2004) Angiogenesis and vasculogenesis are impaired in the precocious-aging klotho mouse. Circulation 110:1148–1155

Siminovitch L, Till JE, McCulloch EA (1964) Decline in colony-forming ability of marrow cells subjected to serial transplantation into irradiated mice. J Cell Physiol 64:23–31

Smith JR, Pereira-Smith OM (1996) Replicative senescence: implications for in vivo aging and tumor suppression. Science 273:63–67

Staller P, Sulitkova J, Lisztwan J, Moch H, Oakeley EJ, Krek W (2003) Chemokine receptor CXCR4 downregulated by von Hippel-Lindau tumour suppressor pVHL. Nature 425:307–311

Stanulis-Praeger B (1987) Cellular senescence revisited: a review. Mech Ageing Dev 38:1–48

Stein GH, Drullinger LF, Soulard A, Dulic V (1999) Differential roles for cyclin-dependent kinase inhibitors p21 and p16 in the mechanisms of senescence and differentiation in human fibroblasts. Mol Cell Biol 19:2109–2117

Stenderup K, Justesen J, Clausen C, Kassem M (2003) Aging is associated with decreased maximal life span and accelerated senescence of bone marrow stromal cells. Bone 33:919–926

Stepanova L, Sorrentino BP (2005) A limited role for p16Ink4a and p19Arf in the loss of hematopoietic stem cells during proliferative stress. Blood 106:827–832

Sudo K, Ema H, Morita Y, Nakauchi H (2000) Age-associated characteristics of murine hematopoietic stem cells. J Exp Med 192:1273–1280

Testa NG, Hendry JH, Molineux G (1985) Long-term bone marrow damage in experimental systems and in patients after radiation or chemotherapy. Anticancer Res 5:101–110

Van Zant G, Holland BP, Eldridge PW, Chen JJ (1990) Genotype-restricted growth and aging patterns in hematopoietic stem cell populations of allophenic mice. J Exp Med 171:1547–1565

Van Zant G, Scott-Micus K, Thompson BP, Fleischman RA, Perkins S (1992) Stem cell quiescence/activation is reversible by serial transplantation and is independent of stromal cell genotype in mouse aggregation chimeras. Exp Hematol 20:470–475

Wang Y, Schulte BA, LaRue AC, Ogawa M, Zhou D (2006) Total body irradiation selectively induces murine hematopoietic stem cell senescence. Blood 107:358–366

Wei W, Sedivy JM (1999) Differentiation between senescence (M1) and crisis (M2) in human fibroblast cultures. Exp Cell Res 253:519–522

Weng NP, Hodes RJ (2000) The role of telomerase expression and telomere length maintenance in human and mouse. J Clin Immunol 20:257–267

Wright WE, Shay JW (2001) Cellular senescence as a tumor-protection mechanism: the essential role of counting. Curr Opin Genet Dev 11:98–103

Wright WE, Piatyszek MA, Rainey WE, Byrd W, Shay JW (1996) Telomerase activity in human germline and embryonic tissues and cells. Dev Genet 18:173–179

Xing Z, Ryan MA, Daria D, Nattamai KJ, Van Zant G, Wang L, Zheng Y, Geiger H (2006) Increased hematopoietic stem cell mobilization in aged mice. Blood 108:2190–2197

Yilmaz OH, Kiel MJ, Morrison SJ (2006) SLAM family markers are conserved among hematopoietic stem cells from old and reconstituted mice and markedly increase their purity. Blood 107:924–930

Yuan R, Astle CM, Chen J, Harrison DE (2005) Genetic regulation of hematopoietic stem cell exhaustion during development and growth. Exp Hematol 33:243–250

Zhang X, Li J, Sejas DP, Pang Q (2005a) The ATM/p53/p21 pathway influences cell fate decision between apoptosis and senescence in reoxygenated hematopoietic progenitor cells. J Biol Chem 280:19635–19640

Zhang X, Li J, Sejas DP, Pang Q (2005b) Hypoxia-reoxygenation induces premature senescence in FA bone marrow hematopoietic cells. Blood 106:75–85

Zimmermann S, Voss M, Kaiser S, Kapp U, Waller CF, Martens UM (2003) Lack of telomerase activity in human mesenchymal stem cells. Leukemia 17:1146–1149

Involvement of Marrow-Derived Endothelial Cells in Vascularization

B. Larrivée[1] · A. Karsan[2] (✉)

[1]Laboratoire de Médecine expérimentale, INSERM U36, Collège de France,
11 Place Marcelin Berthelot 75005 Paris, France

[2]Pathology and Laboratory Medicine and Medical Biophysics,
British Columbia Cancer Agency Research Centre, 675 West 10th Avenue,
V5Z 1L3 British Columbia, Vancouver, Canada
akarsan@bccrc.ca

1	Embryonic Origins of Hematopoiesis and Vasculogenesis	90
1.1	Yolk Sac Hematopoiesis and Vasculogenesis	90
1.2	The Formation and Maturation of Definitive Hematopoietic Cells . .	91
1.3	The Hemangioblast and Endothelial Progenitor Cells	92
2	Hematopoiesis and Vasculogenesis in the Adult	94
2.1	The Hematopoietic Stem Cell	94
2.2	Phenotype and Origin of Endothelial Progenitor Cells	96
2.3	The Angioblast and Evidence of Vasculogenesis in the Adult	99
3	Functions of Endothelial Progenitor Cells in Physiology and Pathology . .	101
3.1	Endothelial Progenitor Cells and Physiological Neovascularization	101
3.2	Contribution of Endothelial Progenitor Cells to Tumor Vascularization	102
3.3	Endothelial Progenitor Cells for Therapy	105
4	Perspectives	106
	References	107

Abstract Until recently, the adult neovasculature was thought to arise only through angiogenesis, the mechanism by which new blood vessels form from preexisting vessels through endothelial cell migration and proliferation. However, recent studies have provided evidence that postnatal neovasculature can also arise though vasculogenesis, a process by which endothelial progenitor cells are recruited and differentiate into mature endothelial cells to form new blood vessels. Evidence for the existence of endothelial progenitors has come from studies demonstrating the ability of bone marrow-derived cells to incorporate into adult vasculature. However, the exact nature of endothelial progenitor cells remains controversial. Because of the lack of definitive markers of endothelial progenitors, the *in vivo* contribution of progenitor cells to physiological and pathological neovascularization remains unclear. Early studies reported that endothelial progenitor cells actively integrate into the adult vasculature and are critical in the development of many types of vascular-dependent disorders such as neoplastic progression. Moreover, it has been suggested that endothelial progenitor cells can be used as a therapeutic strategy aimed at promoting vascular growth in a variety of ischemic diseases. However, increasing numbers of studies have reported no clear contribution of endothelial progenitors in physiological or pathological angiogenesis. In this chapter, we discuss the origin of the endothelial progenitor cell in the embryo and adult, and we discuss the cell's link to the primitive hematopoietic stem cell. We also review the potential significance of endothelial progenitor cells in the formation of a postnatal vascular network and discuss the factors that may account for the current lack of consensus of the scientific community on this important issue.

Keywords Endothelial progenitors · Vasculogenesis · Tumor blood vessels · Endothelial progenitor cell · Hematopoietic stem cell · Hemangioblast · Angiogenesis

1
Embryonic Origins of Hematopoiesis and Vasculogenesis

1.1
Yolk Sac Hematopoiesis and Vasculogenesis

The development of a functional circulatory system is an early prerequisite for the survival and growth of the mammalian embryo. The first differentiated cells to form in the mammalian embryo are those of the hematopoietic and endothelial lineages, which, along with cardiac components, are the backbone of the developing circulatory system (Ema and Rossant 2003). Cells from the hematopoietic and endothelial lineages are mesodermal in origin and are first generated in yolk sac blood islands beginning at embryonic day (E)7 in the mouse (Haar and Ackerman 1971) and between the second and third week of human gestation (Oberlin et al. 2002). Thereafter, hematopoiesis shifts to the fetal liver and again, around the time of birth, to the bone marrow (Moore and Metcalf 1970). The generation of blood cells in blood islands of the yolk sac is referred to as primitive hematopoiesis and results primarily in the production of large, nucleated erythroblasts (Wong et al. 1986), as well as some megakaryocytes (Xu et al. 2001) and primitive macrophages (Shepard and Zon 2000), as opposed to definitive hematopoiesis, which can generate mature cells of all the blood lineages.

Blood islands, which consist of a central focus of hematopoietic cells surrounded by a layer of endothelial cells, arise in the mouse from proximal mesodermal cells in the visceral yolk sac between E7.0 and E7.5 (Shepard and Zon 2000). Between E8.0 and E9.0, the cells constituting the outer layer of the blood island cell aggregates assume a spindle shape as they differentiate into endothelial cells (Shepard and Zon 2000; Fig. 1). The vast majority of the inner cells of the blood islands progressively lose their intercellular attachments as they differentiate into primitive erythroblasts. The simultaneous spatial and temporal appearance of hematopoietic and endothelial cells in the yolk sac blood islands has led to the concept of the hemangioblast, a common precursor for the hematopoietic and endothelial lineages (Ema and Rossant 2003). Moreover, many known markers of endothelial cells are also expressed on hematopoietic cells (Fina et al. 1990; Kallianpur et al. 1994; Matthews et al. 1991). Studies with cells from chick embryos and with embryonic stem cells have lent credence to this concept (Eichmann et al. 1997; Ema and Rossant 2003).

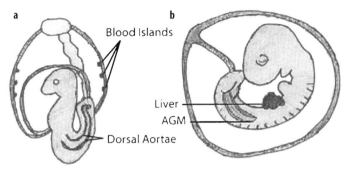

Fig. 1a,b Hematopoietic sites in the developing mouse embryo. Schematic representation of whole mouse embryos at **a** E8.5/9 and **b** E10.5/11. *AGM*, aorta-gonad-mesonephros

While histological studies indicate a restricted hematopoietic potential within the yolk sac, precursor analysis *in vitro* and transplantation studies *in vivo* have provided clear evidence that this tissue is able to generate multiple definitive lineages as well (Yoder and Hiatt 1997). Macrophage precursors were detected in low numbers as early as those of the primitive erythroid lineage (Takahashi and Naito 1993). Definitive erythroid precursors were found at E8.25 and also showed a dramatic increase followed by a general decline of their numbers (Palis et al. 1999). Precursors of the mast cell lineage developed slightly later at E8.5 (Palis et al. 1999). Unlike primitive erythroblasts, definitive erythroid precursors and mast cell precursors do not mature in the yolk sac, suggesting that they are produced for export to other sites, presumably the fetal liver (Palis et al. 1999). These findings suggest a dual role for the yolk sac: the generation of a functional, primitive erythroid lineage and the production of a cohort of definitive precursors that migrate to the fetal liver and establish the initial stage of definitive hematopoiesis in this tissue (Galloway and Zon 2003). Yolk sac hematopoiesis is transient and shows a dramatic decline in activity between E11 and E12. This decline coincides with the onset of activity in the developing liver, which becomes the predominant hematopoietic tissue throughout the remainder of fetal life. In contrast to the restricted program observed in the yolk sac, the fetal liver is a site of multilineage definitive hematopoiesis that includes erythropoiesis, myelopoiesis, and lymphopoiesis (Galloway and Zon 2003).

1.2
The Formation and Maturation of Definitive Hematopoietic Cells

Shortly before the onset of organogenesis the embryo starts to generate a transitory population of embryonic hematopoietic cells that serve its immediate needs. These first hematopoietic cells, consisting mainly of primitive erythroid cells, appear in the embryonic circulation in growing numbers and then colonize the initially inactive fetal liver (Medvinsky and Dzierzak 1996). Definitive hematopoiesis, which results in the production of all hematopoietic lineages, develops slightly later and gradually forms a massive pool in the fetal liver, which becomes the main source of hematopoietic stem cells that subsequently colonize the bone marrow (Dzierzak et al. 1998). Hematopoietic stem cells are defined within the context of stringent transplantation assays used for adult bone marrow cells and are characterized by the following properties: (1) they clonally give rise to all differentiated lineages of the blood cells; (2) they are self-renewing; (3) they possess high proliferative/expansion potential contributing to high-level hematopoietic repopulation of the recipient; and (4) they are active long term (over the lifespan of the individual) (Wognum et al. 2003). Definitive hematopoietic stem cells are produced both in the mature yolk sac and within the paraaortic splanchnopleura (PAS, E8.5–E9.5) and the aorta-gonad-mesonephros region (AGM, E10.5–E11.5) (Galloway and Zon 2003). When AGM, yolk sac, and fetal liver cells are directly compared throughout development for repopulation capabilities by the criteria previously mentioned, the AGM region consistently demonstrates more hematopoietic stem cells than the yolk sac or fetal liver at E10 and E11 (Muller et al. 1994). The definitive hematopoietic stem cells formed in the yolk sac and the AGM region do not mature but instead are believed to migrate and seed the fetal liver, where they undergo terminal differentiation (Palis and Yoder 2001). Thus, within the yolk sac there is temporal overlap between primitive and definitive hematopoiesis.

Although it was previously believed that the definitive hematopoietic stem cells formed in the yolk sac contribute only to primitive hematopoiesis (Muller et al. 1994), it was later shown that yolk sac cells isolated around E9 can engraft and repopulate recipient animals following transplantation into the livers of newborn mice (Yoder and Hiatt 1997; Yoder et al. 1997a, b). Furthermore, yolk sac hematopoietic cells have the ability to repopulate adult bone marrow following coculture on certain types of stromal cells (Matsuoka et al. 2001) or when HoxB4 is ectopically expressed (Kyba et al. 2002). Yolk sac and AGM definitive hematopoietic stem cells appear to circulate (Kumaravelu et al. 2002) and can presumably seed intraembryonic tissues such as the liver and large arteries.

1.3
The Hemangioblast and Endothelial Progenitor Cells

The close spatial and temporal appearance of hematopoietic and endothelial cells in the embryo has led to the postulation of a common progenitor of hematopoietic and vascular cells, the hemangioblast. The term hemangioblast was first introduced to describe discrete cell masses that develop in chick embryo cultures and displayed both hematopoietic and endothelial potential (Haar and Ackerman 1971). Since originally introduced, the concept of the hemangioblast has gained support from studies demonstrating that the hematopoietic and endothelial lineages share expression of a number of different genes such as MB1/QH1 in the quail (Pardanaud et al. 1987; Peault et al. 1983) and CD31, CD34, SCL/Tal-1, and VEGFR-2 in the mouse (Kabrun et al. 1997; Kallianpur et al. 1994; Watt et al. 1995; Young et al. 1995). Histologic observations suggest that clusters of hematopoietic cells are attached to the luminal wall of the dorsal aorta, as if arising from the endothelial cells (Tavian et al. 1996; Mukouyama et al. 1998; Tamura et al. 2002). Gene targeting studies in the mouse have provided further evidence for this bipotential precursor in showing that some of these genes are essential for the development of both lineages (Robb et al. 1995; Shalaby et al. 1995; Shivdasani et al. 1995). Furthermore, overexpression of SCL/Tal-1 in zebrafish embryos results in overproduction of common hematopoietic and endothelial progenitors, perturbation of vasculogenesis, and concomitant loss of pronephric duct and somitic tissue (Gering et al. 1998). Analysis of knock-in embryos in which SCL/Tal-1 is expressed under the control of the VEGFR-2 locus suggests that VEGFR-2 and SCL/Tal-1 act in combination to regulate cell fate decisions for formation of endothelial and hematopoietic cells in early development (Ema et al. 2003). Additional support for the existence of the hemangioblast comes from analysis of a specific mutation in zebrafish known as *cloche*, which affects the development of hematopoietic cells and endocardium (Stainier et al. 1995).

Direct and compelling evidence for the presence of the hemangioblast comes from recent studies utilizing the *in vitro* embryonic stem (ES) cell system (Choi et al. 1998). Embryoid bodies differentiated for 2.5–3.5 days contain a unique cell population, the blast colony-forming cell (BL-CFC). BL-CFCs form blast colonies in the presence of vascular endothelial growth factor (VEGF) in methylcellulose cultures (Choi et al. 1998). Gene expression analysis indicates that cells within blast colonies express a number of genes common to both hematopoietic and endothelial lineages, including SCL, CD34, and VEGFR-2 (Kennedy et al. 1997). Most importantly, BL-CFCs give rise to primitive, definitive hematopoietic and endothelial cells when replated in medium with both hematopoietic and endothelial cell growth factors (Choi et al. 1998; Kennedy et al. 1997). Moreover, in the quail embryo, VEGFR-2$^+$ cells isolated from the mesoderm can give rise to cells of the hematopoietic or endothelial lineages depending on the culture conditions (Eichmann et al. 1997). Furthermore, VE-cadherin$^+$ or CD34$^+$CD45$^-$ endothelial cells, sorted from yolk sac and/or PAS/AGM at different stages of development, can generate both hematopoietic and endothelial cells *in vitro*, therefore identifying these cells as a common precursor for both lineages (Yokomizo et al. 2001). Blast colonies have recently been shown to give rise to a third lineage, the smooth muscle cell (Ema et al. 2003). These characteristics of the BL-CFC suggest that it represents the *in vitro* equivalent of the hemangioblast and, as such, one of the earliest stages of hematovascular development described to date.

BL-CFCs have been detected in small numbers in dissected mouse embryos, suggesting that hemangioblasts may form during normal mouse development (Huber et al. 2004). When cells from pools of late streak to neural plate stage concepti (E7.0–7.5) are harvested, they form colonies with the morphology of BL-CFCs within 3–4 days of culture. Cells from these colonies have been found to express markers associated with hematopoietic and vascular development, including VEGFR-2, SCL, vascular endothelial (VE)-cadherin, and GATA1, and can give rise to cells of the hematopoietic, vascular, and smooth muscle lineages in liquid culture (Huber et al. 2004). The technical difficulties encountered in identifying cells with the properties of the hemangioblast in mouse embryos suggests that these cells are produced in very small numbers and for a very short time, indicating that these cells might be short-lived, differentiating soon after their formation.

Recent insights into the role of endothelial progenitor cells in the embryo have been revealed using a quail-chick parabiosis model (Pardanaud and Eichmann 2006). In this study the authors demonstrated that circulating endothelial progenitors are present early in ontogeny. Under normal conditions, they integrate at very low levels into most tissues. When experimental angiogenic responses were induced by wounding or grafts onto the chorioallantoic membrane, endothelial progenitors were rapidly mobilized prior to the establishment of mature bone marrow, and the progenitors were selectively integrated into sites of neoangiogenesis. However, endothelial progenitors were not mobilized during sprouting angiogenesis following VEGF treatment of the chorioallantoic membrane. Therefore, it appears that embryonic endothelial progenitors are efficiently mobilized during the establishment of an initial vascular supply, but they are not involved during classical sprouting angiogenesis in this avian model.

2
Hematopoiesis and Vasculogenesis in the Adult

2.1
The Hematopoietic Stem Cell

The scale of the hematopoietic system is quite remarkable, considering that daily some 2×10^{11} erythrocytes and 5×10^{10} granulocytes—in addition to platelets, lymphocytes, and monocytes—enter the circulation (Finch et al. 1977). Despite the magnitude of this production system, dysregulation is uncommon and external influences can rapidly induce changes in the blood cell count of a specific lineage. For example, hypoxia can induce an increase in erythrocyte production, but does not affect the neutrophil count, whereas the opposite is true in an acute bacterial infection. This prodigious output of mature cells is ultimately dependent on the precise regulation of primitive hematopoietic stem cells, which, in turn, give rise to an ordered series of transit populations of progenitor and precursor cells of progressively more restricted proliferative and differentiation potential. Identifying and characterizing the

Table 1 Commonly used markers for the purification of hematopoietic stem cells in human and mouse. Markers in bold are also expressed on endothelial progenitors

Mouse hematopoietic stem cell markers	CD34$^{low/-}$, **Sca-1$^+$**, Thy1$^{+/low}$, CD38$^+$, c-kit$^+$, Lin$^-$, FGFR$^+$, CD201, **VEGFR-1**
Human hematopoietic stem cell markers	CD34$^+$, CD59$^+$, Thy1$^+$, CD38$^{low/-}$, c-kit^{low}, Lin$^-$, **CD133, VEGFR-1, VEGFR-2**

Lin, lineage markers (CD11b, B220, Gr-1, TER119)

properties of hematopoietic stem cells has been a formidable task, since it is estimated that about 1 in every 10,000 to 15,000 bone marrow cells is thought to be a stem cell (Coulombel 2004). In the bloodstream, the proportion falls to 1 in 100,000 nucleated cells (Coulombel 2004). Despite the burden of producing over 10^{11} cells per day in the human adult, the great majority of stem cells are not dividing at any one time (Lajtha et al. 1971). The current picture of the stem cell pool is that of a small, but potent group of cells, able to maintain tremendous hematopoietic cell supplies through the division of a small fraction of its members, keeping the remainder of the stem cells in reserve. There appear to be two kinds of hematopoietic stem cells. If bone marrow cells from a transplanted mouse can, in turn, be transplanted to another lethally irradiated mouse and restore its hematopoietic system, the source of the regenerated marrow is considered to be derived from long-term stem cells that are capable of self-renewal (Coulombel 2004). Other cells from bone marrow can immediately regenerate all the different types of blood cells, but under normal circumstances cannot renew themselves over the long term, and these are referred to as short-term progenitor or precursor cells (Coulombel 2004). Progenitor or precursor cells are relatively immature cells that are precursors to a fully differentiated cell of the same tissue type. They are capable of proliferating, but they have a limited capacity to differentiate into more than one cell type, as hematopoietic stem cells do (Messner 1998).

Primitive hematopoietic stem cells and their immediate progeny are morphologically indistinguishable (Messner 1998). Despite continued refinement of multiparameter cell separation strategies, unique phenotypic markers that are able to characterize, resolve, and purify hematopoietic stem cells have not been identified. However, several markers have been used in combination in order to separate hematopoietic stem cells. Some of the markers most frequently used to separate human and murine hematopoietic stem cells are listed in Table 1 (Wognum et al. 2003). Several of these markers (CD34, c-kit, CD133, VEGFR-2/KDR) have been used either alone or in combination as putative markers of endothelial progenitor cells. Thus, attempts to quantify endothelial precursor cells—for instance, by enumerating CD34/VEGFR-2 coexpressing cells—should be approached with some caution, as these markers are not restricted to endothelial progenitor cells.

Fluorescent dye staining approaches have also been used to define hematopoietic stem cells based on efflux capacity and staining profile. The combination of both Rhodamine 123 and Hoechst 33342 staining has demonstrated that Hoechst^{low}–Rhodamine^{low} cells are highly enriched for hematopoietic stem cell activity (Goodell et al. 1996). A dual-wavelength flow cytometric analysis of bone marrow based on differential Hoechst 33342 staining that identifies a "side population" (SP) which gives rise to all mature blood lineages in transplanted mice has attracted much interest in the field of hematopoietic stem cells (Goodell et al. 1996).

To detect hematopoietic stem cell functional activity, *in vivo* transplantation is generally considered to be the most appropriate assay because the ability of cells to reconstitute blood cell production in a myeloablated recipient measures the capacity of stem cells for both extensive proliferation and multilineage differentiation over extended time periods (Coulombel 2004).

2.2
Phenotype and Origin of Endothelial Progenitor Cells

Currently it is extremely difficult to differentiate endothelial progenitors from hematopoietic cells or endothelial cells, since the markers used to isolate endothelial progenitors are also expressed on subsets of hematopoietic cells (CD133, CD34, VEGFR-2, VE-cadherin) and mature endothelial cells (CD34, VEGFR-2, VE-cadherin) (Bhatia 2001). Circulating endothelial progenitors are thought to be included in the cell population expressing CD133 and VEGFR-2 within the subset of CD34$^+$ cells (Gill et al. 2001). Moreover, cells included in the SP, which are reported to be enriched for endothelial progenitors, are also enriched for hematopoietic stem cells (Urbich and Dimmeler 2004). When isolated from bone marrow, umbilical cord blood, or mobilized peripheral blood and cultured *in vitro*, a subset of CD133$^+$VEGFR-2$^+$ cells rapidly loses expression of CD133, but acquires features of mature endothelial cells, such as cobblestone morphology, uptake of acetylated low density lipoproteins (AcLDL), and expression of von Willebrand factor (vWF) (Peichev et al. 2000). Thus, the ability of CD133$^+$ cells to differentiate into endothelial cells *in vitro* and *in vivo* provided evidence for the existence of an adult angioblast or endothelial progenitor cell (Gehling et al. 2000).

Fluorescence hybridization analysis of blood samples from bone marrow transplant recipients who had received gender-mismatched transplants combined with endothelial outgrowth assays provided evidence that a transplantable marrow-derived endothelial progenitor cell does exist in humans (Lin et al. 2000). In this study, serial replating of circulating mononuclear cells followed by culture on type I collagen resulted in late outgrowth of endothelial colonies and ensured that endothelial progenitors obtained from the buffy coat were not contaminated by mature endothelial cells that had sloughed off (Lin et al. 2000). Using a similar method with twice-replated blood cells, we have

Table 2 Common cell markers expressed on human endothelial progenitors, mature endothelial cells, and subsets of hematopoietic cells

Hematopoietic cells markers	CD133 (human), CD117 (c-kit), sca-1 (mouse), CD34 (human), VEGFR-1, VEGFR-2, Tie-2, CD31 (PECAM-1), SP
Endothelial cells markers	CD117 (c-kit), CD34, VEGFR-1, VEGFR-2, VEGFR-3 (lymphatic endothelial cells), Tie-1, Tie-2, VE-cadherin, CD31 (PECAM-1), CD105 (endoglin), von Willebrand factor
Endothelial progenitors markers	CD133, CD117 (c-kit), CD34 (human), VEGFR-2, VEGFR-3 (lymphatic endothelial progenitors), VE-cadherin, CD31 (PECAM-1), CD105 (endoglin), von Willebrand factor, SP

PECAM, platelet/endothelial cell adhesion molecule

enumerated endothelial progenitor cells derived from the umbilical cord (Larrivee et al. 2005). Endothelial progenitor cells were estimated to encompass ~1 in 10^7 total mononuclear cells and 1.5 in 10^6 CD133$^+$ cells (Larrivee et al. 2005). Mature endothelial cells derived *in vitro* from endothelial progenitors display higher proliferative potential than endothelial cells such as human umbilical vein endothelial cells (Lin et al. 2000; Quirici et al. 2001). It has also been reported that VEGFR-3 and CD133 identify a population of lymphatic endothelial progenitors in fetal liver, bone marrow, and peripheral blood (Salven et al. 2003). Table 2 displays some of the markers that are expressed on endothelial progenitors, and it shows how these markers overlap with markers found on endothelial cells and subsets of hematopoietic cells.

The origin of adult endothelial progenitors remains unclear. It is only recently that endothelial progenitors have been linked to cells of the blood lineages, and three potential cell types have been suggested to represent the precursor to the endothelial progenitor cell: hematopoietic stem cells and/or hemangioblasts, monocytes, and marrow-derived mesenchymal stem cells (Fig. 2).

As described earlier, in mammalian embryos several lines of evidence support the contention that cells of the endothelial and hematopoietic lineages share a common progenitor, the hemangioblast (Choi et al. 1998). However, the existence of such a cell in the adult has been difficult to demonstrate. When cultured *in vitro* in the presence of VEGF, primitive CD133$^+$ hematopoietic stem cells form colonies, termed late outgrowth colonies (Gehling et al. 2000; Gill et al. 2001). Cells from these colonies express a wide array of endothelial markers, such as VEGFR-2, CD31, VE-cadherin, and vWF, but lose expression of CD133 (Peichev et al. 2000). Moreover, these cells have endothelial pheno-

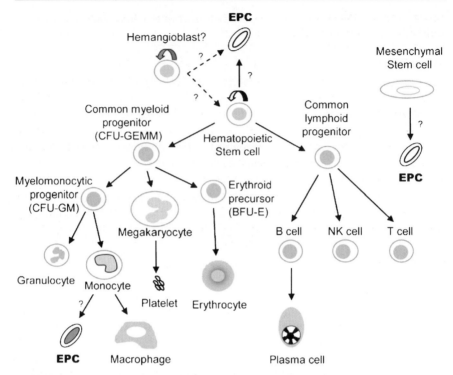

Fig. 2 Schematic representation of the hematopoietic hierarchy and potential origin of endothelial progenitor cells. *EPC*, endothelial progenitor cell

typic properties, and have been reported to integrate into sites of neovascularization *in vivo*. Further evidence of the existence of an adult hemangioblast has been provided by single c-Kit$^+$/Sca1$^+$/Lin$^-$ cell bone marrow transplants in mice, which result in hematopoietic reconstitution as well as integration of rare marrow-derived cells into the endothelium of tumor neovasculature (Grant et al. 2002; Larrivee et al. 2005).

Initially, most attention focused on marrow-derived stem cells as the sole source of endothelial progenitors. However, some studies have suggested that other cells, of hematopoietic and mesenchymal origin, may serve the same purpose (Jiang et al. 2002). Monocytes show some degree of plasticity within the myeloid lineage, and have been shown to differentiate into dendritic cells, macrophages, and Kupffer cells. It has been suggested that a subset of endothelial progenitors may also originate from CD14$^+$ monocytic cells that are capable of assuming morphologic features and phenotypic markers of endothelial cells when cultured in the presence of angiogenic factors (Fernandez Pujol et al. 2001). Endothelial cells derived from cultured monocytes differ from those derived from the differentiation of more primitive hematopoietic stem cells. In the first place, the putative monocytic endothelial progenitors differ from "classical" endothelial progenitors by the expression of monocyte markers (CD14, CD68, CD80, CD45, and CD36) in addition to endothelial markers (CD31, Tie-2, and vWF) (Fernandez Pujol et al. 2000). Furthermore, monocyte-derived endothelial cells display limited proliferative potential and usually die within a few passages, whereas classical $CD14^-$ endothelial cells derived from late outgrowth colonies display high proliferative capacity and can be expanded for numerous passage (Fernandez Pujol et al. 2000; Harraz et al. 2001). However, both subsets of endothelial progenitors have been reported to incorporate into vascular structures of mice after hindlimb ischemia and significantly improve neovascularization (Urbich et al. 2003). It is possible that these progenitors promote neovascularization through different mechanisms. Whereas monocytic $CD14^+$ progenitors may promote angiogenesis through the release of angiogenic cytokines or inflammatory mediators, $CD14^-$ progenitors may accelerate neovascularization by actively incorporating in the endothelium. However, the evidence for this postulate is currently lacking.

Endothelial progenitors have also been shown to originate from a rare cell population, coined the multipotent adult progenitor cell (MAPC) that resides within the mesenchymal stem cell population of human and rodent bone marrow (Reyes and Verfaillie 2001). Unlike hematopoietic stem cells, these mesenchymal stem cells have a high proliferative index *in vitro*, and have been shown to differentiate not only into cells of the mesenchymal lineage, but also to give rise to endothelium, neuroectoderm, and endoderm (Verfaillie et al. 2003). Like hemangioblasts and endothelial progenitors, human MAPCs express VEGFR-2 and CD133 (Reyes et al. 2001). However, unlike endothelial progenitors, they do not express VE-cadherin, CD31, or CD34 (Reyes et al. 2001). In vitro, MAPCs cultured in the presence of VEGF differentiate into cells that express endothelial markers, function *in vitro* as endothelial cells, and contribute *in vivo* to neovessels during tumor angiogenesis and wound healing (Reyes et al. 2001). However, whether these cells can circulate and thereby contribute to angiogenesis at distant sites has never been shown.

2.3
The Angioblast and Evidence of Vasculogenesis in the Adult

The growth of blood vessels through angiogenesis occurs through different mechanisms. Sprouting angiogenesis refers to the process in which activated endothelial cells degrade the basement membrane and migrate toward an inciting factor while laying down a provisional matrix (Papetti and Herman 2002). The endothelial cells then proliferate into the surrounding matrix and form solid sprouts connecting neighboring vessels. These sprouts then develop lumina through a poorly understood morphogenic process. The other means of angiogenesis, intussusception, occurs when the capillary wall extends into the lumen to split a single vessel in two. Intussusception allows the reorganization of existing vessels to permit a vast increase in the number of capillaries without a corresponding increase in the number of endothelial cells (Carmeliet 2005). Until recently, neovascularization in the adult was thought to occur by angiogenesis only (Tonini et al. 2003).

In recent years, another means of neovascularization was described in adults. Vasculogenesis, which refers to the de novo formation of blood vessels from endothelial progenitors, was thought to occur only in the embryo. As described in Sect. 2.2, studies have suggested that cells with endothelial progenitor properties can be isolated from adult sources such as peripheral blood, bone marrow, and umbilical cord blood mononuclear cells (Asahara et al. 1997; Nieda et al. 1997; Peichev et al. 2000; Shi et al. 1998). Likewise hematopoietic cells positive for VEGFR-2 and the hematopoietic stem cell marker CD133 are capable of differentiating into endothelial cells *in vitro* (Gehling et al. 2000; Larrivee et al. 2005). The first studies suggesting the existence of adult circulating endothelial progenitors showed that subsets of bone marrow or peripheral blood mononuclear cells change morphology and start expressing endothelial markers when cultured *in vitro* in the presence of VEGF (Asahara et al. 1997). In vivo, these endothelial progenitors were reported to be incorporated into sites of angiogenesis (Asahara et al. 1999). In the setting of myocardial or hindlimb ischemia, an increase in the mobilization of circulating endothelial progenitors from the bone marrow was reported (Asahara et al. 1999). Likewise, several cytokines such as VEGF, GM-CSF, SDF-1, and erythropoietin have been reported to promote the mobilization of endothelial progenitors (Heeschen et al. 2003; Takahashi et al. 1999; Yamaguchi et al. 2003). In this context, the reported contribution of bone marrow-derived endothelial progenitors to neovessel formation ranges from 5% to 25% in response to granulation tissue formation (Crosby et al. 2000) or growth factor-induced neovascularization (Murayama et al. 2002). The studies suggesting that endothelial progenitors participate in the repair of vascular tissue following ischemic injury led to additional studies suggesting that circulating endothelial progenitor cells also contribute to the formation of the vascular network of tumors (Asahara et al. 1999; Jackson et al. 2001; Kalka et al. 2000). In tumor neovascularization, the initial reported ranges of the contribution of marrow-derived endothelial progenitors to angiogenesis ranged from 35% to 65% (Hilbe et al. 2004; Lyden et al. 2001; Rafii et al. 2002).

While the above studies suggest that circulating endothelial progenitors specifically home to sites of physiological and pathological angiogenesis, and integrate into the neovasculature, more recent studies show minimal to no contribution of marrow-derived endothelial progenitor cells to the endothelium of the neovasculature in adults (Gothert et al. 2004; Larrivee et al. 2005; Machein et al. 2003; Rajantie et al. 2004). Ziegelhoeffer et al. transplanted green fluorescent protein (GFP)-expressing bone marrow cells to investigate whether circulating marrow-derived cells incorporate into collateral arteries after femoral artery ligation (Ziegelhoeffer et al. 2004). After unilateral femoral artery occlusion, the proximal hindlimb and various organs were excised for histological analysis. These investigators failed to detect colocalization of GFP with endothelial or smooth muscle cell markers, but observed accumulations of bone marrow-derived cells around growing collateral arteries and in ischemic distal hindlimbs, which they identified primarily as leukocytes, but also fibroblasts and pericytes. Similar observations were made in the neovasculature of subcutaneously implanted tumors (Larrivee et al. 2005; Ziegelhoeffer et al. 2004).

These findings suggest that in the adult organism, bone marrow-derived cells might not promote vascular growth by incorporating into the endothelium, but rather by functioning as supporting cells. Validation of this position is supported by other studies. For example, Zentilin and colleagues found that while VEGF acts as a potent chemoattractor of bone marrow-derived cells to sites of neovascularization, they were not able to detect direct incorporation of these cells into the newly formed vasculature (Zentilin et al. 2006). Another recent study also demonstrated that marrow-derived cells recruited to angiogenic sites by VEGF enhanced proliferation of endothelial cells by secreting proangiogenic factors distinct from locally derived factors (Grunewald et al. 2006). Similarly, marrow-derived stromal cells were also found to promote neovascularization through the secretion of angiogenic cytokines, rather than by directly integrating into the growing vasculature (Kinnaird et al. 2004). Other studies have also found only rare integration of bone marrow-derived precursors into the endothelium of neovessels using tumor models of experimental angiogenesis (De Palma et al. 2003; Larrivee et al. 2005; Machein et al. 2003; Rajantie et al. 2004; Ziegelhoeffer et al. 2004). In contrast, at least two recent studies have suggested that bone marrow-derived cells may contribute to the neovasculature as pericyte progenitors (De Palma et al. 2005; Song et al. 2005). These latter findings also need to be validated by other groups independently.

3
Functions of Endothelial Progenitor Cells in Physiology and Pathology

3.1
Endothelial Progenitor Cells and Physiological Neovascularization

The adult vasculature is a relatively quiescent organ. The turnover of endothelial cells in healthy tissues is around 3 years (Cines et al. 1998). It has been suggested that under steady-state conditions, endothelial progenitors, which represent less than 1 in 10^7 cells in umbilical cord blood (Larrivee et al. 2005), may contribute to vascular repair (Friedrich et al. 2006). Due to their ability to circulate, certain authors have argued that these cells may represent a reservoir, which may be mobilized during conditions of vascular trauma. Under physiological conditions, it has been reported that endothelial progenitors can be found in a variety of vascular beds. In mice transplanted with bone marrow, 0.2% to 1.4% of endothelial cells in vessels in control tissues were derived from hematopoietic progenitors 4 months after bone marrow transplantation

(Crosby et al. 2000). The rates of endothelial progenitor integration appear to vary depending on the tissue; organs such as the brain show very low levels of endothelial progenitor incorporation, whereas the skin displays relatively high rates of incorporation (up to 2.5% of blood vessels) (Crosby et al. 2000). However, the irradiation used for the bone marrow transplantation may have resulted in higher levels of incorporation of endothelial progenitors than would be observed in physiological conditions, since total body irradiation can result in tissue damage (Chapel et al. 2003). In contrast, endothelial progenitors have been reported to contribute to 8.3% to 11.2% of endothelial cells in vessels that developed in sponge-induced granulation tissue 1 month following marrow transplantation, indicating that rates of incorporation are much higher in models that mimic pathological neovascularization (Kocher et al. 2001). Differences may also relate to the timing of the study following transplantation. In human endothelium, bone marrow-derived endothelial cells were detected at low levels in the skin and gut of sex-mismatched transplant recipients using a combination of immunohistochemistry and XY interphase fluorescent in situ hybridization (FISH) (Jiang et al. 2004).

However, other laboratories have not detected a significant contribution of bone marrow-derived endothelial cells in normal tissues. One study did not detect the presence of bone marrow-derived endothelial cells in pancreatic islets, although they reported the presence of bone marrow-derived epithelial cells (Wang et al. 2006). However, it has been reported that endothelial progenitors are recruited to the pancreas following pancreatic β-cell injury (Mathews et al. 2004). Droetto et al. also reported very low levels of endothelial progenitors in noninjured tissue (Droetto et al. 2004). The low levels of bone marrow differentiation into vascular endothelium seen in these studies may be due to the lack of tissue injury in neonatal transplant models. Even in those reports that did not induce tissue-specific injury, total body irradiation was used to obtain a hematopoietic graft (Lechner and Habener 2003). Irradiation, even at sublethal doses, is known to induce widespread subclinical organ toxicity and thus could result in recruitment of inflammatory cells and/or endothelial progenitors to injured tissues (Chapel et al. 2003). The variation in mechanical or biophysical properties inherent to a specific vascular locus, the absence or presence of inflammatory stimuli, and the unique microenvironment within different organ beds all may significantly affect the recruitment of circulating endothelial progenitors and/or inflammatory cells, thereby explaining some of the apparent differences in the contribution of cells from the bone marrow.

3.2
Contribution of Endothelial Progenitor Cells to Tumor Vascularization

Neovascularization is essential for the growth of solid tumors, and there is some evidence that circulating, bone marrow-derived progenitor cells incorporate into tumor-associated stroma to support tumorigenesis (Zentilin et al. 2006).

The presence of marrow-derived endothelial progenitors in tumor vasculature was first reported by Asahara et al. in murine bone marrow transplantation studies in which endothelial progenitors were identified in highly vascularized lesions of the tumor periphery (Asahara et al. 1997). A possible role for these cells in tumor growth was reported in $Id^{+/-}Id^{-/-}$ mice, which exhibited angiogenic deficits due to an impaired VEGF-driven mobilization of endothelial progenitors as well as an impaired proliferation capacity of VEGFR-1^+ hematopoietic stem cells (Lyden et al. 2001). In these mice, most tumors failed to grow progressively, if at all, and any tumor growth was halted by the lack of an adequate vascularization (Lyden et al. 2001). Following these studies, other laboratories also reported a significant incorporation of endothelial progenitors into the growing tumor vasculature (Garcia-Barros et al. 2003; Hilbe et al. 2004).

The contribution of endothelial progenitors cells to tumor neovessels varies considerably among tumor models (Shaked et al. 2005). Some experimental systems have demonstrated significant but variable integration of endothelial progenitors, whereas in other models bone marrow-derived cells associate, but do not integrate, into the endothelial layer of the vasculature, and promote angiogenesis only indirectly (Hattori et al. 2001; Kopp et al. 2006; Larrivee et al. 2005). For example, in one model of mice transplanted with β-galactosidase-labeled bone marrow, estimates of the levels of marrow-derived endothelial cells integrated in the tumor vasculature were reported to be around 50% (Garcia-Barros et al. 2003). In human tumors, Peters et al. reported that bone marrow-derived endothelial cells contributed to tumor endothelium, averaging 4.9% of the total blood vessels in patients that received bone marrow transplantation with donor cells derived from individuals of the opposite sex (Peters et al. 2005). Other studies show lower but significant levels of endothelial progenitors in the tumor vasculature (Asahara et al. 1997; Crosby et al. 2000). By contrast, other studies show very low to undetectable levels of marrow-derived endothelial progenitor engraftment in the tumor vasculature (De Palma et al. 2003; Larrivee et al. 2005; Machein et al. 2003; Ziegelhoeffer et al. 2004). Whether marrow-derived progenitors integrate into the vasculature, or are merely in close periendothelial association, remains controversial because most analyses are based on histology and are limited by resolution (Zentilin et al. 2006). It is possible that subsets of hematopoietic cells can erroneously be interpreted as endothelial cells, as it has been demonstrated that various types of marrow-derived cells, such as monocytes/macrophages, can reside in close association with endothelial cells in newly formed blood vessels, but do not actually integrate in the blood vessel wall (Gothert et al. 2004; Grunewald et al. 2006; Rajantie et al. 2004; Voswinckel et al. 2003; Ziegelhoeffer et al. 2004).

It is unclear how circulating progenitors home from the bone marrow specifically into the tumor microenvironment, and when during multistep tumorigenesis they are recruited into the preexisting vascular network. Thus, it is likely that the extent to which marrow-derived endothelial progenitors contribute to tumor vascularization is critically regulated by a number of variables, such as tumor type, size, site of implantation, time point of analysis, the type of marker used for their detection and the genetic background of the mouse strain used (Shaked et al. 2005). The type of markers used for the detection of marrow-derived endothelial cells can be especially important, since, as previously mentioned, many cell markers expressed on endothelial cells are also expressed on subsets of hematopoietic cells. Platelet/endothelial cell adhesion molecule (PECAM)-1 and vWF in particular, which are widely used for the detection bone marrow-derived endothelial progenitors, can also be expressed on subsets of hematopoietic cells, such as monocytes and platelets, respectively, which interact with the vessel wall (Newman 1997; Ruggeri 2003). Therefore one must use caution in the interpretation of endothelial progenitor incorporation in the tumor vasculature.

Several studies have suggested that transplanted marrow cells can differentiate into unexpected lineages, including myocytes, hepatocytes, and neurons (Eckfeldt et al. 2005). A potential problem, however, is that reports investigating such differentiation *in vivo* tend to conclude donor origin of transdifferentiated cells on the basis of the existence of donor-specific markers such as lacZ, GFP, or the Y chromosome. Recent work reports the potential of stem cells to fuse with different cell types such as hepatocytes (Grompe 2003) and myocytes. Moreover, it was demonstrated that marrow-derived cells fuse *in vivo* with hepatocytes in liver, Purkinje neurons in the brain, and cardiac muscle in the heart, resulting in the formation of multinucleated cells. No evidence of transdifferentiation without fusion was observed in these tissues (Alvarez-Dolado et al. 2003). This has led to the suggestion that some of the transdifferentiation events reported for tissue-specific stem cells may in fact be a result of cell fusion. Therefore, models of sex-mismatched transplant recipients using XY interphase FISH to detect endothelial progenitors would not be able to discriminate between cell fusion and transdifferentiation (Peters et al. 2005). However, we (and others) have demonstrated that bone marrow-derived endothelial progenitors can arise *in vivo* independently of cell fusion, although the contribution of marrow-derived endothelial progenitors to the tumor endothelium was an exceedingly rare event (Bailey et al. 2004; Jiang et al. 2004; Larrivee et al. 2005). Most studies have used cellular DNA content to determine cell fusion. This assay can be misleading, as fused cells have been shown to lose cellular DNA, which could potentially lead to inaccurate interpretation of results (Pratt et al. 1992). There are also problems with the genetic models which use Cre-expressing mice crossed with reporter mice (floxed reporter allele) to detect fusion, since there may be heterogeneous expression of the Cre recombinase, which would preclude demonstration of cell fusion. Cell fusion cannot therefore be totally excluded when identifying marrow-derived endothelial cells. Furthermore, we have also shown that apoptotic cells can be engulfed by scavenger cells such as macrophages and endothelial cells (Larrivee et al. 2006). When engulfed marrow-derived apoptotic cells (such as neutrophils) express a marker, such as GFP, the scavenger endothelial cell may be misinterpreted as being marrow-derived, therefore adding another level of complexity. Due to these potential reasons for misinterpretation of data and artifactual results, it is possible that levels of endothelial progenitor incorporation into adult blood vessels may be significantly overestimated in some cases.

Whether endothelial progenitors can be used as targets to inhibit tumor blood vessel formation is also unclear. Some have suggested that blockade of endothelial progenitor incorporation into the neovasculature can be used as a target of antitumor therapy. One study has suggested that blocking the mobilization of endothelial progenitors can drastically reduce tumor growth (Lyden et al. 2001). Yet another study demonstrates that spontaneously arising mouse tumors appear to be much less dependent on marrow-derived endothelial progenitors for growth and vascularization than had been reported for subcutaneously grafted tumor models (Sikder et al. 2003). Understanding the factors that regulate contributions from primitive hematopoietic stem cells and their circulating progenitors to new vessel formation may ultimately provide additional ways to influence the process of neovascularization, which may prove beneficial in treating vascular-related diseases. However, the contribution of marrow-derived cells to the tumor vasculature is a highly controversial area. It is therefore clear that further studies are required to determine whether endothelial progenitors can be used as targets for tumor therapies.

More recent studies suggesting that marrow-derived cells may be a source of pericyte progenitors or perivascular cells (macrophages) that contribute proangiogenic cytokines to the tumor neovasculature (De Palma et al. 2005; Grunewald et al. 2006; Song et al. 2005) may provide separate targets for antitumor vascular therapeutics.

3.3
Endothelial Progenitor Cells for Therapy

Therapeutic vasculogenesis refers to the increase of blood vessel growth using cell therapy and has generated interest for a variety of ischemic conditions. Progenitors from bone marrow aspirates or peripheral blood can be reintroduced in the circulation or in the ischemic tissue and have been shown to maintain their ability to participate in blood vessel growth at sites of ischemia or vessel injury (Zammaretti and Zisch 2005). These cells have been shown to have the ability to improve blood circulation in ischemic diseases (Kobayashi et al. 2000). Several studies have shown the therapeutic potential of marrow-derived progenitors. Transplantation of either *ex vivo* expanded endothelial progenitors from $CD133^+VEGFR-2^+$ cells, freshly isolated $CD34^+$ cells, or total bone marrow resulted in increased vessel formation and improved cardiac function (Fuchs et al. 2001; Kawamoto et al. 2001; Kocher et al. 2001). Other researchers have found significant levels of injected endothelial progenitor cells within ischemic tissues, but very low levels in nonischemic tissues, and mature human microvascular endothelial cells injected did not localize to ischemic zones (Park et al. 2004). In one study, CD133$^+$ cells were injected into tissue adjacent to ischemic areas during coronary bypass surgery. All patients were alive and well 3–9 months after surgery: global left-ventricular function was enhanced and infarct tissue perfusion had improved strikingly in most of the patients (Stamm et al. 2003). However, in another study in which patients receiving mononuclear bone marrow cells were compared with those receiving endothelial progenitor cells, there was no difference between treatment groups, suggesting that both cell groups may have the same potential to improve myocardial function and coronary flow reserve (Dzau et al. 2005). However, it is unclear whether the effects observed in these studies come from the integration of endothelial progenitors to sites of injury, or whether they are the result of the mobilization of endothelial progenitors/inflammatory cells that migrate to areas of injury and can participate in vascular recovery by releasing proangiogenic factors (Rajantie et al. 2004; Ziegelhoeffer et al. 2004).

Recently, increased attention has been directed to other populations of marrow-derived cells that are recruited to sites of ongoing angiogenesis, but which do not function as endothelial progenitors. These cells might nevertheless be important to neovessel formation (De Palma et al. 2003; Takakura et al. 2000) by a yet-unknown mechanism, and inhibition of various marrow-derived cells other than endothelial progenitors has also been shown to be of interest for the development of therapies targeting tumor growth. Three types of marrow-derived cells could potentially be used as targets to achieve this goal: (1) endothelial progenitors, (2) mural and stromal cells, and (3) hematopoietic cells. In particular, cytokine-mobilized marrow-derived cells, even though they may not integrate in the vasculature (Zentilin et al. 2006), play a crucial role in the development of new blood vessels through their ability to secrete a variety of angiogenic factors (Grunewald et al. 2006). Since endothelial progenitors appear to integrate in neovessels at a very low frequency in several animal tumor models, it is likely that strategies blocking the mobilization of VEGFR-1 and -2$^+$ cells, which include not only endothelial progenitors but monocytes and other stromal cells, may be useful in inhibiting tumor vascularization by preventing these cells from migrating to tumor sites and releasing inflammatory cytokines. Targeting pericytes has also been shown to be a viable approach to antitumor vascular therapy (Bergers et al. 2003). Thus, the inhibition of recruitment of pericyte progenitors to the developing tumor vasculature may also be an option for targeting the tumor vasculature.

4
Perspectives

Since the discovery of bone marrow-derived endothelial progenitors, there have been remarkable advancements in understanding the mechanisms involved in proliferation, recruitment, mobilization, and incorporation of endothelial progenitors into neoangiogenic vessels. However, their contribution to adult blood vessel formation in physiological and pathological conditions remains very controversial. An increasing number of studies suggest that marrow-derived cells that become associated with the vessel wall represent hematopoietic cells and/or pericytes, rather than true endothelial progenitors that integrate into the neovessel. Emerging trials assessing the role of marrow-derived endothelial progenitors, hematopoietic stem cells, and hematopoietic progenitors in tissue revascularization will open new avenues of research in stem-cell therapeutics. Understanding the mechanisms underlying the role of bone marrow-derived progenitors in physiological and pathological angiogenesis would contribute significantly to our ability to prevent and treat a variety of human pathologies. It is possible that these progenitors, by themselves or in combination with angiogenic factors, or as gene therapy vectors may become important therapeutic tools.

Acknowledgements We thank Ingrid Pollet for figure illustration. Work in the laboratory of A.K. is supported by grants from the Canadian Institutes of Health Research, the Heart and Stroke Foundation of British Columbia and the Yukon, the Canadian Cancer Society, the Stem Cell Network Centres of Excellence, Genome Canada, and Genome British Columbia.

References

Alvarez-Dolado M, Pardal R, Garcia-Verdugo JM, Fike JR, Lee HO, Pfeffer K, Lois C, Morrison SJ, Alvarez-Buylla A (2003) Fusion of bone-marrow-derived cells with Purkinje neurons, cardiomyocytes and hepatocytes. Nature 425:968–973

Asahara T, Murohara T, Sullivan A, Silver M, van der Zee R, Li T, Witzenbichler B, Schatteman G, Isner JM (1997) Isolation of putative progenitor endothelial cells for angiogenesis. Science 275:964–967

Asahara T, Masuda H, Takahashi T, Kalka C, Pastore C, Silver M, Kearne M, Magner M, Isner JM (1999) Bone marrow origin of endothelial progenitor cells responsible for postnatal vasculogenesis in physiological and pathological neovascularization. Circ Res 85:221–228

Bailey AS, Jiang S, Afentoulis M, Baumann CI, Schroeder DA, Olson SB, Wong MH, Fleming WH (2004) Transplanted adult hematopoietic stems cells differentiate into functional endothelial cells. Blood 103:13–19

Bergers G, Song S, Meyer-Morse N, Bergsland E, Hanahan D (2003) Benefits of targeting both pericytes and endothelial cells in the tumor vasculature with kinase inhibitors. J Clin Invest 111:1287–1295

Bhatia M (2001) AC133 expression in human stem cells. Leukemia 15:1685–1688

Carmeliet P (2005) Angiogenesis in life, disease and medicine. Nature 438:932–936

Chapel A, Bertho JM, Bensidhoum M, Fouillard L, Young RG, Frick J, Demarquay C, Cuvelier F, Mathieu E, Trompier F, et al (2003) Mesenchymal stem cells home to injured tissues when co-infused with hematopoietic cells to treat a radiation-induced multi-organ failure syndrome. J Gene Med 5:1028–1038

Choi K, Kennedy M, Kazarov A, Papadimitriou JC, Keller G (1998) A common precursor for hematopoietic and endothelial cells. Development 125:725–732

Cines DB, Pollak ES, Buck CA, Loscalzo J, Zimmerman GA, McEver RP, Pober JS, Wick TM, Konkle BA, Schwartz BS, et al (1998) Endothelial cells in physiology and in the pathophysiology of vascular disorders. Blood 91:3527–3561

Coulombel L (2004) Identification of hematopoietic stem/progenitor cells: strength and drawbacks of functional assays. Oncogene 23:7210–7222

Crosby JR, Kaminski WE, Schatteman G, Martin PJ, Raines EW, Seifert RA, Bowen-Pope DF (2000) Endothelial cells of hematopoietic origin make a significant contribution to adult blood vessel formation. Circ Res 87:728–730

De Palma M, Venneri MA, Roca C, Naldini L (2003) Targeting exogenous genes to tumor angiogenesis by transplantation of genetically modified hematopoietic stem cells. Nat Med 9:789–795

De Palma M, Venneri MA, Galli R, Sergi Sergi L, Politi LS, Sampaolesi M, Naldini L (2005) Tie2 identifies a hematopoietic lineage of proangiogenic monocytes required for tumor vessel formation and a mesenchymal population of pericyte progenitors. Cancer Cell 8:211–226

Droetto S, Viale A, Primo L, Jordaney N, Bruno S, Pagano M, Piacibello W, Bussolino F, Aglietta M (2004) Vasculogenic potential of long term repopulating cord blood progenitors. FASEB J 18:1273–1275

Dzau VJ, Gnecchi M, Pachori AS, Morello F, Melo LG (2005) Therapeutic potential of endothelial progenitor cells in cardiovascular diseases. Hypertension 46:7–18

Dzierzak E, Medvinsky A, de Bruijn M (1998) Qualitative and quantitative aspects of haematopoietic cell development in the mammalian embryo. Immunol Today 19:228–236

Eckfeldt CE, Mendenhall EM, Verfaillie CM (2005) The molecular repertoire of the 'almighty' stem cell. Nat Rev Mol Cell Biol 6:726–737

Eichmann A, Corbel C, Nataf V, Vaigot P, Breant C, Le Douarin NM (1997) Ligand-dependent development of the endothelial and hemopoietic lineages from embryonic mesodermal cells expressing vascular endothelial growth factor receptor 2. Proc Natl Acad Sci U S A 94:5141–5146

Ema M, Rossant J (2003) Cell fate decisions in early blood vessel formation. Trends Cardiovasc Med 13:254–259

Ema M, Faloon P, Zhang WJ, Hirashima M, Reid T, Stanford WL, Orkin S, Choi K, Rossant J (2003) Combinatorial effects of Flk1 and Tal1 on vascular and hematopoietic development in the mouse. Genes Dev 17:380–393

Fernandez Pujol B, Lucibello FC, Gehling UM, Lindemann K, Weidner N, Zuzarte ML, Adamkiewicz J, Elsasser HP, Muller R, Havemann K (2000) Endothelial-like cells derived from human CD14 positive monocytes. Differentiation 65:287–300

Fernandez Pujol B, Lucibello FC, Zuzarte M, Lutjens P, Muller R, Havemann K, Fernandez Pujol B, Lucibello FC, Gehling UM, Lindemann K, et al (2001) Dendritic cells derived from peripheral monocytes express endothelial markers and in the presence of angiogenic growth factors differentiate into endothelial-like cells. Eur J Cell Biol 80:99–110

Fina L, Molgaard HV, Robertson D, Bradley NJ, Monaghan P, Delia D, Sutherland DR, Baker MA, Greaves MF (1990) Expression of the CD34 gene in vascular endothelial cells. Blood 75:2417–2426

Finch CA, Harker LA, Cook JD (1977) Kinetics of the formed elements of human blood. Blood 50:699–707

Friedrich EB, Walenta K, Scharlau J, Nickenig G, Werner N (2006) CD34-/CD133$^+$/VEGFR-2$^+$ endothelial progenitor cell subpopulation with potent vasoregenerative capacities. Circ Res 98:e20–25

Fuchs S, Baffour R, Zhou YF, Shou M, Pierre A, Tio FO, Weissman NJ, Leon MB, Epstein SE, Kornowski R (2001) Transendocardial delivery of autologous bone marrow enhances collateral perfusion and regional function in pigs with chronic experimental myocardial ischemia. J Am Coll Cardiol 37:1726–1732

Galloway JL, Zon LI (2003) Ontogeny of hematopoiesis: examining the emergence of hematopoietic cells in the vertebrate embryo. Curr Top Dev Biol 53:139–158

Garcia-Barros M, Paris F, Cordon-Cardo C, Lyden D, Rafii S, Haimovitz-Friedman A, Fuks Z, Kolesnick R (2003) Tumor response to radiotherapy regulated by endothelial cell apoptosis. Science 300:1155–1159

Gehling UM, Ergun S, Schumacher U, Wagener C, Pantel K, Otte M, Schuch G, Schafhausen P, Mende T, Kilic N, et al (2000) In vitro differentiation of endothelial cells from AC133-positive progenitor cells. Blood 95:3106–3112

Gering M, Rodaway AR, Gottgens B, Patient RK, Green AR (1998) The SCL gene specifies haemangioblast development from early mesoderm. EMBO J 17:4029–4045

Gill M, Dias S, Hattori K, Rivera ML, Hicklin D, Witte L, Girardi L, Yurt R, Himel H, Rafii S (2001) Vascular trauma induces rapid but transient mobilization of $VEGFR2^+AC133^+$ endothelial precursor cells. Circ Res 88:167–174

Goodell MA, Brose K, Paradis G, Conner AS, Mulligan RC (1996) Isolation and functional properties of murine hematopoietic stem cells that are replicating *in vivo*. J Exp Med 183:1797–1806

Gothert JR, Gustin SE, van Eekelen JA, Schmidt U, Hall MA, Jane SM, Green AR, Gottgens B, Izon DJ, Begley CG (2004) Genetically tagging endothelial cells *in vivo*: bone marrow-derived cells do not contribute to tumor endothelium. Blood 104:1769–1777

Grant MB, May WS, Caballero S, Brown GA, Guthrie SM, Mames RN, Byrne BJ, Vaught T, Spoerri PE, Peck AB, Scott EW (2002) Adult hematopoietic stem cells provide functional hemangioblast activity during retinal neovascularization. Nat Med 8:607–612

Grompe M (2003) The role of bone marrow stem cells in liver regeneration. Semin Liver Dis 23:363–372

Grunewald M, Avraham I, Dor Y, Bachar-Lustig E, Itin A, Yung S, Chimenti S, Landsman L, Abramovitch R, Keshet E (2006) VEGF-induced adult neovascularization: recruitment, retention, and role of accessory cells. Cell 124:175–189

Haar JL, Ackerman GA (1971) A phase and electron microscopic study of vasculogenesis and erythropoiesis in the yolk sac of the mouse. Anat Rec 170:199–223

Harraz M, Jiao C, Hanlon HD, Hartley RS, Schatteman GC (2001) CD34– blood-derived human endothelial cell progenitors. Stem Cells 19:304–312

Hattori K, Dias S, Heissig B, Hackett NR, Lyden D, Tateno M, Hicklin DJ, Zhu Z, Witte L, Crystal RG, et al (2001) Vascular endothelial growth factor and angiopoietin-1 stimulate postnatal hematopoiesis by recruitment of vasculogenic and hematopoietic stem cells. J Exp Med 193:1005–1014

Heeschen C, Aicher A, Lehmann R, Fichtlscherer S, Vasa M, Urbich C, Mildner-Rihm C, Martin H, Zeiher AM, Dimmeler S (2003) Erythropoietin is a potent physiologic stimulus for endothelial progenitor cell mobilization. Blood 102:1340–1346

Hilbe W, Dirnhofer S, Oberwasserlechner F, Schmid T, Gunsilius E, Hilbe G, Woll E, Kahler CM (2004) CD133 positive endothelial progenitor cells contribute to the tumour vasculature in non-small cell lung cancer. J Clin Pathol 57:965–969

Huber TL, Kouskoff V, Fehling HJ, Palis J, Keller G (2004) Haemangioblast commitment is initiated in the primitive streak of the mouse embryo. Nature 432:625–630

Jackson KA, Majka SM, Wang H, Pocius J, Hartley CJ, Majesky MW, Entman ML, Michael LH, Hirschi KK, Goodell MA (2001) Regeneration of ischemic cardiac muscle and vascular endothelium by adult stem cells. J Clin Invest 107:1395–1402

Jiang S, Walker L, Afentoulis M, Anderson DA, Jauron-Mills L, Corless CL, Fleming WH (2004) Transplanted human bone marrow contributes to vascular endothelium. Proc Natl Acad Sci U S A 101:16891–16896

Jiang Y, Vaessen B, Lenvik T, Blackstad M, Reyes M, Verfaillie CM (2002) Multipotent progenitor cells can be isolated from postnatal murine bone marrow, muscle, and brain. Exp Hematol 30:896–904

Kabrun N, Buhring HJ, Choi K, Ullrich A, Risau W, Keller G (1997) Flk-1 expression defines a population of early embryonic hematopoietic precursors. Development 124:2039–2048

Kalka C, Masuda H, Takahashi T, Kalka-Moll WM, Silver M, Kearney M, Li T, Isner JM, Asahara T (2000) Transplantation of expanded endothelial progenitor cells for therapeutic neovascularization. Proc Natl Acad Sci U S A 97:3422–3427

Kallianpur AR, Jordan JE, Brandt SJ (1994) The SCL/TAL-1 gene is expressed in progenitors of both the hematopoietic and vascular systems during embryogenesis. Blood 83:1200–1208

Kawamoto A, Gwon HC, Iwaguro H, Yamaguchi JI, Uchida S, Masuda H, Silver M, Ma H, Kearney M, Isner JM, Asahara T (2001) Therapeutic potential of expanded endothelial progenitor cells for myocardial ischemia. Circulation 103:634–637

Kennedy M, Firpo M, Choi K, Wall C, Robertson S, Kabrun N, Keller G (1997) A common precursor for primitive erythropoiesis and definitive haematopoiesis. Nature 386:488–493

Kinnaird T, Stabile E, Burnett MS, Epstein SE (2004) Bone-marrow-derived cells for enhancing collateral development: mechanisms, animal data, and initial clinical experiences. Circ Res 95:354–363

Kobayashi T, Hamano K, Li TS, Katoh T, Kobayashi S, Matsuzaki M, Esato K (2000) Enhancement of angiogenesis by the implantation of self bone marrow cells in a rat ischemic heart model. J Surg Res 89:189–195

Kocher AA, Schuster MD, Szabolcs MJ, Takuma S, Burkhoff D, Wang J, Homma S, Edwards NM, Itescu S (2001) Neovascularization of ischemic myocardium by human bone-marrow-derived angioblasts prevents cardiomyocyte apoptosis, reduces remodeling and improves cardiac function. Nat Med 7:430–436

Kopp HG, Ramos CA, Rafii S (2006) Contribution of endothelial progenitors and proangiogenic hematopoietic cells to vascularization of tumor and ischemic tissue. Curr Opin Hematol 13:175–181

Kumaravelu P, Hook L, Morrison AM, Ure J, Zhao S, Zuyev S, Ansell J, Medvinsky A (2002) Quantitative developmental anatomy of definitive haematopoietic stem cells/long-term repopulating units (HSC/RUs): role of the aorta-gonad-mesonephros (AGM) region and the yolk sac in colonisation of the mouse embryonic liver. Development 129:4891–4899

Kyba M, Perlingeiro RC, Daley GQ (2002) HoxB4 confers definitive lymphoid-myeloid engraftment potential on embryonic stem cell and yolk sac hematopoietic progenitors. Cell 109:29–37

Lajtha LG, Gilbert CW, Guzman E (1971) Kinetics of haemopoietic colony growth. Br J Haematol 20:343–354

Larrivee B, Niessen K, Pollet I, Corbel SY, Long M, Rossi FM, Olive PL, Karsan A (2005) Minimal contribution of marrow-derived endothelial precursors to tumor vasculature. J Immunol 175:2890–2899

Larrivee B, Olive PL, Karsan A (2006) Tissue distribution of endothelial cells *in vivo* following intravenous injection. Exp Hematol 12:1741–1745

Lechner A, Habener JF (2003) Bone marrow stem cells find a path to the pancreas. Nat Biotechnol 21:755–756

Lin Y, Weisdorf DJ, Solovey A, Hebbel RP (2000) Origins of circulating endothelial cells and endothelial outgrowth from blood. J Clin Invest 105:71–77

Lyden D, Hattori K, Dias S, Costa C, Blaikie P, Butros L, Chadburn A, Heissig B, Marks W, Witte L, et al (2001) Impaired recruitment of bone-marrow-derived endothelial and hematopoietic precursor cells blocks tumor angiogenesis and growth. Nat Med 7:1194–1201

Machein MR, Renninger S, de Lima-Hahn E, Plate KH (2003) Minor contribution of bone marrow-derived endothelial progenitors to the vascularization of murine gliomas. Brain Pathol 13:582–597

Mathews V, Hanson PT, Ford E, Fujita J, Polonsky KS, Graubert TA (2004) Recruitment of bone marrow-derived endothelial cells to sites of pancreatic beta-cell injury. Diabetes 53:91–98

Matsuoka S, Tsuji K, Hisakawa H, Xu M, Ebihara Y, Ishii T, Sugiyama D, Manabe A, Tanaka R, Ikeda Y, et al (2001) Generation of definitive hematopoietic stem cells from murine early yolk sac and paraaortic splanchnopleures by aorta-gonad-mesonephros region-derived stromal cells. Blood 98:6–12

Matthews W, Jordan CT, Gavin M, Jenkins NA, Copeland NG, Lemischka IR (1991) A receptor tyrosine kinase cDNA isolated from a population of enriched primitive hematopoietic cells and exhibiting close genetic linkage to c-kit. Proc Natl Acad Sci U S A 88:9026–9030

Medvinsky A, Dzierzak E (1996) Definitive hematopoiesis is autonomously initiated by the AGM region. Cell 86:897–906

Messner HA (1998) Human hematopoietic progenitor in bone marrow and peripheral blood. Stem Cells 16 [Suppl 1]:93–96

Moore MA, Metcalf D (1970) Ontogeny of the haemopoietic system: yolk sac origin of *in vivo* and *in vitro* colony forming cells in the developing mouse embryo. Br J Haematol 18:279–296

Mukouyama Y, Hara T, Xu M, Tamura K, Donovan PJ, Kim H, Kogo H, Tsuji K, Nakahata T, Miyajima A (1998) In vitro expansion of murine multipotential hematopoietic progenitors from the embryonic aorta-gonad-mesonephros region. Immunity 8:105–114

Muller AM, Medvinsky A, Strouboulis J, Grosveld F, Dzierzak E (1994) Development of hematopoietic stem cell activity in the mouse embryo. Immunity 1:291–301

Murayama T, Tepper OM, Silver M, Ma H, Losordo DW, Isner JM, Asahara T, Kalka C (2002) Determination of bone marrow-derived endothelial progenitor cell significance in angiogenic growth factor-induced neovascularization *in vivo*. Exp Hematol 30:967–972

Newman PJ (1997) The biology of PECAM-1. J Clin Invest 100:S25–S29

Nieda M, Nicol A, Denning-Kendall P, Sweetenham J, Bradley B, Hows J (1997) Endothelial cell precursors are normal components of human umbilical cord blood. Br J Haematol 98:775–777

Oberlin E, Tavian M, Blazsek I, Peault B (2002) Blood-forming potential of vascular endothelium in the human embryo. Development 129:4147–4157

Palis J, Yoder MC (2001) Yolk-sac hematopoiesis: the first blood cells of mouse and man. Exp Hematol 29:927–936

Palis J, Robertson S, Kennedy M, Wall C, Keller G (1999) Development of erythroid and myeloid progenitors in the yolk sac and embryo proper of the mouse. Development 126:5073–5084

Papetti M, Herman IM (2002) Mechanisms of normal and tumor-derived angiogenesis. Am J Physiol Cell Physiol 282:C947–970

Pardanaud L, Eichmann A (2006) Identification, emergence and mobilization of circulating endothelial cells or progenitors in the embryo. Development 133:2527–2537

Pardanaud L, Altmann C, Kitos P, Dieterlen-Lievre F, Buck CA (1987) Vasculogenesis in the early quail blastodisc as studied with a monoclonal antibody recognizing endothelial cells. Development 100:339–349

Park S, Tepper OM, Galiano RD, Capla JM, Baharestani S, Kleinman ME, Pelo CR, Levine JP, Gurtner GC (2004) Selective recruitment of endothelial progenitor cells to ischemic tissues with increased neovascularization. Plast Reconstr Surg 113:284–293

Peault BM, Thiery JP, Le Douarin NM (1983) Surface marker for hemopoietic and endothelial cell lineages in quail that is defined by a monoclonal antibody. Proc Natl Acad Sci U S A 80:2976–2980

Peichev M, Naiyer AJ, Pereira D, Zhu Z, Lane WJ, Williams M, Oz MC, Hicklin DJ, Witte L, Moore MA, Rafii S (2000) Expression of VEGFR-2 and AC133 by circulating human $CD34^+$ cells identifies a population of functional endothelial precursors. Blood 95:952–958

Peters BA, Diaz LA, Polyak K, Meszler L, Romans K, Guinan EC, Antin JH, Myerson D, Hamilton SR, Vogelstein B, et al (2005) Contribution of bone marrow-derived endothelial cells to human tumor vasculature. Nat Med 11:261–262

Pratt CI, Wu SQ, Bhattacharya M, Kao C, Gilchrist KW, Reznikoff CA (1992) Chromosome losses in tumorigenic revertants of EJ/ras-expressing somatic cell hybrids. Cancer Genet Cytogenet 59:180–190

Quirici N, Soligo D, Caneva L, Servida F, Bossolasco P, Deliliers GL (2001) Differentiation and expansion of endothelial cells from human bone marrow $CD133^+$ cells. Br J Haematol 115:186–194

Rafii S, Heissig B, Hattori K (2002) Efficient mobilization and recruitment of marrow-derived endothelial and hematopoietic stem cells by adenoviral vectors expressing angiogenic factors. Gene Ther 9:631–641

Rajantie I, Ilmonen M, Alminaite A, Ozerdem U, Alitalo K, Salven P (2004) Adult bone marrow-derived cells recruited during angiogenesis comprise precursors for periendothelial vascular mural cells. Blood 104:2084–2086

Reyes M, Verfaillie CM (2001) Characterization of multipotent adult progenitor cells, a subpopulation of mesenchymal stem cells. Ann N Y Acad Sci 938:231–233; discussion 233–235

Reyes M, Lund T, Lenvik T, Aguiar D, Koodie L, Verfaillie CM (2001) Purification and expansion of postnatal human marrow mesodermal progenitor cells. Blood 98:2615–2625

Robb L, Lyons I, Li R, Hartley L, Kontgen F, Harvey RP, Metcalf D, Begley CG (1995) Absence of yolk sac hematopoiesis from mice with a targeted disruption of the scl gene. Proc Natl Acad Sci U S A 92:7075–7079

Ruggeri ZM (2003) Von Willebrand factor. Curr Opin Hematol 10:142–149

Salven P, Mustjoki S, Alitalo R, Alitalo K, Rafii S (2003) VEGFR-3 and CD133 identify a population of $CD34^+$ lymphatic/vascular endothelial precursor cells. Blood 101:168–172

Shaked Y, Bertolini F, Man S, Rogers MS, Cervi D, Foutz T, Rawn K, Voskas D, Dumont DJ, Ben-David Y, et al (2005) Genetic heterogeneity of the vasculogenic phenotype parallels angiogenesis; implications for cellular surrogate marker analysis of antiangiogenesis. Cancer Cell 7:101–111

Shalaby F, Rossant J, Yamaguchi TP, Gertsenstein M, Wu XF, Breitman ML, Schuh AC (1995) Failure of blood-island formation and vasculogenesis in Flk-1-deficient mice. Nature 376:62–66

Shepard JL, Zon LI (2000) Developmental derivation of embryonic and adult macrophages. Curr Opin Hematol 7:3–8

Shi Q, Rafii S, Wu MH, Wijelath ES, Yu C, Ishida A, Fujita Y, Kothari S, Mohle R, Sauvage LR, et al (1998) Evidence for circulating bone marrow-derived endothelial cells. Blood 92:362–367

Shivdasani RA, Mayer EL, Orkin SH (1995) Absence of blood formation in mice lacking the T-cell leukaemia oncoprotein tal-1/SCL. Nature 373:432–434

Sikder H, Huso DL, Zhang H, Wang B, Ryu B, Hwang ST, Powell JD, Alani RM (2003) Disruption of Id1 reveals major differences in angiogenesis between transplanted and autochthonous tumors. Cancer Cell 4:291–299

Song S, Ewald AJ, Stallcup W, Werb Z, Bergers G (2005) PDGFRbeta[+] perivascular progenitor cells in tumours regulate pericyte differentiation and vascular survival. Nat Cell Biol 7:870–879

Stainier DY, Weinstein BM, Detrich HW, Zon LI, Fishman MC (1995) Cloche, an early acting zebrafish gene, is required by both the endothelial and hematopoietic lineages. Development 121:3141–3150

Stamm C, Westphal B, Kleine HD, Petzsch M, Kittner C, Klinge H, Schumichen C, Nienaber CA, Freund M, Steinhoff G (2003) Autologous bone-marrow stem-cell transplantation for myocardial regeneration. Lancet 361:45–46

Takahashi K, Naito M (1993) Development, differentiation, and proliferation of macrophages in the rat yolk sac. Tissue Cell 25:351–362

Takahashi T, Kalka C, Masuda H, Chen D, Silver M, Kearney M, Magner M, Isner JM, Asahara T (1999) Ischemia- and cytokine-induced mobilization of bone marrow-derived endothelial progenitor cells for neovascularization. Nat Med 5:434–438

Takakura N, Watanabe T, Suenobu S, Yamada Y, Noda T, Ito Y, Satake M, Suda T (2000) A role for hematopoietic stem cells in promoting angiogenesis. Cell 102:199–209

Tamura H, Okamoto S, Iwatsuki K, Futamata Y, Tanaka K, Nakayama Y, Miyajima A, Hara T (2002) In vivo differentiation of stem cells in the aorta-gonad-mesonephros region of mouse embryo and adult bone marrow. Exp Hematol 30:957–966

Tavian M, Coulombel L, Luton D, Clemente HS, Dieterlen-Lievre F, Peault B (1996) Aorta-associated CD34[+] hematopoietic cells in the early human embryo. Blood 87:67–72

Tonini T, Rossi F, Claudio PP (2003) Molecular basis of angiogenesis and cancer. Oncogene 22:6549–6556

Urbich C, Dimmeler S (2004) Endothelial progenitor cells: characterization and role in vascular biology. Circ Res 95:343–353

Urbich C, Heeschen C, Aicher A, Dernbach E, Zeiher AM, Dimmeler S (2003) Relevance of monocytic features for neovascularization capacity of circulating endothelial progenitor cells. Circulation 108:2511–2516

Verfaillie CM, Schwartz R, Reyes M, Jiang Y (2003) Unexpected potential of adult stem cells. Ann N Y Acad Sci 996:231–234

Voswinckel R, Ziegelhoeffer T, Heil M, Kostin S, Breier G, Mehling T, Haberberger R, Clauss M, Gaumann A, Schaper W, Seeger W (2003) Circulating vascular progenitor cells do not contribute to compensatory lung growth. Circ Res 93:372–379

Wang X, Ge S, Gonzalez I, McNamara G, Rountree CB, Xi KK, Huang G, Bhushan A, Crooks GM (2006) Formation of pancreatic duct epithelium from bone marrow during neonatal development. Stem Cells 24:307–314

Watt SM, Gschmeissner SE, Bates PA (1995) PECAM-1: its expression and function as a cell adhesion molecule on hemopoietic and endothelial cells. Leuk Lymphoma 17:229–244

Wognum AW, Eaves AC, Thomas TE (2003) Identification and isolation of hematopoietic stem cells. Arch Med Res 34:461–475

Wong PM, Chung SW, Chui DH, Eaves CJ (1986) Properties of the earliest clonogenic hemopoietic precursors to appear in the developing murine yolk sac. Proc Natl Acad Sci U S A 83:3851–3854

Xu MJ, Matsuoka S, Yang FC, Ebihara Y, Manabe A, Tanaka R, Eguchi M, Asano S, Nakahata T, Tsuji K (2001) Evidence for the presence of murine primitive megakaryocytopoiesis in the early yolk sac. Blood 97:2016–2022

Yamaguchi J, Kusano KF, Masuo O, Kawamoto A, Silver M, Murasawa S, Bosch-Marce M, Masuda H, Losordo DW, Isner JM, Asahara T (2003) Stromal cell-derived factor-1 effects on expanded endothelial progenitor cell recruitment for ischemic neovascularization. Circulation 107:1322–1328

Yoder MC, Hiatt K (1997) Engraftment of embryonic hematopoietic cells in conditioned newborn recipients. Blood 89:2176–2183

Yoder MC, Hiatt K, Dutt P, Mukherjee P, Bodine DM, Orlic D (1997a) Characterization of definitive lymphohematopoietic stem cells in the day 9 murine yolk sac. Immunity 7:335–344

Yoder MC, Hiatt K, Mukherjee P (1997b) In vivo repopulating hematopoietic stem cells are present in the murine yolk sac at day 9.0 postcoitus. Proc Natl Acad Sci U S A 94:6776–6780

Yokomizo T, Ogawa M, Osato M, Kanno T, Yoshida H, Fujimoto T, Fraser S, Nishikawa S, Okada H, Satake M, et al (2001) Requirement of Runx1/AML1/PEBP2alphaB for the generation of haematopoietic cells from endothelial cells. Genes Cells 6:13–23

Young PE, Baumhueter S, Lasky LA (1995) The sialomucin CD34 is expressed on hematopoietic cells and blood vessels during murine development. Blood 85:96–105

Zammaretti P, Zisch AH (2005) Adult 'endothelial progenitor cells'. Renewing vasculature. Int J Biochem Cell Biol 37:493–503

Zentilin L, Tafuro S, Zacchigna S, Arsic N, Pattarini L, Sinigaglia M, Giacca M (2006) Bone marrow mononuclear cells are recruited to the sites of VEGF-induced neovascularization but are not incorporated into the newly formed vessels. Blood 107:3546–3554

Ziegelhoeffer T, Fernandez B, Kostin S, Heil M, Voswinckel R, Helisch A, Schaper W (2004) Bone marrow-derived cells do not incorporate into the adult growing vasculature. Circ Res 94:230–238

Part II
Therapeutic Implication and Clinical Experience

Comparison of Intracardiac Cell Transplantation: Autologous Skeletal Myoblasts Versus Bone Marrow Cells

A. G. Zenovich[1] · B. H. Davis[2] · D. A. Taylor[1] (✉)

[1]Center for Cardiovascular Repair, 312 Church Street SE, NHH 7-105A,
Minneapolis MN, 55455, USA
dataylor@umn.edu

[2]Department of Medicine, Duke University Medical Center, Durham NC, 27708, USA

1	Introduction		118
2	The Goals of Cell-Based Therapies		119
3	Autologous Skeletal Myoblasts: Future Tools for Repair of Failing Myocardium?		120
3.1	Overview of Preclinical Data		120
3.2	Initial Clinical Experience		121
3.3	Important Issues of SKMB Transplantation to Be Resolved		122
	3.3.1	Expansion of Cells	122
	3.3.2	Risk of Arrhythmias	122
	3.3.3	Location of Transplantation: Does It Matter?	126
	3.3.4	Role of the Environment	127
	3.3.5	Inflammation and SKMBs	128
	3.3.6	Autologous Versus Allogeneic Cells	128
3.4	Latest Information		130
4	Bone Marrow Mononuclear Cells: Recent Studies Show Positive Effects in Ischemic Injury		131
4.1	Bone Marrow Mononuclear Cells: Brief Overview		132
4.2	Endothelial Progenitor Cells		133
4.3	Mesenchymal Stem Cells		134
4.4	Cardiac Progenitor Cells		135
4.5	Clinical Studies		136
4.6	Umbilical Cord Blood Cells		144
5	Skeletal Myoblasts Versus Bone-Marrow Cells: How Far to Go to Reach the Best Cell for Cardiac Repair?		145
5.1	Creating a Centralized Registry for the Results of Trials and Biorepository for Blood Samples to Examine Accumulated Data and Set Direction for the Future of the Field		145
5.2	Increased Mechanistic Understanding Should Allow Us to Create the Best Cell-Based "Clinical Product"		146
5.3	The Two Important Steps in Defining "Best Cell"		148
5.4	Evaluating the Best Delivery Route for a Cell-Based Clinical Product Will Be Beneficial for Clinicians and Patients		150
5.5	Arriving at a Consensus Regarding Trial Design and Outcome Measurements		151
5.6	Testing Cell-Based Models in Drug Development to Accelerate Design of Therapies Targeted at Repair		153
6	Summary		154
	References		155

Abstract An increasing number of patients living with cardiovascular disease (CVD) and still unacceptably high mortality created an urgent need to effectively treat and prevent disease-related events. Within the past 5 years, skeletal myoblasts (SKMBs) and bone marrow (or blood)-derived mononuclear cells (BMNCs) have demonstrated preclinical efficacy in reducing ischemia and salvaging already injured myocardium, and in preventing left ventricular (LV) remodeling, respectively. These findings have been translated into clinical trials, so far totaling over 200 patients for SKMBs and over 800 patients for BMNCs. These safety/feasibility and early phase II studies showed promising but somewhat conflicting symptomatic and functional improvements, and some safety concerns have arisen. However, the patient population, cell type, dose, time and mode of delivery, and outcome measures differed, making comparisons problematic. In addition, the mechanisms through which cells engraft and deliver their beneficial effects remain to be fully elucidated. It is now time to critically evaluate progress made and challenges encountered in order to select not only the most suitable cells for cardiac repair but also to define appropriate patient populations and outcome measures. Reiterations between bench and bedside will increase the likelihood of cell therapy success, reduce the time to development of combined of drug- and cell-based disease management algorithms, and offer these therapies to patients to achieve a greater reduction of symptoms and allow for a sustained improvement of quality of life.

Keywords Acute myocardial infarction · Bone marrow · Cell therapy · Heart failure · Stem cells

1
Introduction

Cardiovascular disease (CVD) has become a major health issue throughout the world, exceeding infection and cancer as the leading cause of death in the Western world and in many developing countries (LeGrand 2000; Thom et al. 2006). Although CVD mortality has decreased because of advances in therapies for atherosclerosis, hypercholesterolemia, hypertension, diabetes, and post-acute myocardial infarction (post-AMI) left ventricular (LV) remodeling (Pearson et al. 2002; Smith et al. 2006), CVD still accounts for 1 in every 2.7 deaths in the United States, translating into approximately 2.5 million deaths each year (Thom et al. 2006). In addition, the prevalence of the risk factors for CVD, such as hypertension, obesity, and type 2 diabetes, has been on the rise in recent years (Appel et al. 2006; Haffner 2002; Pearson et al. 2002; Wyatt et al. 2006). Current data show that the incidence of clinical CVD in the 30- to 50-year-old age group is increasing (Juonala et al. 2006; Yan et al. 2006). Moreover, as a result of improved prevention, recognition, and treatment of AMI, the percentage of patients surviving AMI has grown, but unfortunately so has the prevalence of post-AMI heart failure (HF)—with at least a third of patients manifesting HF symptomatology in the first year following AMI (Miller and Missov 2001). Currently, the causes are attributed to both the fairly limited efficacy of pharmacological agents at reducing LV remodeling and hospitalizations for HF exacerbations (Bertrand 2004; Cohn 2002; Doggrell 2005; Hernandez et al. 2005; Jong et al. 2003; Jost et al. 2005; Reiffel 2005; Thattassery and Gheorghiade 2004; Torp-Pedersen et al. 2005), as well as to underutilization of these drugs, which precludes translating the successes observed in trials into clinical settings (Lenzen et al. 2005; Levy et al. 2002). In addition to the increasing number of patients, survival of HF patients has also increased following wider recommendations for clinical use of implantable cardioverter-defibrillators (ICDs) (Moss et al. 2002). The number of patients with an unmet medical need is likely to continue to increase, as the number of people over 65 years of age in the United States doubles in the next 25 years because of aging of the "baby-boomers," with nearly 15% of this population projected to develop HF due to aging, CVD, and type 2 diabetes (Thom et al. 2006).

The urgency of this growing problem has created an unmet need for a more advanced understanding of the entire continuum of CVD so that we can design therapies to treat the entire spectrum of disease. New therapies are needed to prevent LV remodeling after acute injury and to stop the progressive loss of cardiac function in a chronically failing myocardium. Finally, therapies should be designed to halt the CVD process, beginning with improvement of vascular health.

All these needs have fueled research directed at cell-based therapies. The main cell types that have been evaluated clinically are skeletal myoblasts (SKMBs) and bone-marrow derived mononuclear cells (BMNCs), or subsets thereof. In this chapter, we provide a brief overview of the therapeutic effects of each of these types of cells, and also of several important challenges associated with the development of cell-based therapies.

2
The Goals of Cell-Based Therapies

As with every new therapy, be it drug-, device-, or cell-based, potential applications drive the conception and progression of the idea. Cell therapy was envisioned for use after AMI to prevent LV remodeling and onset of HF. Ultimately, cell therapy should be applicable in a broader CVD context: from the beginning stages of atherosclerosis to advanced HF. Such versatility could be afforded by creating "clinical products" in which cell type, purity, dose, route, and criteria for optimal administration (timing of injection relative to injury and degree of injury, frequency of therapeutic application), are based on the continuum of disease. Potential adverse effects will also have to be well-characterized. Taking a multi-faceted approach should generate products with a significant capability of repairing underlying cardiac injury and thus promoting functional recovery to a degree better than current drug therapies can offer; that result alone would represent a paradigm shift in the treatment of CVD.

A realistic goal of cell therapy is restoring at least some degree of function and perfusion to the injured and remodeled myocardium. Full myocardial regeneration is currently not yet achievable, although progress has been made in developing cell-based patches and sheets that could be applied to injured areas of the LV—again, to promote repair (Hata et al. 2006; Liu et al. 2004; Miyagawa et al. 2005). The two primary cell types that have shown capabilities for repair in the heart to date are SKMBs and BMNCs, mostly studied in HF and AMI, respectively.

3
Autologous Skeletal Myoblasts: Future Tools for Repair of Failing Myocardium?

SKMBs, derived from muscle "satellite cells," can expand and form neofibers after muscle injury, thereby regenerating skeletal muscle (Mauro 1961). Because of those properties and because SKMBs express contractile proteins very similar to those in the heart, SKMBs were the first candidate for cardiac repair. The idea muscle-based repair emerged in 1987 and was translated into dynamic cardiomyoplasty, when previously paced latissimus dorsi muscle was surgically wrapped around the failing heart in an attempt to provide some contractile support to the LV (Chachques et al. 1987). Although dynamic cardiomyoplasty did not deliver the results hoped for, cellular cardiomyoplasty (transplantation of SKMBs into the heart) did (Chiu et al. 1995; Murry et al. 1996; Scorsin et al. 1997; Zibaitis et al. 1994). Transplanted cells survived and formed striated muscle grafts within the damaged cardiac tissue, which at the time was considered a success.

3.1
Overview of Preclinical Data

In 1998, we demonstrated for the first time that engraftment of SKMBs into injured myocardium improved LV function and attenuated remodeling (Taylor et al. 1998). In that study, SKMBs improved the contractility of scarred segments of the heart without strict differentiation into cardiomyocytes. Rather, SKMBs yielded myogenin-positive SKMB-like cells (situated in the center of the scar) and myogenin-negative more primitive cardiac muscle-like cells (found around the scar periphery) (Atkins et al. 1999c). The transplanted SKMBs adapted to the surrounding myocardium by forming myofibers that were electrically isolated from host cardiomyocytes and yet improved LV performance (Atkins et al. 1999a).

However, the mechanism(s) of that improvement presented a puzzle that remains unresolved. Numerous pathways have been suggested, from modulation of LV wall stress to active contraction of the injected cells (Ott and Taylor 2006). It is more likely that the improvement of LV function comes from a combination of both a direct effect of the transplanted cells on LV geometry and performance, and a "paracrine" effect of exogenous cells on LV remodeling and angiogenesis (Van Den Bos and Taylor 2003). Because these mechanisms are still not fully understood, there is a considerable variance of opinion with regards to the long-term effect of SKMB transplantation: do autologous myoblasts improve contractility or only prevent further deterioration of the injured myocardial segments? Whether SKMBs are better or worse than other cells to do the former or the latter may not matter, as both SKMBs and BMNCs could be beneficial in patients with HF. It could, however, have implications for the timing of cell therapy. Preclinical data substantiate the claims that autologous SKMBs improve both diastolic and systolic myocardial performance after both acute and chronic injury (Agbulut et al. 2004; Fuchs et al. 2001; Hiasa et al. 2004; Horackova et al. 2004; Hutcheson et al. 2000; Ohno et al. 2003; Ott et al. 2004, 2006; Taylor et al. 1998; Thompson et al. 2003).

3.2
Initial Clinical Experience

The advantages of autologous SKMBs in CVD/HF extend beyond benefits observed in animal models. Their autologous nature (Koh et al. 1993) overcomes the two major limitations of cardiac transplantation in CVD: a shortage of donor tissue and the complexities of immunosuppression. Their capacity for myogenesis increases the likelihood of improved contractility. Finally, their relatively high resistance to ischemia may be crucial for survival in infarcted regions, where ischemia dominates (Reffelmann et al. 2003).

The first observational clinical study using cell therapy to treat CVD was initiated by Menasche and colleagues (2003) in 2000. In this trial, an average of 871×10^6 cells (at least 85% SKMBs) was injected into a nonrevascularizable LV segment as an adjunct to coronary artery bypass grafting (CABG). Significant improvements in LV ejection fraction (EF) and regional wall thickening in the treated segments were observed, suggesting anti-remodeling effects of SKMBs. More recently, Dib et al. (2005) administered SKMBs concurrently with CABG, or as an adjunct to an LV assist device (LVAD) implanted as a bridge to transplantation. Following the combined (with cells) procedure, myocardial perfusion improved and LVEF increased. Several explanted hearts were examined post-LVAD at the time of cardiac transplantation, and engrafted SKMBs were seen in 4 of the 5 specimens within the infarcted regions. In another clinical study, a lower dose of SKMBs (mean of 196×10^6) was injected—as a sole therapy—into infarcted myocardium [via a catheter system capable of percutaneous transluminal non-fluoroscopic LV electromechanical mapping—e.g., NOGA system (Biosense Webster, Inc., California, USA)]) yielding improved regional wall motion and a mildly increased LVEF over 3–6 months (Smits et al. 2003). Data continue to emerge (Chachques et al. 2004; Gavira et al. 2006; Herreros et al. 2003; Ince et al. 2004; Siminiak et al. 2004, 2005; Table 1) showing that SKMBs can be delivered in the context of HF (reduced LVEF, ongoing ischemia, neurohormonal activation, potential hemodynamic instability, risk of arrhythmias, etc.) and not only survive within the infarcted regions of myocardium, but most importantly attenuate LV remodeling (Pagani et al. 2003). The degree of functional improvement may not only depend on the baseline LVEF, but also on the route of delivery. Overall, patients who received SKMBs as an adjunct to CABG demonstrated a mean increase in LVEF ranging from 6% to 18%, while those who received the cells without the concomitant surgical procedure showed 6% to 24% improvement. However, upon a close examination of the data, it is apparent that in those patients whose baseline LVEF is quite low (mean of 24%), the presence or absence of CABG may not matter clinically because of the predominance of irreversibly damaged myocytes, scar, or both. With higher baseline LVEFs (around 35%), CABG could increase blood flow and augment tissue perfusion to improve engraftment so that a larger proportion of cells would contribute to repair. A comparison of patients' treatment regimens and the extent of revascularization of the cell-treated area should offer insights into of whether or not the differences in outcome simply represent variations in the use of SKMBs or geographical and institutional differences in the treatment of HF.

3.3
Important Issues of SKMB Transplantation to Be Resolved

3.3.1
Expansion of Cells

The first hurdle associated with any autologous cells is the need for cell expansion in the clinical setting. This process necessitates a sufficient time between injury, harvest of cells, and therapeutic application. In many patients (as well as healthy donors), this window of time ranges from several days to several weeks, which does not seem to pose a problem in the context of chronic injury (typical for HF). However, in the case of AMI, a treatment without delay may be significantly more beneficial than a postponed intervention with regards to its effects on the acutely ischemic area. Therefore, alternative cells could be employed, such as BMNCs, or—if myoblasts are truly superior—allogeneic cells from a healthy appropriately matched donor may offer a solution. Randomized studies comparing SKMBs and BMNCs in this context would bring substantial clarity to this issue.

3.3.2
Risk of Arrhythmias

The reports of electrical adverse events in patients after autologous SKMB transplantation have generated broad skepticism within the clinical community about the safety of this potential treatment option. Despite the appropriateness of these concerns, these events should be placed in the pathophysiological context of systolic HF, where arrhythmias are inherent to the disease process

Table 1 Published cell therapy trials with SKMBs to date. Included are studies with five or more patients enrolled and for which results were published prior to 2 November 2006, where methodological and outcome details were available. For details, please refer to original publications

Investigator (Country)	No. of patients	Dx	Average SKMB dose[a] ×10^6	Delivery route	Length of follow-up (months)	Baseline LVEF, %	Follow-up LVEF, %	Functional outcomes; change in symptomatology (NYHA class)
Dib et al. 2005 (USA)	30	Post-AMI HF	2.2–300	Transepi+CABG or LVAD	24	28	35 (y 1) 36 (y 2)	Myocardial viability (by PET) increased. NYHA class significantly improved (baseline: mean of 2.1; 1-y follow-up: mean of 1.4) but increased at 2-y follow-up (no differences with baseline).
Chachques et al. 2004 (France)	20	Post-AMI	300	Transepi w/o CABG	14±5	28	52	Wall motion score improved, glucose uptake (by PET) increased NYHA class significantly improved (baseline: mean of 2.5; follow-up: mean of 1.2).
Herreros et al. 2003; Gavira et al. 2006 (Spain)	12 (+14 historical controls)	Ischemic HF+prior AMI	221	Transepi+CABG	3 and 12	36; controls: 36	54 (3 m); 55 (12 m); Controls: 39	Regional wall motion (by E) and viability (glucose uptake by PET) improved 7/12 patients improved by 1 NYHA class.
Menasche et al. 2003 (France)	10	Post-AMI HF	871	Transepi w/o CABG	10.9	24	32	14/22 segments demonstrated increased systolic thickening NYHA functional class significantly improved (baseline: 2.7±0.2; follow-up: 1.6±0.1).

Table 1 (continued)

Investigator (Country)	No. of patients	Dx	Average SKMB dose[a] ×10⁶	Delivery route	Length of follow-up (months)	Baseline LVEF, %	Follow-up LVEF, %	Functional outcomes; change in symptomatology (NYHA class)
Siminiak et al. 2004 (Poland)	10	>3 m post-AMI	0.4–50 (range)	Transepi+CABG	4 and 12	35 (25–40)	42 (29–47)	Of 9 previously dyskinetic segments, 5 became akinetic, and of 10 akinetic segments, 4 became hypokinetic at 4 m, this effect was maintained at 12 m Reduction in symptomatology not assessed.
Siminiak et al. 2005 (Poland)	9	Post-AMI HF	17–106 (range)	Transcoronary	10 wks	41 (30–49)	44 (33–49)	Symptoms improved in all patients by at least 1 NYHA class Symptoms improved in all patients by at least 1 NYHA class.
Ince et al. 2004 (Germany)	6 (+6 controls)	Ischemic HF	210	Transendo	12	24; controls: 24	32; controls: 21	Walking distance and NYHA class significantly improved.
Smits et al. 2003 (Netherlands)	5	Post-AMI (anterior) HF	196	Transendo EMM-guided	3 and 6	36	41	Regional contractility (by MRI) significantly increased.

Abbreviations: AMI, acute myocardial infarction; CABG, coronary artery bypass grafting; Dx, diagnosis; E, echocardiography; EMM, electromagnetic mapping; HF, heart failure; ICM, ischemic cardiomyopathy; LV, left ventricle; LVEF, left ventricular ejection fraction; m, month(s); MRI, magnetic resonance imaging; NYHA class, New York Heart Association functional class; PET, positron emission tomography; SKMB, skeletal myoblast; Transendo, transendocardial; Transepi, transepicardial; wks, weeks; y, year(s)

[a]"(Range)" denotes dose-escalation or variation was used

(Moss 2003). In fact, many of the patients in recent trials (Table 1) met the Multicenter Automatic Defibrillator Implantation Trial (MADIT)-II and now Sudden Cardiac Death in Heart Failure Trial (SCD-HeFT) criteria (Mark et al. 2006; Moss et al. 2002), which were presented after those cell therapy trials had begun. These criteria suggest that patients with systolic HF of ischemic etiology (post-AMI) with LVEF less than or equal to 30%–35% may significantly benefit from ICDs in terms of survival, as these devices are highly successful in terminating lethal arrhythmias. Whether SKMBs augmented arrhythmogenesis in these patients is difficult to discern. Dib and colleagues (2005) have not reported a dramatically increased incidence of electrical instability after SKMB administration. In fact, there were only 2 patients with ventricular tachycardia (VT) post-cell transplantation with no clear-cut connection to cellular cardiomyoplasty. Specifically, 1 patient displayed a VT of ischemic etiology (attributable to stenoses of the bypass grafts), and another patient displayed VT, bigeminy, and junctional rhythm—all of which disappeared after discontinuation of digoxin and initiation and up-titration of carvedilol. Supplementation with low-dose amiodarone has shown beneficial effects: a reduction of VT from 40% to 10% (Siminiak et al. 2004, 2005). It is also possible that some of the patients that exhibited arrhythmogenesis could have had a suboptimally treated HF.

Alternatively, arrhythmias as much may have been avoided in more recent studies as a result of the shift in clinical practice owing to low-dose amiodarone and the inclusion of patients who met the MADIT-II and SCD-HeFT criteria and had already received ICDs. However, this strategy may be problematic in terms of quantifying efficacy of cell-based treatments. In currently published studies (Table 1), patients with a mean baseline LVEF of 30% or lower exhibited symptomatic benefits following the SKMB procedure, but the anti-remodeling effects were not as pronounced as in patients whose LVEF were in the 30% to 40% range. The case might very well be that the patients who are eligible for an ICD under the MADIT-II or SCD-HeFT criteria may have a degree of injury too large to be repaired by SKMBs any other cell type, or the ischemic process (or both) and the chronic downregulation of blood flow could create conditions that are relatively harsh for transplanted cells to survive, even taking into consideration that SKMBs are considered relatively resistant to ischemia.

The presence of arrhythmias highlights another potential limitation to cell-based therapy: the unclear (so far) ability of any transplanted cells to electrically integrate with native tissue (Abraham et al. 2005). Because SKMBs are the best-studied and most myogenic cell type, more experience exists with these cells than others. Scorsin et al. (2000) did not observe any electrical integration of myoblasts preclinically. Similarly, Suzuki and co-authors (2001) reported that in the absence of connexin-43 overexpression, SKMBs did not couple very well with surrounding myocardium but the treated area synchronously contracted with surrounding tissue and contributed to overall cardiac performance. Our group has reported similar findings (Thompson et al. 2003).

So is coupling important? Are connexin-43 and N-cadherin (Reinecke et al. 2000, 2002) the only requirements for electrical coupling of cells? Or does the coupling involve additional specific molecular factors? To what extent is electrical integration of the graft with the surrounding myocardium related to the function of the myocardial segment? Some answers to these questions are emerging, but more research on mechanistic insights into this process is much needed. In this regard, Marban's group had made progress on antiarrhythmic engineering of SKMBs by genetically modifying SKMBs to express gap junctions (and connexin 43) (Abraham et al. 2005). Interestingly, cocultures of human SKMBs with rat cardiomyocytes produced spiral reentry waves—similar to VT/ventricular fibrillation in humans—that were terminated by nitrendipine, L-type Ca^{2+} channel blocker, but not by lidocaine (standard treatment for reentry). The genetically modified SKMBs had a much higher proportion of cells that did not exhibit arrhythmias in culture. Therefore, there is much to be learned with regards to mechanisms of arrhythmias after cell transplantation.

3.3.3
Location of Transplantation: Does It Matter?

The location of transplanted cells (center of scar or its periphery), the homogeneity of the scar and its contractile properties relative to the border zone, the environment of the scar and the border zone, and the number of cells engrafted play major roles in electrical outcome. However, we have yet to dissect the variables involved and to assign primary and secondary order of importance. For example, there are data showing increased incidence of arrhythmias in animals who receive SKMBs into the border zone versus the center of the infarct (Atkins et al. 1999a, b), and there are preliminary observations showing the exact opposite (J. McCue, C. Swingen, T. Feldberg, C. Caron, S. Prabhu, R. Motillal, D.A. Taylor, unpublished data).

Notwithstanding the heterogeneity of the data, the importance of location in terms of functional outcome has recently been highlighted by several studies where delivery of cells into revascularized versus nonrevascularized scar augmented outcome. More recently, Ott et al. demonstrated that the close proximity of small injections of transplanted SKMBs (microdepots) led to a more uniform contractile improvement compared with larger volume injections with greater distances from each other (macrodepots), likely due to lesser environmental stress per cell and possibly paracrine influences (Ott et al. 2005a). Recently, we have developed several approaches to increase the ability to direct cell delivery to specific locations. First, we developed a thoracoscopic approach (Thompson et al. 2004), and more recently a robotic approach (Ott et al. 2006). Specifically, SKMBs have been transplanted into the embolization-induced HF myocardium using the da Vinci robotic system (Ott et al. 2006). A high degree of precision has been reached in the placement of the cells into apical, anterior, and lateral segments. LVEF, wall thickening, regional wall motion, and LV end-diastolic volume have improved following the procedure as demonstrated by magnetic resonance imaging (MRI). This approach may bring better outcomes due to a higher capacity to control the placement of the cells in a hypo/akinetic versus dyskinetic myocardium. Clearly, the location of SKMB transplantation into the myocardium does matter. As more preclinical and clinical studies emerge, we will be moving closer to a more complete understanding of why cell location matters from a mechanistic standpoint.

3.3.4
Role of the Environment

The environment of the infarcted myocardium consists of the border zone with viable or partially viable cells and a scarred center. These two areas have distinctly different oxygen concentrations, which are primarily dependant on the blood flow. The border zone and the scar also possess different diffusion characteristics. Recent experiments from our group (B.H. Davis, T. Schroeder, M.D. Dewhirst, K. Olbrich, D.A. Taylor, unpublished data) performed in C_2C_{12} myoblasts placed in an artificial three-dimensional infarction construct have shown that the survival of transplanted cells decreased toward the center of the scar, where the milieu becomes more ischemic (Fig. 1). Availability of oxygen remains one of the main limiting factors in cell survival, even with SKMBs that are believed to be relatively ischemia-resistant. An attempt to improve cell survival by increasing available glucose did not rescue the cells from hypoxia-induced apoptosis. However, improving biochemical alterations (such as amino acids) may have a better impact on cell viability rather than blood flow alone. In this regard, glutamine deprivation reduced oxygen consumption rate (OCR) in the myoblasts within the infarct (Fig. 1). The myoblasts with reduced OCR survived better in an ischemic milieu because of increased oxygen penetration depth. Together, these observations suggest that submersion of cells into the glutamine-free media could improve survival. Alternatively, glutamine antagonists could be used to pretreat the cells prior to transplantation. Unfortunately, the use of glutamine antagonists in patients to precondition the ischemic myocardial tissue is not a viable option due to a high rate of central nervous system toxicity (lethargy, confusion, and decreased mental status) (Hidalgo et al. 1998), an adverse reactions profile that would not be suitable for CVD patients undergoing cell therapy. However, treating cells in vitro could overcome these limitations. Esterified L-cysteine-*S*-*N*-methylcarbamate, which showed reduction of glutamine concentration in several tumors, could also be potentially examined as an adjunct glutamine reducer in SKMB media (Jayaram et al. 1990).

3.3.5
Inflammation and SKMBs

Inflammation is likely a cue for endogenous recruitment and repair. Cytokines and inflammatory mediators may also be important in the survival of cells. More than half of HF patients have atherosclerosis (Thom et al. 2006), which has now been established as an inflammatory disease (Hansson 2005). As cytokines mediate inflammatory responses and are largely responsible for the progression of atherosclerotic plaque (Hansson 2005), they could also negatively affect SKMB survival, most likely either via upregulation of NF-κB or downregulation of Akt (Tan et al. 2006). A considerable amount of work needs to be done to understand the interactions of SKMBs (or any transplanted cell type) with proinflammatory factors and the importance of those interactions in relationship to outcome. So far, it is unclear if cytokines that increase smooth muscle cell proliferation in atherosclerosis, such as interleukin (IL)-10 and IL-18 (Raines and Ferri 2005), would impact cells' fate, and whether monocyte chemoattractant protein (MCP)-1, IL-1β, fractalkine, interferon (IFN)-γ and/or stromal-derived factor (SDF)-1α participate in cell survival.

3.3.6
Autologous Versus Allogeneic Cells

As the development of therapeutic SKMB transplantation continues, the use of allogeneic (instead of autologous) cells may advance into clinical studies. The main reason for the use of allogeneic cells is the availability of an off-shelf product with known potency and defined characteristics. These factors are likely to be important both for use in an acute injury setting and in patients

Fig. 1a–c a Penetration depth of oxygen (*solid black bars*) and glucose (*ascending bars*) in a tissue-engineered construct seeded with primary SKMBs. Oxygen (55 mmHg) and glucose (5 mM) concentrations at the surface of the gel were designed to approximate values seen proximal to capillaries in mature myocardial tissue. As seen in this panel, oxygen concentration drops to zero in a fraction of the distance of glucose, indicating that oxygen (and not glucose) may be the limiting factor for survival of transplanted SKMBs. **b** Oxygen penetration depth in tissue-engineered constructs seeded with primary SKMBs maintained in control conditions (*solid black bars*) (5 mM glucose and 4 mM glutamine at the construct surface), high glucose conditions (*ascending bars*) (20 mM glucose and 4 mM glutamine at the construct surface), and glutamine deprivation conditions (*gray bars*) (5 mM glucose and 0.05 mM glutamine at the construct surface). As shown in this panel, constructs under glutamine deprivation allowed oxygen to penetrate much deeper into the constructs. **c** Cell survival in constructs described in panel b as measured by nitro blue tetrazolium chloride (NBT) staining. Cell survival was statistically equal ($p=0.589$) under control (*solid bars*) and high glucose conditions (*ascending bars*). Cell survival was significantly increased in constructs under maintained glutamine deprivation by 1,500 µm depth (*, $p=0.009$). Improved viability was secondary to improved oxygen penetration

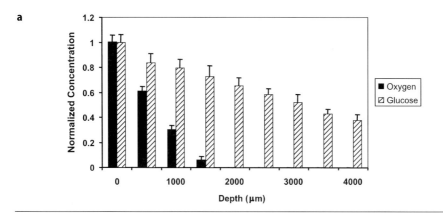
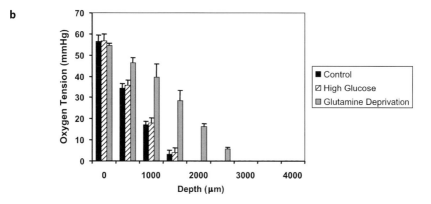
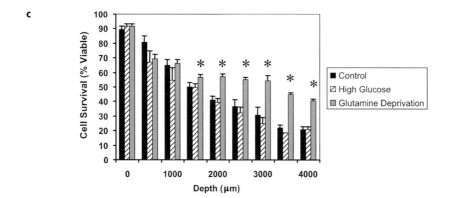

with advanced HF, where SKMBs have been subjected to a prolonged period of stress: globally reduced cardiac output, alterations in levels of oxygen, amino acids and other metabolites, neurohormonal activation, etc. These cells are likely to be suboptimal and may potentially offer less functional benefits. In this regard, important progress has been made by Skuk et al. who have carried out allogeneic SKMB transplantation in nonhuman primates (Skuk et al. 2002). Obviously, the use of allogeneic cells will require carefully optimized immunosuppression. The bad news is that HF patients are already receiving extensive pharmacological regimens, and therefore, additional medications could impose the risks of drug interactions and synergistic adverse reactions. The good news is that immunomodulatory strategies are being actively explored, and so far the combination of tacrolimus and mycophenolate mofetil has been able to achieve efficient immunosuppression with reduced side effects following allogeneic SKMB transplantation. We can also learn from allogeneic myoblast transplantation in Duchenne muscular dystrophy (Camirand et al. 2004), where insights into SKMB survival and transplant tolerance will determine the potential of the use of allogeneic SKMBs in patients with CVD.

3.4
Latest Information

The Myoblast Autologous Grafting in Ischemic Cardiomyopathy (MAGIC) trial, a phase II randomized clinical trial to examine the efficacy and safety of CABG plus SKMBs versus CABG alone in approximately 300 patients in North America and Europe, has been recently reported at the Scientific Sessions of the American Heart Association in November 2006 (Menasche 2006). The trial was halted in the first part of 2006 because the design of the trial was no longer considered state-of-the-art (as the number of CABG cases declined), and recruitment fell much below projected targets. France, Belgium, the United Kingdom, Germany, Italy, and other European countries actively participated in this trial.

The inclusion criteria were LVEF between 15% and 35%, and a history of AMI prior to screening with residual akinesia, affecting at least three contiguous LV segments, that was unresponsive to administration of dobutamine. In addition, patients had to have a clinical indication for CABG. This trial utilized two doses of SKMBs: a mean of 400×10^6 and 800×10^6 cells. The cells were harvested and expanded under guidelines Good Manufacturing Practice in Paris, France, and Cambridge, MA, USA, achieving viabilities of 95% and a purity of 89%. Thirty patients received high dose of SKMBs, 33 received the low-dose, and 34 a placebo. CABG was performed in noncell-transplanted LV segments in all patients who received cell therapy, and the mean cross-clamp time ranged from 59 to 64 min without significant differences between the three groups.

Kaplan-Meier analysis of survival free of major adverse cardiac events (MACE) showed no differences between high-dose and placebo and low-dose versus placebo at 30 days and 6 months following the procedure ($p=0.12$ and

0.87 for the high-dose group and $p=0.43$ and 0.09 for the low-dose group, respectively). MACE curves separated early in the course of follow-up (at 1 month), although the trial was not powered to detect these differences statistically. Nevertheless, patients in the high-dose SKMB group appeared to have survived equivalently or slightly better compared to placebo, indicating no increase in MACE attributable to SKMB transplantation. Time to first ventricular arrhythmia was not different among the three groups at both 30 days and 6 months ($p=0.30$ and 0.12 for the high-dose group, and $p=0.20$ and 0.23 for the low-dose group). These data appear to significantly decrease if not dissolve concerns about potential augmentation of arrhythmogenesis with SKMBs.

Importantly, administration of high-dose SKMBs reduced LV end-diastolic volume by the mean of 23 ml (range: −42 to 0 ml), $p=0.006$, and decreased LV systolic volume by the mean of 18 ml (−34–6.0 ml), $p=0.008$ versus placebo. These changes translated into a mean 3.0% change in LVEF (3.0–14.0), $p=0.04$ compared with the placebo group. Of course, the limitations of this trial are in the small number of patients, relatively short length of follow-up, and the inability to perform MRI for a more precise quantification (versus echocardiography) of the anti-remodeling effects of SKMB therapy. Another important remark is that no patient in either the low- or high-dose group exhibited overt acute HF.

Overall, the MAGIC trial showed that SKMB administration has favorable effects on the LV remodeling process, which is the culprit of HF progression. This outcome was achieved without compromising safety of patients. The results of the MAGIC trial in conjunction with earlier studies in the United States and Europe suggest, at the very least, that SKMBs deserve a more in-depth evaluation.

In summary, autologous SKMB administration for patients with HF has the potential of being a relatively efficacious treatment, if such holds true in definitive phase III trials. The results of the MAGIC trial have certainly brought back the enthusiasm for SKMB-based repair of myocardial damage and the possibility of SKMB therapy being a part of the treatment paradigm in patients with HF. The role of SKMBs in the treatment of AMI remains to be determined.

4
Bone Marrow Mononuclear Cells: Recent Studies Show Positive Effects in Ischemic Injury

BMNCs have received attention recently following presentation of the latest clinical trials showing reduction in a composite endpoint of death, recurrence of myocardial infarction and revascularization in patients post-AMI (Schachinger et al. 2006b). In the past bone marrow cells were thought to give rise only to hematopoietic cells. We now serow that BMNCs are a heterogeneous population of hematopoietic precursors, containing endothelial progenitor cells (EPCs) and their subsets (e.g., AC 133$^+$, or VEGF R2$^+$, or CD34$^+$ progenitors), mesenchymal stem cells (MSCs), and multiple other populations including monocyte precursors, T and B cell precursors, CD14 cells, etc. (Saulnier et al. 2005; Verfaillie et al. 2003). Similar cell populations have been isolated from umbilical cord bloodw (Zhai et al. 2004). As these cell types show satisfactory results in treatment of AMI, several aspects merit a discussion.

4.1
Bone Marrow Mononuclear Cells: Brief Overview

Several BMNCs populations have been targeted for cardiac repair. These include c-kit$^+$ cells, EPCs, mixed BMNC fractions, and other subsets. At issue is their potential for cardiac-related angiogenesis and myogenesis. Bone marrow-derived progenitors by their nature respond to the microenvironment and develop a correspondent phenotype (Orlic et al. 2001). The differentiation of BMNCs into cardiomyocyte-like cells has been demonstrated (Makino et al. 1999; Pittenger et al. 1999). Tomita et al. have substantiated that finding showing that BMNCs transplanted into cryoinjured myocardium differentiate into myogenic cells expressing myosin heavy chain and troponin I—hallmarks of muscle cells (Tomita et al. 1999). Therapeutic agents, such as dexamethasone (Grigoriadis et al. 1988) and 5-azacytidine (Wakitani et al. 1995), accelerated the formation of myotube-like structures in vitro, and the cells then started beating spontaneously. Wang and colleagues (2000) demonstrated differentiation of BMNCs (in the environment of a normal myocardium) into cardiomyocytes that contained not only myosin heavy chain but also gap junctions. Orlic and coauthors confirmed that a BMNC population (lin$^-$ c-kit$^+$ cells) could repair myocardial scar when delivered subcutaneously together with granulocyte-colony stimulating factor (Orlic et al. 2001). Other cell populations isolated from BMNCs also achieved positive results (i.e., reduction of LV remodeling and lessening the degree of cardiomyocyte apoptosis) (Kocher et al. 2001). In that regard, our group has also shown that when injected into the center and the peripheral regions of the scar, some BMNCs differentiated into striated muscle and improved LV function (Thompson et al. 2003). Whether BMNC populations do or do not produce functioning cardiomyocytes may not be clinically relevant, as Murry et al. (2004) have shown that lin$^-$ c-kit$^+$ cells did not produce cardiomyocytes after transplantation into the ischemic myocardium, but instead differentiated into hematopoietic cells; however, despite the lack of differentiation into cardiomyocytes, these cells did prevent LV remodeling. These and other (Jackson et al. 2001) findings indicated that bone marrow administration may deliver beneficial results when transplanted into a patient with CVD, without ethical dilemmas associated with the use of embryonic progenitors. Moreover, the techniques of collection and expansion of bone marrow cells have been established because of the use of bone marrow transplantation in hematology and oncology as a therapeutic procedure.

4.2
Endothelial Progenitor Cells

EPCs are bone marrow-derived cells that express CD133 (AC133), CD34, and vascular endothelial growth factor (VEGF)-R2 (KDR) markers on their surface at various times in their maturation process. EPCs are considered to play an important role in maintaining vascular integrity and mediating angiogenesis (Hill et al. 2003; Kalka et al. 2000). Recent data have shown associative evidence between the quantity and function of circulating EPCs and the risk for CVD. For example, the number and the functional capacity of $CD34^+KDR^+$ EPCs was inversely related to the level of risk for CVD in 519 patients and also correlated with a composite measure of events (AMI, hospitalization, revascularization, or CVD-related death) at follow-up after adjustments for age, gender, and other risk factors and covariates (Hill et al. 2003; Werner et al. 2005). Event-free survival increased proportionately to the baseline level of EPCs. Fadini et al. showed that the same population of EPCs was an independent predictor of early atherosclerosis measured by carotid intima-media thickness in 137 subjects (Fadini et al. 2006). However, we believe that the risk for CVD could be better reflected by a combined assessment of several BMNC-derived, EPC-related populations—"reparative" (such as $CD34^+$, $AC133^+$ EPCs) and "pro-inflammatory" (such as $CD45^+$, $CD14^+$, and $CD3^+$, and the like) cells. Our group recently showed that a reduction in $CD31^+CD45^-$ vascular progenitor cells, thought to be related to EPCs, is associated with aging and disease state in the mouse $ApoE^{-/-}$ model of atherosclerosis (Rauscher et al. 2003). In our hands, the delivery of functionally viable cells could prevent the progression of atherosclerosis and reduce inflammation, as reflected by decreasing circulating IL-6.

Recent research has shown that the number of circulating EPCs is increased in patients following AMI, most likely representing an attempt for endogenous repair (Shintani et al. 2001). EPCs are presumed to be mobilized by the ischemic damage in the heart (and other tissues) and migrate to damaged areas to induce neovascularization. When EPCs were injected into the rats tail vein or LV cavity after a period of ischemia, more than a twofold increase in the homing of infused EPCs was observed when compared to animals undergoing sham surgery (Aicher et al. 2003). LV dimensions, fractional shortening, and regional wall motion improved only in rats that received EPCs and not in the controls (Kawamoto et al. 2001). It is likely that the improvements seen in part depended on increased myocardial perfusion and decreased inflammation following administration of EPCs. Indeed, after an intravenous infusion of EPCs into an infarcted region, a marked increase in capillary density in the infarcted area and its borders occurred (Kocher et al. 2001). That effect has been attributed to a combination of vasculogenesis and angiogenesis, and several paracrine properties have recently been attributed to these cells (Kinnaird et al. 2004), although the mechanism of these beneficial action on the ischemic myocardium remain to be investigated. Currently, $CD34^+$ cells are being evaluated for effects in patients with refractory angina, chronic ischemia, and intermittent claudication.

4.3
Mesenchymal Stem Cells

MSCs are multipotent progenitors derived from the marrow stroma. These cells are negative for the $CD34^+$ marker that is characteristic of EPCs, but express a series of other distinguishing markers, such as CD29, CD44, CD71, CD90, CD105, CD106, CD120a, CD124, and Src homology domains (SH) (Haynesworth et al. 1992; Pittenger and Martin 2004). MSCs can differentiate into most cell types of mesodermal origin including fat, bone, cartilage, and skeletal muscle precursors (Jiang et al. 2002). There is also conflicting evidence on the capacity of MSCs to differentiate into cardiomyocyte-like and endothelial cells after intramyocardial injections (Kawada et al. 2004; Shake et al. 2002; Toma et al. 2002). The prevailing thought is that such differentiation can only happen when these cells are in contact with native cardiomyocytes and does not happen within the scar (Strauer et al. 2002), where instead cells can give rise to other mesodermal cells including fibroblasts, osteoblasts, chondrocytes, and adipocytes. If this holds true, the optimal time for therapeutic MSC administration is more likely to be from early after injury (within days), when surviving cardiomyocytes are still present in the infarcted territory, than when the scar has fully matured. Functionally, MSCs engraft in relatively high numbers, and appear to increase neovascularization and improve regional contractility and diastolic function (Schuster et al. 2004). Several other studies have suggested that MSCs can home to sites of injury following injection into the coronary or peripheral vasculature, reduce the size of the infarcted territory, and restore functional characteristics of the injured myocardium (Amado et al. 2005; Bittira et al. 2003; Strauer et al. 2002). However, it has also been reported that intracoronary administration of MSCs can cause microinfarctions and induce damage of otherwise healthy myocardium (Vulliet et al. 2004). These safety aspects need to be carefully evaluated in the ongoing clinical trials.

MSCs are the only allogeneic progenitor population in clinical trials for treatment of CVD. Recently, MSCs have been defined as "immunoprivileged" (Amado et al. 2005; Jiang et al. 2005) because they do not express MHC-class II and B-7 molecules, which prevents their engaging in the usual T cell responses to produce soluble mediators of rejection (Zimmet and Hare 2005). Although a certain degree of skepticism about that fact remains, such a property definitely increases the attractiveness of these cells for future clinical use. Observing functional improvement of the myocardium with an off-the-shelf cell that lacks negative immunological effects could accelerate the development of a commercial cell therapy product. However, whether MSCs are going to hold up to their promise in clinical studies remains to be seen. A safety/feasibility study is underway, and the early data appear promising, but whether MSCs will equal or outperform any other cell type remains to be seen. A recent preclinical study from our group suggests that MSCs and SKMBs both improve function after ischemia-induced cardiac injury to a similar degree (Thompson et al. 2003).

4.4
Cardiac Progenitor Cells

Recently, several undifferentiated cell populations have been isolated from neonatal and acutely infarcted, failing hearts by their expression of c-kit, multidrug resistant gene (MRD)-1, isl-1, or sca-1 stem cell markers and by concomitant lack of expression of hematopoietic markers (Anversa and Nadal-Ginard 2002; Oh et al. 2003; Urbanek et al. 2005). Interestingly, the number of some these cells was increased after AMI, but was very low in failing hearts, suggesting that these cells take part in ongoing minor repair, which becomes insufficient in HF (Beltrami et al. 2001). Mouquet et al. have recently identified a similar side population within the bone marrow (Mouquet et al. 2005). More recently, we have isolated an upstream progenitor population in neonatal and adult hearts, which appears to give rise to these downstream (more mature) cardiac progenitor populations. These stage-specific embryonic antigen-1-positive uncommitted cardiac progenitor cells (UPCs) could be expanded in vitro and differentiated down myocyte, smooth muscle, and endothelial cell pathways (Ott et al. 2007). To date, methods for harvest, expansion, and in vitro growth of all cardiac progenitor cell (CPC) populations are limited. Smith and colleagues demonstrated that CPCs could be isolated from biopsy specimens obtained from humans and grown under in vitro conditions (Smith et al. 2005). We have shown that we can expand UPCs to large numbers in vitro over several weeks, providing numbers sufficient for cardiac repair (Ott et al. 2007).

Functional repair is the ultimate aim of cell therapy and should at least theoretically be best initiated with cardiac-derived cells. We have recently shown that expanded UPCs were capable of functional repair when injected into infarcted rat heart at 2 weeks following the ligation of the left anterior descending artery (Ott et al. 2005b). LVEF improved in the treated animals (baseline: 34.8±4.2%, week 5:56.5±6.5%, $p=0.001$), and, as expected, decreased in control animals (baseline: 36.5±3.7%, week 5:28.2±3.8%, $p<0.001$). Overall, LV remodeling was attenuated in UPC-treated animals compared to controls. At follow-up, maximal $+dP/dt$ was higher in UPC-treated animals and the relaxation time was shorter compared with controls. As predicted by the hemodynamic improvement and positive anti-remodeling effects, the infarct size was reduced with UPCs. Engraftment of UPCs within scars was histologically verified (Ott et al. 2005b). Similarly to the UPC population, c-kit$^+$ cells are involved in repair after being injected into an ischemic myocardium (Beltrami et al. 2003). Endogenous Sca-1$^+$ CPCs possess the ability to differentiate into functional cardiomyocytes (Oh et al. 2003). And when isl-1$^+$ cells were co-cultured with neonatal cardiomyocytes, those cells were able to electrically integrate with the myocardial cells in vitro by forming gap junctions (Laugwitz et al. 2005).

Overall, the biology of CPCs and their capacity for repair are creating interest, and the results of preclinical use are intriguing. It is possible that future of cardiac repair may involve endogenous mobilization or recruitment of these cells.

4.5
Clinical Studies

Trials performed to date (Table 2) have focused on the use of BMNCs, EPCs, MSCs, and cardiac-derived progenitor cells (CPCs) for a broad range of CVD—from advanced coronary atherosclerosis to end-stage HF. As shown in Table 2, the outcomes of clinical studies are quite divergent—from no effect on patients' symptoms and/or objective measures of LV performance to a reduction of CVD-related events at follow-up. Whether this discrepancy represents a difference in disease context (advanced atherosclerotic CVD versus acute ischemic insult versus chronic downregulation of blood flow and contractility), patient population, cell type, and dose or some other factors remains to be resolved.

Overall, the data on the use of BMNCs in advanced atherosclerotic disease, AMI and ST-elevation myocardial infarction (STEMI) to date are encouraging. Although a small number of patients with advanced atherosclerosis (and no AMI or HF) have been studied (Tse et al. 2003, 2006), cell therapy substantially reduced anginal episodes per week to an extent that appears greater than the reduction seen with ranolazine, a new antianginal agent (Chaitman 2006). The improvement in symptomatology with BMNCs correlated with increased myocardial perfusion. Unfortunately, BMNCs in the context of reperfused and/or stented AMI were not as beneficial. Although a trend toward a reduction in the size of the infarcted area was observed, no functional improvement was gained (Lunde et al. 2006). This apparent lack of a positive effect might have occurred due to an efficient reperfusion and prompt restoration of coronary flow that precluded the potential for BMNC-based repair (Kuethe et al. 2004). When reperfusion and stenting were not uniformly utilized, BMNCs and other cell types (AC133$^+$ EPCs and MSCs) improved myocardial viability (or reduced infarct size), wall motion, coronary flow, and LVEF. However, some disappointments taint the otherwise bright picture: (1) compared to BMNCs, CPCs did not perform very well, and (2) several patients who received AC133$^+$ EPCs showed either restenosis or *de novo* lesions. Whether the cells were the primary suspects or innocent bystanders is quite difficult to discern from the small number of patients studied (Bartunek et al. 2005). It is possible that the success of BMNCs lies in the administration of unfractionated cells, which then allows for the cell–cell and cell–tissue signaling interactions in vivo (the extent of which are not entirely known at this time) that are otherwise absent when

Table 2 Cell therapy trials with BMNCs, MSCs, EPCs, and CPCs. Included are studies with 5 or more patients enrolled for which results were published prior to or on 1 November 2006, where methodological and outcome details were available. Controls comprise historical and active randomized participants and those patients who received placebo. Dose-escalation/variation was used in studies where range is provided. For details, please refer to original publications

Investigator (country)	No. of patients	Cell type	Dx	Mean cell dose, ×10^6	Delivery route	Follow-up time, m	Functional outcomes
Tse et al. 2003, 2006 (China)	12	BMNCs	Advanced CAD	12–16	Transendo EMM-guided	3, 6 and 44±10	At 3 m, regional wall motion, thickening improved, hypoperfused areas lessened (by MRI); no significant changes in LVEF at 3 or 6 m (baseline LVEF 60%). Angina reduced from 26.5 to 16.4 episodes/w. At long-term follow-up, 2 patients died, 1 patient received CABG.
Hamano et al. 2001 (Japan)	5	BMNCs	Advanced CAD	300–2,200	Transepi +CABG	12	Perfusion improved in 3 out of 5 patients (by S).
ASTAMI: Lunde et al. 2006 (Norway)	47 (+50 controls)	BMNCs	AMI treated with PCI	54–130	IC	6	No differences in LVEF (by MRI), or perfusion (by SPECT); trend toward infarct size reduction (by MRI).
Kuethe et al. 2004 (Germany)	5	BMNCs	AMI (reperfused, stented)	39	IC, 6.3±0.4 d post PCI	3 and 12	LVEF and regional wall motion did not change (by E). Coronary flow (by IC Doppler) and contractility indexes (by DE) remained similar at follow-up.

Table 2 (continued)

Investigator (country)	No. of patients	Cell type	Dx	Mean cell dose, $\times 10^6$	Delivery route	Follow-up time, m	Functional outcomes
TOPCARE-AMI: Assmus et al. 2002; Schachinger et al. 2004, 2006a; Britten et al. 2003 (Germany)	30 / 29	CPCs / BMNCs	AMI	CPCs: 13 BMNCs: 238	IC, 4.9 d after AMI	4 and 12	At 4 m, wall motion in the infarction zone improved (all by V and DE); myocardial viability increased (by PET); CPCs and BMNCs behaved similarly. Migratory capacity predicted LV remodeling in multivariate analysis. LVEF improved from 51.6% to 60%; ESV reduced At 1 y, 1 patient in each cell group died due to cardiogenic shock, no other MACE or malignant arrhythmias; MRI at 1 y showed maintenance of LVEF and reduction of infarct size, no reactive LV hypertrophy. Coronary flow normalized in infarct-related arteries.
Chen et al. 2004 (China)	34 (+35 controls)	MSCs	AMI	8,000–10,000	IC, 8±4 h after AMI	6	LVEF and regional wall motion increased (by V); perfusion defects decreased (by PET); real-time EMM showed improvements in mechanical capabilities, electrical properties and functional indexes of LV.
Strauer et al. 2002 (Germany)	10 (+10 controls)	BMNCs	AMI	28	IC, 5–9 d post AMI	3	Infarcted region decreased to 12±7% from 30±13% (by V); LV contractility and EDV improved, perfusion increased (by DE, RV, RC).
Bartunek et al. 2005 (Belgium)	19 (+16 controls)	AC133	AMI	12.6	IC, 11.6±1.4 d post AMI	4	LVEF increased from 45% to 52.1% similar to controls (by E), perfusion improved (by SPECT). 7 patients in cell therapy group developed restenosis (vs 4 in control group), 2 patients in cell therapy group had de novo lesions.

Table 2 (continued)

Investigator (country)	No. of patients	Cell type	Dx	Mean cell dose, ×10^6	Delivery route	Follow-up time, m	Functional outcomes
Stamm et al. 2003 (Germany)	6	AC 133 (purified)	AMI	1.02–1.57	Transepi +CABG	3	LVEF, EDV, EDD improved (by E); perfusion (area at-risk) improved in 5 out of 6 patients.
Fernandez-Aviles et al. 2004 (Spain)	20	BMNCs	Extensive AM	78	IC, 13.5±5.5 d post-AMI	6	LVEF improved by mean of 6%, contractile reserve increased; ESV decreased in (by MRI, DE).
Galinanes et al. 2004 (UK)	14	BMNCs	Extensive AMI	Aspirated BMNCs	+CABG	1.5 and 10	Improvement in segmental and global LV function only in segments that received cells+CABG (by DE). Effect persisted through follow-up.
REPAIR-AMI: Schachinger et al. 2006b (Germany)	101 (+103 controls)	BMNCs	STEMI	236	IC, 3–6 days after AMI	4 and 12	LVEF increased by a mean of 5.5% (by V), patients with LVEF<49% benefited most At 1 y, BMNC-treated patients exhibited reduction in a combined MACE endpoint (death, AMI recurrence, revascularization).
Janssens et al. 2006 (Belgium)	33 34	BMNCs Controls	STEMI	172	IC with 3 ischemia-reperfusion cycles, 24 h post Dx	4	Contractility enhanced at follow-up in segments with severe and complete hyperenhancement (markedly reduced viability) the cell-treated group, a trend toward reduction in LV mass (by MRI); patients with larger infarctions exhibited restoration of myocardial viability (by PET).

Table 2 (continued)

Investigator (country)	No. of patients	Cell type	Dx	Mean cell dose, $\times 10^6$	Delivery route	Follow-up time, m	Functional outcomes
BOOST: Meyer et al. 2006; Schaefer et al. 2006; Wollert et al. 2004 (Germany)	30 (+30 controls)	BMNCs	STEMI	2,460	IC, 4.8±1.3 days post-PCI	12	LVEF increased by 6.7% mostly due to improved regional wall motion in the peripheral area (by MRI); LVEDV and infarct size decreased. Diastolic function improved (by E). LV functional benefits did not persist at 1 y.
Katritsis et al. 2005 (Greece)	11 (+11 controls)	MSCs +EPCs	STEMI (6 pts <1 m; 6 pts >1 m)	1–2 (MSCs: 66±11%; EPCs: 28±11%)	IC, with ischemia-reperfusion cycle	4	Myocardial contractility improved in 5/11 patients but not in controls (by DE); perfusion improved in 6/11 patients (by SPECT).
Fuchs et al. 2006 (Israel)	27	BMNCs	Refractory angina +ischemia	28	Transendo EMM-guided	3 and 12	Exercise duration increased from 418±136 s to 489±142 s; ischemia lessened (by SPECT).CCS angina class improved from 3.2±0.5 to 2.0±0.9 At 1 year, 5 patients had revascularization procedures, functional and symptomatic improvements were maintained in other patients.
Fuchs et al. 2003 (USA)	10	BMNCs	Refractory angina	78	Transendo EMM-guided	3	Perfusion improved (by SPECT); CCS angina score decreased from 3.1±0.3 to 2.0±0.9.
TOPCARE-CHD: Assmus et al. 2006 (Germany)	34 35 (+23 controls)	CPCs BMNCs	Prior AMI (>3 m)	CPCs: 22 BMNCs: 205	IC, with cross-over to the other cell type	3	LVEF increased in patients that crossed over to BMNCs, absolute increase=2.9% (by MRI). No changes in LVEF with CPCs. NYHA class improved in BMNC group—reduction to of 2.0±0.7 from 2.2±0.6, no improvement with CPCs.

Table 2 (continued)

Investigator (country)	No. of patients	Cell type	Dx	Mean cell dose, ×10^6	Delivery route	Follow-up time, m	Functional outcomes
IACT: Strauer et al. 2005 (Germany)	18	BMNCs	Prior AM	15–22 per ea infusion, 4–6 total	IC, 5 m–8.5 y	3	LVEF increased by 15% (by V), infarct size fell by 30% (by SPECT), myocardial viability of infarcted zone increased by 15% (by PET).
Perin et al. 2003; Dohmann et al. 2005 (study conducted in Brazil)	14 (+7 controls)	BMNCs	Severe CAD+HF	25.5	Transendo EMM-guided	2, 4, and 6	Mean LVEF increased from 30% to 35.5% (by E); perfusion improved (by SPECT); NYHA class decreased from 2.2±0.9 to 1.1±0.4; CCS angina reduced from 2.6±0.8 to 1.3±0.6 class.
Blatt et al. 2005 (Israel)	6	BMNCs	ICM	50 ml of aspirated BMNCs	IC, after induction of ischemia by balloon inflation for 3 min	4	LVEF improved from mean of 25% to 28% (by E); wall motion (by DE) increased but only in segments with baseline hibernation. NYHA class fell to mean of 2.3 from 3.5; one patient developed postprocedure hypotension and troponin increase.
Perin et al. 2004 (study conducted in Brazil)	11 (+9 controls)	BMNCs	End-stage ICM	15 injections, 0.2 ml/ea (50 ml aspirated)	Transendo EMM-guided	2, 6 and 12	LVEF increased at 2 m, did not change at 6 and 12 m; perfusion improved (by SPECT); NYHA class decreased from a mean of 2.2 to 1.4, CCS fell from 2.6 to 1.2 class. Exercise capacity improved (by treadmill).

Table 2 (continued)

Investigator (country)	No. of patients	Cell type	Dx	Mean cell dose, ×10^6	Delivery route	Follow-up time, m	Functional outcomes
Silva et al. 2004 (USA)	5	BMNCs	Pre-Trnsp HF	15 injections, 0.2 ml/ea (50 ml aspirated)	Transendo EMM-guided	2 and 6	Exercise capacity (by treadmill oxygen consumption) improved in 4/5 patients, disqualifying them from listing for Trnsp.

CABG, coronary artery bypass graft; CAD, coronary artery disease; CCS, Canadian Cardiovascular Society; d, days; DE, dobutamine echocardiography; Dx, diagnosis; ea, each; EDD, end-diastolic (LV) dimension; EDV, end-diastolic (LV) volume; EMM, electromechanical mapping of LV; h, hours; HF, heart failure; IC, intracoronary; ICM, ischemic cardiomyopathy; LV, left ventricular; LVEDV, left ventricular end-diastolic volume; LVEF, left ventricular ejection fraction; m, month(s); MACE, major adverse cardiac events; NA, not available; NYHA class, New York Heart Association functional class; PCI, percutaneous coronary intervention; PET, positron emission tomography; pts, patients; RC, right heart catheterization; RV, radionuclide ventriculography; S, scintigraphy; sec, seconds; SPECT, single photon emission tomography; STEMI, acute ST-elevation myocardial infarction; Transendo, transendocardial; Transepi, transepicardial; Trnsp, transplantation; Tx, therapy/treatment; V, ventriculography; w, week; y, years an isolated cell population is administered. In other words, unfractionated BMNCs have both mature and immature EPCs along with other progenitors, and it is quite possible that a combination of these cells may be the best choice, although specific CPC populations have not been clinically tested.

The REPAIR-AMI (Schachinger et al. 2006b) study highlighted several important aspects with regards to efficacy of cell therapy. First, administration of the mean number of 236×10^6 BMNCs in patients with ST-elevation myocardial infarction resulted in a higher event-free survival at 1 year than patients who received placebo. That was the first randomized study showing that exogenous BMNCs do in fact participate in tissue repair and can withstand the rigorous test of clinically driven endpoints, at least in a phase II study. Second, it is becoming apparent that there needs to be a sufficient degree of tissue injury for the cells to show efficacy. For instance, in the REPAIR-AMI study, those patients that had a baseline LVEF at 48.5% or lower benefited the most from BMNCs, and those above this cut-off showed little or no benefit.

Unfortunately, despite the best efforts to reduce the progression of the pathophysiological process post-AMI, up to 50% of patients manifest symptomatic HF by year 7 post-AMI (Abbate et al. 2006; Miller and Missov 2001). HF notwithstanding, refractory angina is a growing problem post-AMI, with approximately 20% of patients remaining symptomatic despite best efforts to tailor pharmacological therapy and interventional approaches (Yang et al. 2004). BMNCs have exhibited a considerable improvement of symptomatic status and functional outcomes in the latter group, which persisted at 1 year follow-up. The reduction of anginal episodes was paralleled by increased exercise treadmill time and improved myocardial perfusion. In the former group (HF), successes of BMNCs varied, most likely dependent on the baseline LVEF and, quite possibly, on the delivery technique. Clearly, patients enjoyed a reduction of shortness of breath, angina, and other symptoms; however, the effect on LV contractility was not always very pronounced. Possibly, administration of cells can prevent HF from worsening, at least for a period of time. The data from Silva et al. (2004) showing delisting of patients from transplantation because of the increase in exercise capacity, albeit in a very small number of patients, supports these beneficial effects on BMNCs in HF. The TOPCARE-HF and BOOST-2 studies have been initiated to gain a more systematic insight into the response of the myocardium to BMNCs, when HF pathophysiology predominates. Given the reduced number and migratory capacity of EPCs shown in preclinical studies and the deficits in EPC quantity seen in patients with advanced CVD and HF (Werner et al. 2005), it will be interesting to see if BMNCs are capable of improving cardiac function or if the HF milieu only allows ischemia-resistant cells, such as SKMBs or MSCs, to survive. So far, unlike in the SKMB trials, symptomatic and functional improvements in HF patients treated with BMNCs occurred without the additional burden of electrical events. These large trials will provide more data to definitively answer the question of safety with BMNCs.

BMNCs, CPCs, and AC133[+] cells have been given intracoronarily and as an adjunct to CABG with granulocyte colony-stimulating factor (filgrastim and PEG-filgrastim) mostly in the context of safety and feasibility studies. These studies have been reviewed elsewhere (Boyle et al. 2006; Korbling et al. 2003).

4.6
Umbilical Cord Blood Cells

A relatively new source for progenitor cells is umbilical cord blood, which contains fetal-derived populations identical to those found in bone marrow (Erices et al. 2000). Umbilical cord blood cells (UCBCs) are easily obtained, albeit not in large volumes, have the potential to develop into multiple lineages, do not pose as many ethical questions as embryonic cells, and are less immunogenic than allogeneic bone marrow counterparts, which means a larger proportion of the population could receive cells from appropriately matched donors. If UCBCs are stored at birth, they could provide an autologous source of stem cells to treat myocardial damage later in life. Current studies in animal models show that UCBCs injected directly into the infarcted myocardium improve LV ejection fraction, anteroseptal wall thickening, and dP/dt (max), while decreasing infarct size (Henning et al. 2004). In addition, intravenous injection of UCBCs in mice following ligation-induced injury resulted in an approximately 20% higher capillary density in the border zones of the infarction—a finding not observed in untreated animals (Ma et al. 2005). Recent data have suggested that human UCBC-derived CD34[+] cells may be capable both of preventing injury progression and of partially reversing systolic and diastolic dysfunction, if administered shortly after AMI (Leor et al. 2006). Several other investigators published studies showing similar functional outcomes with varying populations of UCBCs (Hirata et al. 2005; Kim et al. 2005). However, no evidence to date suggests that cord blood cells injected into the infarcted myocardium are able to produce mature cardiomyocytes in humans, or that the functional benefits seen in animal models could be replicated in patients with AMI, HF, or both. Overall, it appears that UCBCs may appear to be an interesting cell of choice to be used in further studies of treatment of myocardial injury.

In summary, the data with regards to BMNCs across the continuum of CVD are not uniform, but nonetheless encouraging. Taken together, the results show a great deal of evidence toward the efficacy of BMNC therapy in settings of AMI, refractory angina, and HF. At this time, it is clear that at least in AMI patients, a phase III trial will take place soon. Current ongoing clinical trials are listed on the Internet (http://www.clinicaltrials.gov; http://www.thescientist.com/supplementary/html/24104). There are reasons to be optimistic realizing there is much yet to be learned in the process of defining a cell-based clinical product.

5
Skeletal Myoblasts Versus Bone-Marrow Cells: How Far to Go to Reach the Best Cell for Cardiac Repair?

Translating research findings from bench experiments to bedside efficacy to develop a new therapeutic product is a process of multiple interrelated steps. The first stage is the idea, which comes from the basic science of the pathophysiology of a disease. Next, that idea must be tested in clinically relevant animal models. If the data indicate a potential benefit, further testing takes place in consecutive clinical studies according to a regulatory framework. However, if unexpected issues arise or new pieces of the puzzle emerge (for example, delivery-related issues in the case of cell therapy), the process needs to move back to the bench, and only when those issues are resolved can the process return back to bedside. Cell therapy with either SKMBs or BMNCs and its subsets is in the iterative stage between bedside and bench at the present time. The success of the "clinical product" rests on these re-iterations, as these refinement cycles address issues that may hinder or even preclude clinical utilization. Ultimately, selection of the cell to exercise the fullest capacity for repair delivered via the route that is easiest to operate and least dangerous for the patient, thereby reaching the most suitable environment will define the best cell. At this time, however, we still have several things to accomplish before we can assign that status to a particular cell type. In this regard, there are several important prerequisites that merit discussion.

5.1
Creating a Centralized Registry for the Results of Trials and Biorepository for Blood Samples to Examine Accumulated Data and Set Direction for the Future of the Field

Decades of CVD research have highlighted the importance of centralized databases in advancing our understanding of a disease process. The field of cardiovascular medicine, in particular, would not have advanced as far as it has in the past 25–30 years if the Framingham Study or the Thrombolysis in Myocardial Infarction (TIMI) trials had not been initiated and executed in a centralized matter. Large databases provide the power to examine the data retrospectively while being able to control for numerous covariates—a step not possible to accomplish in a review or even a meta-analysis. Centralized databases also act as testing grounds for new hypotheses, often before clinical trials commence.

The field of cell therapy has arrived at the point where the next advancement should be creating and employing a large database of all results of clinical trials to serve as a filter for the hypotheses. With the aid of such a tool, ideas could be segregated before hundreds of thousands of dollars are spent only to find out, for example, that a specific unforeseen factor interfered with the outcome. Centralized data collecting efforts in acute HF, such as the Acute

Decompensated Heart Failure National Registry (Yancy and Fonarow 2004), have brought extremely valuable data with regards to the outcomes of clinical management of HF patients. We believe that a centralized registry of cell therapy trials could not only advance the field but could generate the next set of questions to ask, which, in turn, will greatly advance the science and will move the field closer to creating a cell-based "clinical product." With the powers of the Internet and data sharing, it is possible to design the registry in such a way that it is not only compliant with appropriate regulations on the conduct of research and on patient privacy but is also effective in reducing the work load of users, and, most importantly, could be accessible from multiple points, similar to a Web page.

The second aspect of the centralized registry is pairing it with a biorepository for blood samples. Although such an initiative requires committed funding and resources, centralizing sample collection, storage, flow cytometry, and assays makes a great deal of sense and could help greatly advance the science. The goal would be to use the progenitor cell characterization in conjunction with clinical data and examine the dynamics within multiple populations of progenitors in varying states of disease when different types of cells are given. Understanding the mechanisms of repair and regeneration is the next obstacle of the field today—and the biorepository may substantially advance that knowledge. In that regard, the ability to go back to the samples when new markers, receptors, and pathways emerge is far more cost-efficient than the currently available approaches. Combining the registry for the clinical trials data and the biorepository for the blood and tissue samples seems to be exactly what the field needs to make another decade of major progress and help shape future cell therapy products.

5.2
Increased Mechanistic Understanding Should Allow Us to Create the Best Cell-Based "Clinical Product"

Nine years after the first report suggested efficacy of SKMBs in treatment of failing myocardium, cell therapy has been applied, albeit only in clinical studies, in almost 800 patients (sum of patients in all published studies, and including the recently presented MAGIC trial). In addition, many trials are still actively recruiting patients. However even after over 800 patients have been treated, we do not possess a solid understanding of the mechanism of cell engraftment, survival, and tissue repair. While a complete definition of the mechanisms of beneficial action as well as adverse effects has not been required for the regulatory approval and clinical use of pharmacological treatments [e.g., levonorgestrel for emergency contraception (Gemzell-Danielsson 2006), immunomodulatory drugs in myelodysplasia (Galili and Raza 2006), farnesyl transferase inhibitors in cancer (Appels et al. 2005), mitoxantrone in multiple sclerosis (Neuhaus et al. 2005), adverse effects of bisphosphonates on the bone

(Pickett 2006), behavioral side effects of certain triptans in the treatment of migraine headaches (Lambert 2005), etc.], optimizing of cell-based products cannot proceed without additional mechanistic insights into cell-mediated repair. Knowing critical components of how the SKMB- or BMNC-driven repair works should enable us to target these (and other) types of cells to the right pathophysiological contexts, to achieve efficacy comparable or better than that of pharmacological therapies, especially when measured by long-term follow-up studies. These insights can only be obtained when clinical and basic science work together, so that the process can balance between clinically relevant questions and scientifically important observations.

Even though our understanding of repair at this point is limited, progress has been made and pathways as well as individual components are being identified. For example, we now understand that survival of BMNCs and their actions are mediated (at least to some extent) by Akt, (Gnecchi et al. 2006) and that VEGF, (Wang et al. 2006; Zen et al. 2006) stromal-derived factor-1 (Misao et al. 2006; Ratajczak et al. 2006) and several cytokines (Takahashi et al. 2006) play a role as well. But what about thrombopoietin, erythropoietin, hypoxia-induced factor-1-α, tissue growth factor-β, and other molecules implicated in other aspects of CVD treatment or angiogenesis (Kirito et al. 2005; Vandervelde et al. 2005)? To what extent do any of these mediators act similarly on SKMBs? There is evidence that supplying VEGF decreases the amount of SKMBs undergoing necrotic/apoptotic process following transplantation into the myocardium (Yau et al. 2005). There is also initial evidence that treatment of C_2C_{12} SKMBs with tumor necrosis factor-α (TNF-α) not only suppresses morphological and biochemical differentiation, but induces apoptosis following its initial stimulation of proliferation and survival (Stewart et al. 2004). Such influence of TNF-α makes sense, since increased concentrations of this cytokine in HF correlate with reduced systolic LV function (Kaur et al. 2006). However, we have yet to define how the molecular cascades act in synergism to promote repair. This state of fragmented knowledge about signaling is very reminiscent of the early days of understanding of the clotting cascade and the mechanisms involved into coagulation and hemostasis. Eventually, however, a more or less complete understanding of pathways emerged and intrinsic and extrinsic pathways were brought together. That effort serves well even today; therapies continue to be developed targeting specific components (such as factor X versus thrombin inhibition) and the management of such serious conditions as AMI continues to evolve as new targets are being refined and introduced into clinical practice. There is no doubt that cell therapy will take a similar path. Although, because the field of cell therapy lies on the crossroads of cardiology, vascular biology, hematology, and immunology, deciphering the pathways involved in repair will take a lot more effort than the abovementioned coagulation cascade. However, borrowing the knowledge from oncology, hematology, and immunology and applying it in the context of cardiac repair will accelerate acquiring of new knowledge and will lead to refining of the eventual therapeutic product.

5.3
The Two Important Steps in Defining "Best Cell"

At present, discrepant clinical trial outcomes exist for the similar cells in different patients and different cells in the similar patients. Comparisons are difficult because timing after injury, dosing, and route of cell administration also differ. Yet some generalities are emerging. For example, SKMBs seem to engraft into the myocardium and result in functional improvement in HF, and BMNCs show positive results in treating acutely injured myocardium. What is clear from a basic science point of view is that different environments in the myocardium at the time of injury likely generate different milieus, and therefore the cells that engraft in one environment may not survive in another one. Whether the discrepant clinical results are a result of a rush to be first clinically to apply various cells types in various contexts, or if segregation of the types of cells (at least between SKMBs and BMNCs) for the appropriate types of injury (AMI and HF) has already happened inadvertently, remains to be understood.

Because developing a successful therapy, one that is based on a biological understanding of the human body and the pathogenesis of disease, requires multiple re-iterations between bench and beside, we need to go back to bench research now to compare various cell types side by side in various types of ischemic injury *in appropriate animal models*. This may seem to be a simple process, but in reality all available cells, delivery routes, and injury models taken together would result in approximately 2,400 comparisons to be done. This clearly is a prohibitive number for a promising therapy. Therefore, the field needs to come to a consensus on the clinical relevance and conduct comparisons accordingly (i.e., concentrate on the most common types of injury, such as reperfused AMI at up to 4 h from the onset of symptoms, ischemic HF with a mild-to-moderate ischemic process). As the data become available, we can then build additional hypotheses as to what may or may not work in other types of pathology and models. Such experiments should also bring additional insights into our understanding of how and when repair happens. This suggestion may sound contradictory to reality, considering the number of preclinical studies and clinical studies that have been published in the field (i.e., 1,347 Medline hits on keyword searches for "heart" and "cell transplantation" as of 19 November 2006). However, only a few side-by-side comparisons of different cell populations have been performed. We clearly lack direct comparisons of different cell types in clearly defined clinically relevant models of disease.

Comparing different cell types in various contexts of disease will also help us definite how to improve survival of transplanted cells, which is currently one of the largest hurdles of cell therapy. Most reports suggest that 70%–90% of all transplanted cells die within the first few days of transplantation into infarct scar. Studies have shown that a subset of the transplanted cells survive and multiply, but the question is how to promote survival either by genetic expression or by pharmacological means, or by modification of the media. Preconditioning of cells before transplantation via heat shock or transfecting cells with pro-survival factors (Akt, heat shock proteins, growth factors) (Kohin et al. 2001; Zhang et al. 2001) or glutamine deprivation (or antagonists) as we described earlier may help increase their survival rate in vivo. More work will need to be done in this area to better define the relationship between the microenvironment of the infarct scar and the adjacent myocardial segments and outcome of the transplanted cells. In addition, we have to look beyond the scar and evaluate the impact of hypertension (HTN), valvular insufficiency (VI), and nocturnal dips in oxygen saturation and relate them with the ability of cells to repair the myocardium. HTN and VI are important causal factors of LV hypertrophy and remodeling (Udelson et al. 2003) and therefore will impact the signaling and quite possibly the ability to promote engraftment of exogenous cells. This may mean that adequate antihypertensive therapy and correction of valvular abnormalities may positively impact the engraftment and survival of the delivered cells. Also, correction of valve abnormalities is now feasible with percutaneously placed devices—a positive development for HF patients, many of whom are not considered to be ideal candidates for surgical procedures. Some of these devices, such as the Coapsys platform for mitral insufficiency repair developed by Myocor, positively impact LV remodeling in early clinical trials (TRACE data, OUS clinical trial; J. Price, personal communication). It is possible that correction of VI prior to cellular cardiomyoplasty will provide a better environment for cell survival and ability to perform repair, as the effects of the cells will not be counteracted by VI-associated remodeling. Nocturnal dips in tissue oxygen saturation are the result of sleep apnea, which is present in many CVD patients mainly due to HTN and obesity. In HF, however, a large proportion of patients is suffering from sleep apnea (Ferreira et al. 2006). Therefore, tissue hypoxia may be augmented by sleep apnea, thus decreasing the chances of the transplanted cells surviving. Although this aspect has not been evaluated, it intuitively makes sense. It is possible that correction of sleep apnea may be required to optimize the potential of exogenous cells to repair the myocardium, decrease the apnea-associated diastolic dysfunction (Sidana et al. 2005), and allow the cells to augment impaired systolic properties of myocardial segments.

Insights from previous research endeavors are important. A confounding inflammatory response to needle punctures during cell administration is very reminiscent of early percutaneous or transmyocardial revascularization studies where the creating channels with laser promoted inflammation (Fleischer et al. 1996). The possibility that needle-based cell delivery is proinflammatory should be explored further; and if so, we need to define specific cytokines that might be involved. If that suggestion holds, the proinflammatory action of the delivery vehicle should be evaluated against the etiology of the disease, as atherosclerosis is now considered an inflammatory disorder (Hansson 2005) and cytokine abnormalities have been found in HF (Kotlyar et al. 2006; Toth et al. 2006).

In addition to surviving in the ischemic environment at the time of implantation, the ideal cell for myocardial repair will be able to, become, despite ischemia, a fully functioning cardiomyocyte or an endothelial cell. However, none of the progenitor cells currently used satisfies both of these criteria at the numbers sufficient for maximal repair or recovery of function. Therefore, it is important to continue working toward understanding the differentiation of progenitor cells in a cardiomyocyte phenotype at the bench level, even if it does not seem clinically relevant. The goal, then, might be to design specific pharmacological/molecular tools to induce a differentiation pathway prior to implantation so that a specific phenotype would manifest slowly enough to allow neovascularization to become functional.

Moreover, injected cells have significantly different electrical properties than cardiomyocytes. These differences have led to VT observed in some of the clinical trials as previously described. For cardiovascular cell therapy to reach its potential, it will be critical to electrically integrate transplanted cells into the surviving myocardium and to ensure the absence of negative electrical consequences. This problem may be approached by genetically altering transplanted cells (to promote electrical coupling), developing new adjunctive safety measures (such as co-administration of antiarrhythmics), delivery of cells only in patients who meet the MADIT-II or SCD-HeFT criteria and have ICDs, or modifying the transplanted cells to become true cardiomyocytes that can survive in a harsh milieu.

Lastly, and perhaps most importantly, there is an urgent need for a taskforce to define the nomenclature of progenitor cells to arrive to a consensus of which cells we are going to call "progenitor cells." Similar taxonomy efforts have been recently accomplished by Krumholz et al. (2006) for clarification of nomenclature in CVD. A writing group consisting of experts in cell biology, taxonomy, cell differentiation, cell therapy, hematology, and translational research could very rapidly accomplish this task. Efforts in this direction will advance the field—and may help avoid unfortunate outcomes. Even though a task force of the European Society of Cardiology released a guiding document on the clinical investigations of autologous adult stem cells for cardiac repair (Bartunek et al. 2006), taxonomic questions were not covered.

5.4
Evaluating the Best Delivery Route for a Cell-Based Clinical Product Will Be Beneficial for Clinicians and Patients

It is clear that choosing the best delivery route is an important prerequisite for success, closely following the choice of cell for the right environment. A major obstacle to achieving efficacy is a rather poor engraftment seen when cells are administered by intracoronary, intravenous, and intracardiac routes. This limitation is likely to have emerged due to multiple factors, of which the technical difficulties of injecting the cells exactly into the center or the periphery of the scar or precise catheter manipulations in the coronary tree cannot be overemphasized. Therefore, training of the operators gains a pivotal importance. Recently, concerns of operator error halted the GENASIS trial (Genetic Angiogenic Stimulation Investigational Study—Corautus Genetics phase IIb clinical trial to evaluate safety and efficacy of VEGF-2 for treatment of patients with severe angina). As we go forward, creating a specialized network of centers for cell therapy, as recently proposed by the National Heart, Lung, and Blood Institute (NHLBI), would allow for training of interventional cardiologists by experts in delivery techniques. Alternatively, it may also make sense to restrict the number of centers per region that act as referral centers to utilize cell therapy—at least until the techniques come to solid maturity. We have learned that operator volume and experience was a critical determinant of success in CABG and percutaneous coronary intervention (PCI) clinical trials and also in routine clinical practice. As the field of cell therapy goes forward, we cannot ignore the importance of appropriately trained specialists.

The data thus far have suggested that intracoronary delivery, at least in the context of AMI, can provide a comparable level of engraftment of cells to surgical delivery. This outcome again highlights the need for side-by-side comparisons of various delivery methods in controlled, well-designed experiments. Understanding of the biology involved in the delivery route–engraftment interaction could translate into the development of optimal situation-specific delivery systems. No doubt this process will take some time. However, the technological progress in the past 10–15 years has been so rampant that it will not at all be surprising if the next 5–10 years brings major advancements in this regard.

5.5
Arriving at a Consensus Regarding Trial Design and Outcome Measurements

At the present time, clinical trials in cell therapy suffer from several major shortcomings primarily involving design and selection of endpoints. For example, most studies have been accompanied by an additional revascularization procedure, either by percutaneous coronary intervention or CABG, making any functional improvement due to exogenous cells nearly impossible to distinguish from the standard of care. In addition, patient characteristics at study entry need to be matched more carefully in prospective trials, which would include baseline comparisons beyond the standard regimen of demographic and basic clinical disease-defining parameters, such as assessments of biomarkers, cytokine profiles, and levels of circulating progenitors to characterize the milieu and relate the impact of exogenous cells to outcome appropriately. Furthermore, medication regimens need to be tracked more carefully throughout the course of the trial, as illustrated by the recently published trial (Lunde et al. 2006) where the patients who received BMNCs were prescribed more diuretics (40% in the cell therapy group versus 26% in the control group), which might have negatively impacted the engraftment of the cells and the overall outcome of the study. We also need to account for the stimulatory effects of drugs, such as statins, peroxisome proliferator-activated receptor agonists, erythropoietin, estrogen, angiotensin receptor blockers, and possibly others in various disease states. Right now, it is completely unclear which, if any, of these combinations of drugs alter the number and the function of progenitor cells available for repair. Clearly, if cell therapy is to be adequately evaluated as a therapy, we will need this type of information for the design of definitive phase III trials and also going forward with clinical applications. The best chance to obtain this information prospectively is through a centralized registry, as the number of covariates to discern the drug–cell effect is going to be disproportionately large and may require a large number of patients. In addition, we lack data that evaluate time in disease progression as well as time in dynamics of transplanted cells as an additional factor in treatment. Over all, there is a lack of standardization in the current preclinical approach to cell therapy; for example, cell types, doses, preclinical models, and endpoints all differ. Attempts to standardize these parameters and to decide on a consensus will move us forward.

What we select to be an "endpoint" in cell therapy likely matters. So far, clinical trials have been geared toward measuring functional improvement of the LV by assessing global EF. As we know from the HF trials, improvement of regional contractility may not always translate into better HF numbers because of differences in loading conditions. Since the data with both SKMBs and BMNCs so far suggest that exogenous cells are capable of anti-remodeling effects, measuring those as an endpoint in prospective trials will require using a technique with a high sensitivity and specificity in measurements of regional contractile parameters. However, we have begun to appreciate observer-dependence of those measurements. Even though cardiac magnetic resonance imaging (CMR) offers the best topographic assessment of the heart, the variability is minimized by conducting clinical trials with centralized core laboratories where the personnel undergo regular inter- and intra-observer reproducibility assessments. More attention needs to be paid to peri-infarct zone and scar volume and myocardial perfusion quantification. Over the last 10 years, CMR has matured to offer quantitative assessment of myocardial perfusion (Jerosch-Herold et al. 1998; Zenovich et al. 2001). Measuring changes in blood flow was proposed as an endpoint for angiogenesis studies (Wilke et al. 2001), and it is now becoming apparent that cell therapy will need a sensitive measure of blood flow as well. CMR has been used to detect the presence of exogenous cells in the myocardium, as well as to characterize the myocardium prior to transplantation of cells to delineate the areas of myocardial damage (Zhou et al. 2006).

Concurrently, we need to critically evaluate the endpoints that are used at the present time and come to an agreement, most likely through an AHA/ACC-sponsored consensus document, similarly to available data standards for AMI, HF, and atrial fibrillation, that would outline the standard sets of data to be captured in the cell therapy trials. As that process goes along, some endpoints with high subject variability, such as exercise treadmill time, will be critically evaluated and new, biologically relevant, and clinically translatable endpoints will be introduced. Such a process will also enormously aid acceptance of new endpoints by the Food and Drug Administration and will over time accelerate bringing cell therapies to market.

5.6
Testing Cell-Based Models in Drug Development to Accelerate Design of Therapies Targeted at Repair

Recognition of involvement of progenitor cells in tissue repair and investigation of its mechanisms may provide a foundation for a new approach in drug development—testing the effects of candidate molecules on BMNCs, EPCs, MSCs, and other cell types. At the time when the cost of bringing a new drug to market ranges from US $500 million to US $2 billion, dependent on the therapeutic application (Adams and Brantner 2006), new avenues must be thought to reduce the price tag of drug discovery research as well as clinical trials to allow more candidate molecules to reach phase I and II trials. It is clear that the drugs that impact major pathways in CVD, such as renin-angiotensin-aldosterone or HMG-co-α reductase, have already been discovered, tested, and brought to market, and have shown their clinical abilities to save patients' lives. Therefore, the next generation of therapeutics will be directed at initiation of atherosclerotic lesions and arresting their development via reduction of inflammation and targeting specific receptors. We propose that along with those avenues, new compounds should also be evaluated for their ability to aid exogenous cells engraft and survive. In HF, the increasing number of patients creates an opportunity to design new therapies to reduce symptomatology and reverse remodeling. Recently, trials of endothelin antagonists (darusentan, tezosentan) (Anand et al. 2004; O'Connor et al. 2003), TNF-α antagonists (e.g., etanercept) (Mann et al. 2004), and a Ca^{2+} sensitizer (levosimendan) (Cleland et al. 2006) have been disappointing either due to a lack of beneficial effect or because of adverse reactions that created a prospective application to a wide patient cohort problematic. SKMBs, on the other hand, have a great potential to be a part of a therapeutic armamentarium in HF, as the evidence for efficacy, at least so far, points in a positive direction. Pharmacological approaches to aid engraftment, survival, and electrical integration of SKMBs would be beneficial to the development of cell-based therapies.

In addition to testing new drugs for their effects on exogenous repair and seeking approaches to improve survival of cells, we propose that models involving cells with reparative potential could be employed in drug discovery science. If a new drug is targeting repair, it makes sense that models that utilize reparative cells are used early in the process to reduce costs of further development in case of a negative outcome. Progenitor cells could also be employed as tools for toxicogenomics and safety evaluations. After all, if a progenitor cell dies, then the efficacy of a compound as well as its potential safety should be reexamined. Here, the use of biomarkers together with evaluations of progenitor cells may bridge bench science and human pharmacology, providing additional insights into biological mechanisms of disease and advancing the science at the same time.

Overall, the use of SKMBs, BMNCs, EPCs, MSCs, and other cell types may be extremely helpful both early in drug development and also in the human testing of new candidate molecules.

In summary, we believe we have outlined the major issues in cell therapy today. As the field develops further and products moves closer to market, addressing each prerequisite will increase the likelihood of a successful outcome. The ultimate success, however, will be the prevention of atherosclerosis and CVD, the reduction of hospitalization and major adverse cardiac events, and in prolonging a healthy life for patients who currently have limited options available to them.

6
Summary

Cell transplantation has opened a new frontier in CVD. The concept of repairing or regenerating ischemic cardiac tissue creates a real possibility, and while many questions still remain, it has an excellent chance of eventually becoming a clinical reality. Cell-based therapies have the potential of providing physicians with alternatives that extend beyond revascularization and medical management to reverse damage that, in many cases, has already been done and may not be truly controllable for a large patient population.

To further advance cell therapy for CVD, we now have to come to a field-wide consensus and standardize future studies. The diversity of cell types, application techniques, and disease stages can be a hurdle and an opportunity, and only collaboration will allow us to move forward as a field instead of expanding information that cannot be combined or compared. Recent clinical trials have shown that cell therapy with SKMBs and BMNCs is able to demonstrate clinical benefits in AMI and HF. Promising results evoked the scientific enthusiasm to warrant large-scale controlled clinical trials to determine the best and safest application of this technology, and to gain a better understanding of its mechanism(s).

As the field progresses, we have a responsibility to promise patients (and the press) only what we can deliver; that is, to tell the truth about cardiac repair. BMNCs, MSCs, SKMBs, or other types of cells hold a great promise to modify pathophysiological process in specific ways. It is crucial to understand for clinicians, patients, and the press that specificity precludes a panacea. As we go forward, some applications will succeed and some will fail. Cells may not be found guilty of failures. On the contrary, the disease contexts may come to be the primary determinants of efficacy. We have already experienced a similar process with angiogenic growth factors in CVD, and we now realize that those trials should have more carefully targeted the disease process, as the results uniformly showed that sicker patients had larger therapeutic benefits. As investigators, we need to be realistic of the expectations we place on cell therapy, and ultimately we need to under-promise and over-deliver, based on rigorous science; otherwise, the great potential will eventually be destroyed. Cell therapy is, however, a new and very promising alternative that warrants much further exploration, inspiration, and investment of our time and resources.

Acknowledgements This work has been supported in part by NHLBI/NIH award to Dr. Doris A. Taylor (R01-HL-063346), Minnesota Partnership for Biotechnology and Medical Genomics award, and by funding from the Center for Cardiovascular Repair, University of Minnesota. The authors sincerely thank Harald C. Ott, MD, for his continuous contributions to the ongoing success of the Center for Cardiovascular Repair.

References

Abbate A, Bussani R, Amin MS, et al (2006) Acute myocardial infarction and heart failure: role of apoptosis. Int J Biochem Cell Biol 38:1834–1840
Abraham M, Henrikson C, Tung L, et al (2005) Antiarrhythmic engineering of skeletal myoblasts for cardiac transplantation. Circ Res 97:159–167
Adams C, Brantner V (2006) Estimating the cost of new drug development: is it really 802 million dollars? Health Aff 25:420–428
Agbulut O, Vandervelde S, Al Attar N, et al (2004) Comparison of human skeletal myoblasts and bone marrow-derived CD133+ progenitors for the repair of infarcted myocardium. J Am Coll Cardiol 44:458–463
Aicher A, Brenner W, Zuhayra M, et al (2003) Assessment of the tissue distribution of transplanted human endothelial progenitor cells by radioactive labeling. Circulation 107:2134–2139
Amado L, Saliaris A, Schuleri K, et al (2005) Cardiac repair with intramyocardial injection of allogeneic mesenchymal stem cells after myocardial infarction. Proc Natl Acad Sci USA 102:11474–11479
Anand I, McMurray J, Cohn J, et al (2004) Long-term effects of darusentan on left-ventricular remodelling and clinical outcomes in the Endothelin A Receptor Antagonist Trial in Heart Failure (EARTH): randomised, double-blind, placebo-controlled trial. Lancet 364:347–354
Anversa P, Nadal-Ginard B (2002) Myocyte renewal and ventricular remodelling. Nature 415:240–243
Appel LJ, Brands MW, Daniels SR, Karanja N, Elmer PJ, Sacks FM (2006) Dietary approaches to prevent and treat hypertension: a scientific statement from the American Heart Association. Hypertension 47:296–308
Appels N, Beijnen J, Schellens J (2005) Development of farnesyl transferase inhibitors: a review. Oncologist 18:565–578

Assmus B, Schachinger V, Teupe C, et al (2002) Transplantation of progenitor cells and regeneration enhancement in acute myocardial infarction (TOPCARE-AMI). Circulation 106:3009–3017

Assmus B, Honold J, Schachinger V, et al (2006) Trancoronary transplantation of progenitor cells after myocardial infarction. N Engl J Med 355:1222–1232

Atkins B, Hutcheson K, Hueman M, et al (1999a) Reversing post-MI dysfunction: improved myocardial performance after autologous skeletal myoblast transfer to infarcted rabbit heart. Circ Suppl 100:I-838

Atkins BZ, Hueman MT, Meuchel JM, et al (1999b) Myogenic cell transplantation improves in vivo regional performance in infarcted rabbit myocardium. J Heart Lung Transplant 18:1173–1180

Atkins BZ, Lewis CW, Kraus WE, et al (1999c) Intracardiac transplantation of skeletal myoblasts yields two populations of striated cells in situ. Ann Thorac Surg 67:124–129

Bartunek J, Vanderheyden M, Vandekerckhove B, et al (2005) Intracoronary injection of CD133-positive enriched bone marrow progenitor cells promotes cardiac recovery after recent myocardial infarction: feasibility and safety. Circulation 112:I178-183

Bartunek J, Dimmeler S, Drexler H, et al (2006) The consensus of the task force of the European Society of Cardiology concerning the clinical investigation of the use of autologous adult stem cells for repair of the heart. Eur Heart J 27:1338–1340

Beltrami AP, Urbanek K, Kajstura J, et al (2001) Evidence that human cardiac myocytes divide after myocardial infarction. N Engl J Med 344:1750–1757

Beltrami AP, Barlucchi L, Torella D, et al (2003) Adult cardiac stem cells are multipotent and support myocardial regeneration. Cell 114:763–776

Bertrand ME (2004) Provision of cardiovascular protection by ACE inhibitors: a review of recent trials. Curr Med Res Opin 20:1559–1569

Bittira B, Shum-Tim D, Al-Khaldi A, et al (2003) Mobilization and homing of bone marrow stromal cells in myocardial infarction. Eur J Cardiothorac Surg 24:393–398

Blatt A, Cotter G, Leitman M, et al (2005) Intracoronary administration of autologous bone marrow mononuclear cells after induction of short ischemia is safe and may improve hibernation and ischemia in patients with ischemic cardiomyopathy. Am Heart J 150:981–987

Boyle JA, Schulman SP, Hare JM (2006) Stem cell therapy for cardiac repair: ready for the next step. Circulation 114:339–352

Britten MB, Abolmaali ND, Assmus B, et al (2003) Infarct remodeling after intracoronary progenitor cell treatment in patients with acute myocardial infarction (TOPCARE-AMI): mechanistic insights from serial contrast-enhanced magnetic resonance imaging. Circulation 108:2212–2218

Camirand G, Rousseau J, Ducharme M, et al (2004) Novel Duchenne muscular dystrophy treatment through myoblast transplantation tolerance with anti-CD45RB, anti-CD154 and mixed chimerism. Am J Transplant 4:1255–1265

Chachques JC, Grandjean PA, Tommasi JJ, et al (1987) Dynamic cardiomyoplasty: a new approach to assist chronic myocardial failure. Life Support Syst 5:323–327

Chachques JC, Herreros J, Trainini J, et al (2004) Autologous human serum for cell culture avoids the implantation of cardioverter-defibrillators in cellular cardiomyoplasty. Int J Cardiol 95 [Suppl 1]:S29–S33

Chaitman B (2006) Ranolazine for the treatment of chronic angina and potential use in other cardiovascular conditions. Circulation 113:2462–2472

Chen SL, Fang WW, Ye F, et al (2004) Effect on left ventricular function of intracoronary transplantation of autologous bone marrow mesenchymal stem cell in patients with acute myocardial infarction. Am J Cardiol 94:92–95

Chiu RC, Zibaitis A, Kao RL (1995) Cellular cardiomyoplasty: myocardial regeneration with satellite cell implantation. Ann Thorac Surg 60:12–18

Cleland JG, Freemantle N, Coletta AP, et al (2006) Clinical trials update from the American Heart Association: REPAIR-AMI, ASTAMI, JELIS, MEGA, REVIVE-II, SURVIVE, and PROACTIVE. Eur J Heart Fail 8:105–110

Cohn JN (2002) Lessons learned from the valsartan heart failure trial (Val-HeFT): angiotensin receptor blockers in heart failure. Am J Cardiol 90:992–993

Dib N, Michler R, Pagani F, et al (2005) Safety and feasibility of autologous myoblast transplantation in patients with ischemic cardiomyopathy: four-year follow-up. Circulation 112:1748–1755

Doggrell SA (2005) CHARMed—the effects of candesartan in heart failure. Expert Opin Pharmacother 6:513–516

Dohmann H, Perin E, Borojevic R, et al (2005) Sustained improvement in symptoms and exercise capacity up to six months after autologous transendocardial transplantation of bone marrow mononuclear cells in patients with severe ischemic heart disease (in Portuguese). Arq Bras Cardiol 84:360–366

Erices A, Conget P, Minguell J (2000) Mesenchymal progenitor cells in human umbilical cord blood. Br J Haematol 109:235–242

Fadini G, Coracina A, Baesso I, et al (2006) Peripheral blood CD34+KDR+ endothelial progenitor cells are determinants of subclinical atherosclerosis in a middle-aged general population. Stroke 37:2277–2282

Fernandez-Aviles F, San Roman JA, Garcia-Frade J, et al (2004) Experimental and clinical regenerative capability of human bone marrow cells after myocardial infarction. Circ Res 95:742–748

Ferreira S, Winck J, Bettencourt P, et al (2006) Heart failure and sleep apnoea: to sleep perchance to dream. Eur J Heart Fail 8:227–236

Fleischer K, Goldschmidt-Clermont P, Fonger J, et al (1996) One-month histologic response of transmyocardial laser channels with molecular intervention. Ann Thorac Surg 62:1051–1058

Fuchs S, Baffour R, Zhou Y, et al (2001) Transendocardial delivery of autologous bone marrow enhances collateral perfusion and regional function in pigs with chronic experimental myocardial ischemia. J Am Coll Cardiol 37:1726–1732

Fuchs S, Satler LF, Kornowski R, et al (2003) Catheter-based autologous bone marrow myocardial injection in no-option patients with advanced coronary artery disease: a feasibility study. J Am Coll Cardiol 41:1721–1724

Fuchs S, Kornowski R, Weisz G, et al (2006) Safety and feasibility of transendocardial autologous bone marrow cell transplantation in patients with advanced heart disease. Am J Cardiol 97:823–829

Galili N, Raza A (2006) Immunomodulatory drugs in myelodysplastic syndromes. Expert Opin Investig Drugs 15:805–813

Galinanes M, Loubani M, Davies J, et al (2004) Autotransplantation of unmanipulated bone marrow cells into scarred myocardium is safe and enhances cardiac function in humans. Cell Transplant 13:7–13

Gavira J, Herreros J, Perez A, et al (2006) Autologous skeletal myoblast transplantation in patients with nonacute myocardial infarction: 1-year follow-up. J Thorac Cardiovasc Surg 131:799–804

Gemzell-Danielsson K (2006) Effects of levonorgestrel on ovarian function when used for emergency contraception. Minerva Ginecol 58:205–207

Gnecchi M, He H, Noiseux N, et al (2006) Evidence supporting paracrine hypothesis for Akt-modified mesenchymal stem cell-mediated cardiac protection and functional improvement. FASEB J 20:661–669

Grigoriadis A, Heersche J, Aubin J (1988) Differentiation of muscle, fat, cartilage, and bone from progenitor cells present in a bone-derived clonal cell population: effect of dexamethasone. J Cell Biol 106:2139–2151

Haffner SM (2002) Can reducing peaks prevent type 2 diabetes: implication from recent diabetes prevention trials. Intl J Clin Pract Suppl 129:33–39

Hamano K, Nishida M, Hirata K, et al (2001) Local implantation of autologous bone marrow cells for therapeutic angiogenesis in patients with ischemic heart disease: clinical trial and preliminary results. Jpn Circ J 65:845–847

Hansson GK (2005) Inflammation, atherosclerosis, and coronary artery disease. N Engl J Med 352:1685–1695

Hata H, Matsumiya G, Miyagawa S, et al (2006) Grafted skeletal myoblast sheets attenuate myocardial remodeling in pacing-induced canine heart failure model. J Thorac Cardiovasc Surg 132:918–924

Haynesworth SE, Baber MA, Caplan AI (1992) Cell surface antigens on human marrow-derived mesynchemal cells are detected by monoclonal antibodies. Bone Marrow Transplant 13:69–80

Henning RJ, Abu-Ali H, Balis JU, et al (2004) Human umbilical cord blood mononuclear cells for the treatment of acute myocardial infarction. Cell Transplant 13:729–739

Hernandez AF, Velazquez EJ, Solomon SD, et al (2005) Left ventricular assessment in myocardial infarction: the VALIANT registry. Arch Intern Med 165:2162–2169

Herreros J, Prosper F, Perez A, et al (2003) Autologous intramyocardial injection of cultured skeletal muscle-derived stem cells in patients with non-acute myocardial infarction. Eur Heart J 24:2012–2020

Hiasa K, Egashira K, Kitamoto S, et al (2004) Bone marrow mononuclear cell therapy limits myocardial infarct size through vascular endothelial growth factor. Basic Res Cardiol 99:165–172

Hidalgo M, Rodriguez G, Kuhn J, et al (1998) A Phase I and pharmacological study of the glutamine antagonist acivicin with the amino acid solution aminosyn in patients with advanced solid malignancies. Clin Cancer Res 4:2763–2770

Hill JM, Zalos G, Halcox JP, et al (2003) Circulating endothelial progenitor cells, vascular function, and cardiovascular risk. N Engl J Med 348:593–600

Hirata Y, Sata M, Motomura N, et al (2005) Human umbilical cord blood cells improve cardiac function after myocardial infarction. Biochem Biophys Res Commun 327:609–614

Horackova M, Arora R, Chen R, et al (2004) Cell transplantation for treatment of acute myocardial infarction: unique capacity for repair by skeletal muscle satellite cells. Am J Physiol Heart Circ Physiol 287:H1599–1608

Hutcheson KA, Atkins BZ, Hueman MT, et al (2000) Comparing the benefits of cellular cardiomyoplasty with skeletal myoblasts or dermal fibroblasts on myocardial performance. Cell Transplant 9:359–368

Ince H, Petzsch M, Rehders T, et al (2004) Transcatheter transplantation of autologous skeletal myoblasts in postinfarction patients with severe left ventricular dysfunction. J Endovasc Ther 11:695–704

Jackson KA, Majka SM, Wang H, et al (2001) Regeneration of ischemic cardiac muscle and vascular endothelium by adult stem cells. J Clin Invest 107:1395–1402

Janssens S, Dubois C, Bogaert J, et al (2006) Autologous bone marrow-derived stem-cell transfer in patients with ST-segment elevation myocardial infarction: double-blind, randomised controlled trial. Lancet 367:113–121

Jayaram H, Lui M, Plowman J, et al (1990) Oncolytic activity and mechanism of action of a novel L-cysteine derivative, L-cysteine, ethyl ester, S-(N-methylcarbamate) monohydrochloride. Cancer Chemother Pharmacol 26:88–92

Jerosch-Herold M, Wilke N, Stillman A (1998) Magnetic resonance quantification of the myocardial perfusion reserve with a Fermi function model for constrained deconvolution. Med Phys 25:73–84

Jiang XX, Zhang Y, Liu B, et al (2005) Human mesenchymal stem cells inhibit differentiation and function of monocyte-derived dendritic cells. Blood 105:4120–4126

Jiang Y, Jahagirdar BN, Reinhardt RL, et al (2002) Pluripotency of mesenchymal stem cells derived from adult marrow. Nature 418:41–49

Jong P, Yusuf S, Rousseau MF, et al (2003) Effect of enalapril on 12-year survival and life expectancy in patients with left ventricular systolic dysfunction: a follow-up study. Lancet 361:1843–1848

Jost A, Rauch B, Hochadel M, et al (2005) Beta-blocker treatment of chronic systolic heart failure improves prognosis even in patients meeting one or more exclusion criteria of the MERIT-HF study. Eur Heart J 26:2689–2697

Juonala M, Viikari J, Rasanen L, et al (2006) Young adults with family history of coronary heart disease have increased arterial vulnerability to metabolic risk factors: the Cardiovascular Risk in Young Finns Study. Arterioscler Thromb Vasc Biol 26:1376–1382

Kalka C, Masuda H, Takahashi T, et al (2000) Transplantation of ex vivo expanded endothelial progenitor cells for therapeutic neovascularization. Proc Natl Acad Sci U S A 97:3422–3427

Katritsis DG, Sotiropoulou PA, Karvouni E, et al (2005) Transcoronary transplantation of autologous mesenchymal stem cells and endothelial progenitors into infarcted human myocardium. Catheter Cardiovasc Interv 65:321–329

Kaur K, Sharma A, Singal P (2006) Significance of changes in TNF-alpha and IL-10 levels in the progression of heart failure subsequent to myocardial infarction. Am J Physiol Heart Circ Physiol 291:H106-113

Kawada H, Fujita J, Kinjo K, et al (2004) Nonhematopoietic mesenchymal stem cells can be mobilized and differentiate into cardiomyocytes after myocardial infarction. Blood 104:3581–3587

Kawamoto A, Gwon HC, Iwaguro H, et al (2001) Therapeutic potential of ex vivo expanded endothelial progenitor cells for myocardial ischemia. Circulation 103:634–637

Kim B, Tian H, Prasongsukarn K, et al (2005) Cell transplantation improves ventricular function after a myocardial infarction: a preclinical study of human unrestricted somatic stem cells in a porcine model. Circulation 112 [9 Suppl]:I96-I104

Kinnaird T, Stabile E, Burnett M, et al (2004) Local delivery of marrow-derived stromal cells augments collateral perfusion through paracrine mechanisms. Circulation 109:1543–1549

Kirito K, Fox N, Komatsu N, et al (2005) Thrombopoietin enhances expression of vascular endothelial growth factor (VEGF) in primitive hematopoietic cells through induction of HIF-1alpha. Blood 105:4258–4263

Kocher AA, Schuster MD, Szabolcs MJ, et al (2001) Neovascularization of ischemic myocardium by human bone-marrow-derived angioblasts prevents cardiomyocyte apoptosis, reduces remodeling and improves cardiac function. Nat Med 7:430–436

Koh GY, Klug MG, Soonpaa MH, et al (1993) Differentiation and long-term survival of C2C12 myoblast grafts in heart. J Clin Invest 92:1548–1554

Kohin S, Stary CM, Howlett RA, et al (2001) Preconditioning improves function and recovery of single muscle fibers during severe hypoxia and reoxygenation. Am J Physiol Cell Physiol 281:C142-146

Korbling M, Estrov Z, Champlin R (2003) Adult stem cells and tissue repair. Bone Marrow Transplant 32 [Suppl 1]:S23–24

Kotlyar E, Vita J, Winter M, et al (2006) The relationship between aldosterone, oxidative stress, and inflammation in chronic, stable human heart failure. J Card Fail 12:122–127

Krumholz HM, Currie PM, Riegel B, et al (2006) AHA Scientific statement. A taxonomy for disease management. A scientific statement from the American Heart Association Disease Management Taxonomy Writing Group. Circulation 114:1432–1445

Kuethe F, Richartz BM, Sayer HG, et al (2004) Lack of regeneration of myocardium by autologous intracoronary mononuclear bone marrow cell transplantation in humans with large anterior myocardial infarctions. Int J Cardiol 97:123–127

Lambert G (2005) Preclinical neuropharmacology of naratriptan. CNS Drug Rev 11:289–316

Laugwitz KL, Moretti A, Lam J, et al (2005) Postnatal isl1+ cardioblasts enter fully differentiated cardiomyocyte lineages. Nature 433:647–653

LeGrand C (2000) Fifty-third World Health Assembly: impact on cardiovascular disease. Can J Cardiol 16:857–859

Lenzen MJ, Boersma E, Reimer WJ, et al (2005) Under-utilization of evidence-based drug treatment in patients with heart failure is only partially explained by dissimilarity to patients enrolled in landmark trials: a report from the Euro Heart Survey on Heart Failure. Eur Heart J 26:2706–2713

Leor J, Guetta E, Feinberg M, et al (2006) Human umbilical cord blood-derived CD133+ cells enhance function and repair of the infarcted myocardium. Stem Cells 24:772–780

Levy D, Kenchaiah S, Larson MG, et al (2002) Long-term trends in the incidence of and survival with heart failure. N Engl J Med 347:1397–1402

Liu J, Hu Q, Wang Z, et al (2004) Autologous stem cell transplantation for myocardial repair. Am J Physiol Heart Circ Physiol 287:H501–511

Lunde K, Solheim S, Aakhus S, et al (2006) Intracoronary injection of mononuclear bone marrow cells in acute myocardial infarction. N Engl J Med 355:1199–1209

Ma N, Stamm C, Kaminski A, et al (2005) Human cord blood cells induce angiogenesis following myocardial infarction in NOD/SCID-mice. Cardiovasc Res 66:45–54

Makino S, Fukuda K, Miyoshi S, et al (1999) Cardiomyocytes can be generated from marrow stromal cells in vitro. J Clin Invest 103:697–705

Mann D, McMurray J, Packer M, et al (2004) Targeted anticytokine therapy in patients with chronic heart failure: results of the Randomized Etanercept Worldwide Evaluation (RENEWAL). Circulation 109:1594–1602

Mark D, Nelson C, Anstrom K, et al (2006) Cost-effectiveness of defibrillator therapy or amiodarone in chronic stable heart failure: results from the Sudden Cardiac Death in Heart Failure Trial (SCD-HeFT). Circulation 114:135–142

Mauro A (1961) Satellite cell of skeletal muscle fibers. J Biophys Biochem Cytol 9:493–495

Menasche P (2006) Late-breaking clinical trials III. First randomized placebo-controlled myoblast autologous grafting in ischemic cardiomyopathy (MAGIC) trial. Presented at the Scientific Sessions of the American Heart Association, 15 November 2006, Chicago

Menasche P, Hagege AA, Vilquin JT, et al (2003) Autologous skeletal myoblast transplantation for severe postinfarction left ventricular dysfunction. J Am Coll Cardiol 41:1078–1083

Meyer G, Wollert K, Lotz J, et al (2006) Intracoronary bone marrow cell transfer after myocardial infarction: eighteen months' follow-up data from the randomized, controlled BOOST (BOne marrOw transfer to enhance ST-elevation infarct regeneration) trial. Circulation 113:1287–1294

Miller LW, Missov ED (2001) Epidemiology of heart failure. Cardiol Clin 19:547–555

Misao Y, Takemura G, Arai M, et al (2006) Importance of recruitment of bone marrow-derived CXCR4+ cells in post-infarct cardiac repair mediated by G-CSF. Cardiovasc Res 71:455–465

Miyagawa S, Sawa Y, Sakakida S, et al (2005) Tissue cardiomyoplasty using bioengineered contractile cardiomyocyte sheets to repair damaged myocardium: their integration with recipient myocardium. Transplantation 80:1586–1595

Moss AJ (2003) MADIT-I and MADIT-II. J Cardiovasc Electrophysiol 14:S96–S98

Moss AJ, Zareba W, Hall WJ, et al (2002) Prophylactic implantation of a defibrillator in patients with myocardial infarction and reduced ejection fraction. N Engl J Med 346:877–883

Mouquet F, Pfister O, Jain M, et al (2005) Restoration of cardiac progenitor cells after myocardial infarction by self-proliferation and selective homing of bone marrow-derived stem cells. Circ Res 97:1090–1092

Murry CE, Kay MA, Bartosek T, et al (1996) Muscle differentiation during repair of myocardial necrosis in rats via gene transfer with MyoD. J Clin Invest 98:2209–2217

Murry CE, Soonpaa MH, Reinecke H, et al (2004) Haematopoietic stem cells do not transdifferentiate into cardiac myocytes in myocardial infarcts. Nature 428:664–668

Neuhaus O, Wiendl H, Kieseier B, et al (2005) Multiple sclerosis: mitoxantrone promotes differential effects on immunocompetent cells in vitro. J Neuroimmunol 168:128–137

O'Connor C, Gattis W, Adams KJ, et al (2003) Tezosentan in patients with acute heart failure and acute coronary syndromes: results of the Randomized Intravenous TeZosentan Study (RITZ-4). J Am Coll Cardiol 41:1452–1457

Oh H, Bradfute SB, Gallardo TD, et al (2003) Cardiac progenitor cells from adult myocardium: homing, differentiation, and fusion after infarction. Proc Natl Acad Sci U S A 100:12313–12318

Ohno N, Fedak P, Weisel R, et al (2003) Transplantation of cryopreserved muscle cells in dilated cardiomyopathy: effects on left ventricular geometry and function. J Thorac Cardiovasc Surg 126:1537–1548

Orlic D, Kajstura J, Chimenti S, et al (2001) Bone marrow cells regenerate infarcted myocardium. Nature 410:701–705

Ott HC, Taylor DA (2006) From cardiac repair to cardiac regeneration—ready to translate? Expert Opin Biol Ther 6:867–878

Ott HC, Bonaros N, Marksteiner R, et al (2004) Combined transplantation of skeletal myoblasts and bone marrow stem cells for myocardial repair in rats. Eur J Cardiothorac Surg 25:627–634

Ott HC, Kroess R, Bonaros N, et al (2005a) Intramyocardial microdepot injection increases the efficacy of skeletal myoblast transplantation. Eur J Cardiothorac Surg 27:1017–1021

Ott HC, Matthiesen T, Brechtken J, et al (2005b) A novel population of adult derived cardiac progenitor cells is capable of functional myocardial repair. Circulation 112:II-332

Ott HC, Brechtken J, Swingen C, et al (2006) Robotic minimally invasive cell transplantation for heart failure. J Thorac Cardiovasc Surg 132:170–173

Ott HC, Matthiesen T, Brechtken J, et al (2007) The adult human heart as a source for stem cells: repair strategies with embryonic like progenitor cells. Nat Clin Pract Cardiovasc Med 4 suppl 1:S27–S39

Pagani FD, DerSimonian H, Zawadzka A, et al (2003) Autologous skeletal myoblasts transplanted to ischemia-damaged myocardium in humans. Histological analysis of cell survival and differentiation. J Am Coll Cardiol 41:879–888

Pearson TA, Blair SN, Daniels SR, et al (2002) AHA Guidelines for Primary Prevention of Cardiovascular Disease and Stroke: 2002 Update: consensus panel guide to comprehensive risk reduction for adult patients without coronary or other atherosclerotic vascular diseases. American Heart Association Science Advisory and Coordinating Committee. Circulation 106:388–391

Perin EC, Dohmann HF, Borojevic R, et al (2003) Transendocardial, autologous bone marrow cell transplantation for severe, chronic ischemic heart failure. Circulation 107:2294–2302

Perin EC, Dohmann HF, Borojevic R, et al (2004) Improved exercise capacity and ischemia 6 and 12 months after transendocardial injection of autologous bone marrow mononuclear cells for ischemic cardiomyopathy. Circulation 110:II213–218

Pickett F (2006) Bisphosphonate-associated osteonecrosis of the jaw: a literature review and clinical practice guidelines. J Dent Hyg 80:10

Pittenger MF, Martin BJ (2004) Mesenchymal stem cells and their potential as cardiac therapeutics. Circ Res 95:9–20

Pittenger MF, Mackay AM, Beck SC, et al (1999) Multilineage potential of adult human mesenchymal stem cells. Science 284:143–147

Raines E, Ferri N (2005) Thematic review series: the immune system and atherogenesis. Cytokines affecting endothelial and smooth muscle cells in vascular disease. J Lipid Res 46:1081–1092

Ratajczak M, Reca R, Wysoczynski M, et al (2006) Modulation of the SDF-1-CXCR4 axis by the third complement component (C3)—implications for trafficking of CXCR4+ stem cells. Exp Hematol 34:986–995

Rauscher FM, Goldschmidt-Clermont PJ, Davis BH, et al (2003) Aging, progenitor cell exhaustion, and atherosclerosis. Circulation 108:457–463

Reffelmann T, Leor J, Muller-Ehmsen J, et al (2003) Cardiomyocyte transplantation into the failing heart-new therapeutic approach for heart failure? Heart Fail Rev 8:201–211

Reiffel J (2005) Practical algorithms for pharmacologic management of the post myocardial infarction patient. Clin Cardiol 28:I28–37

Reinecke H, MacDonald GH, Hauschka SD, et al (2000) Electromechanical coupling between skeletal and cardiac muscle. Implications for infarct repair. J Cell Biol 149:731–740

Reinecke H, Poppa V, Murry CE (2002) Skeletal muscle stem cells do not transdifferentiate into cardiomyocytes after cardiac grafting. J Mol Cell Cardiol 34:241–249

Saulnier N, Di Campli C, Zocco M, et al (2005) From stem cell to solid organ. Bone marrow, peripheral blood or umbilical cord blood as favorable source? Eur Rev Med Pharmacol Sci 9:315–324

Schachinger V, Assmus B, Britten MB, et al (2004) Transplantation of progenitor cells and regeneration enhancement in acute myocardial infarction: final one-year results of the TOPCARE-AMI Trial. J Am Coll Cardiol 44:1690–1699

Schachinger V, Assmus B, Honold J, et al (2006a) Normalization of coronary blood flow in the infarct-related artery after intracoronary progenitor cell therapy: intracoronary Doppler substudy of the TOPCARE-AMI trial. Clin Res Cardiol 95:13–22

Schachinger V, Erbs S, Elsasser A, et al (2006b) Intracoronary bone marrow-derived progenitor cells in acute myocardial infarction. N Engl J Med 355:1210–1221

Schaefer A, Meyer G, Fuchs M, et al (2006) Impact of intracoronary bone marrow cell transfer on diastolic function in patients after acute myocardial infarction: results from the BOOST trial. Eur Heart J 27:929–935

Schuster MD, Kocher AA, Seki T, et al (2004) Myocardial neovascularization by bone marrow angioblasts results in cardiomyocyte regeneration. Am J Physiol Heart Circ Physiol 287:H525–532

Scorsin M, Hagege AA, Marotte F, et al (1997) Does transplantation of cardiomyocytes improve function of infarcted myocardium? Circulation 96:II-188–193

Scorsin M, Hagege A, Vilquin JT, et al (2000) Comparison of the effects of fetal cardiomyocyte and skeletal myoblast transplantation on postinfarction left ventricular function. J Thorac Cardiovasc Surg 119:1169–1175

Shake JG, Gruber PJ, Baumgartner WA, et al (2002) Mesenchymal stem cell implantation in a swine myocardial infarct model: engraftment and functional effects. Ann Thorac Surg 73:1919–1925; discussion 1926

Shintani S, Murohara T, Ikeda H, et al (2001) Mobilization of endothelial progenitor cells in patients with acute myocardial infarction. Circulation 103:2776–2779

Sidana J, Aronow W, Ravipati G, et al (2005) Prevalence of moderate or severe left ventricular diastolic dysfunction in obese persons with obstructive sleep apnea. Cardiology 104:107–109

Silva G, Perin E, Dohmann H, et al (2004) Catheter-based transendocardial delivery of autologous bone-marrow-derived mononuclear cells in patients listed for heart transplantation. Tex Heart Inst J 31:214–219

Siminiak T, Kalawski R, Fiszer D, et al (2004) Autologous skeletal myoblast transplantation for the treatment of postinfarction myocardial injury: phase I clinical study with 12 months of follow-up. Am Heart J 148:531–537

Siminiak T, Fiszer D, Jerzykowska O, et al (2005) Percutaneous trans-coronary-venous transplantation of autologous skeletal myoblasts in the treatment of post-infarction myocardial contractility impairment: the POZNAN trial. Eur Heart J 26:1188–1195

Skuk D, Goulet M, Roy B, et al (2002) Efficacy of myoblast transplantation in nonhuman primates following simple intramuscular cell injections: toward defining strategies applicable to humans. Exp Neurol 175:112–126

Smith R, Barlie L, Cho HC, et al (2005) Unique phenotype of cardiospheres derived from human endomyocardial biopsies. Circulation 112:II-334

Smith SC Jr, Allen J, Blair SN, et al (2006) AHA/ACC guidelines for secondary prevention for patients with coronary and other atherosclerotic vascular disease: 2006 update: endorsed by the National Heart, Lung, and Blood Institute [erratum appears in Circulation. 2006 Jun 6;113 (22):e847]. Circulation 113:2363–2372

Smits PC, van Geuns RJ, Poldermans D, et al (2003) Catheter-based intramyocardial injection of autologous skeletal myoblasts as a primary treatment of ischemic heart failure: clinical experience with six-month follow-up. J Am Coll Cardiol 42:2063–2069

Stamm C, Westphal B, Kleine HD, et al (2003) Autologous bone-marrow stem-cell transplantation for myocardial regeneration. Lancet 361:45–46

Stewart C, Newcomb P, Holly J (2004) Multifaceted roles of TNF-alpha in myoblast destruction: a multitude of signal transduction pathways. J Cell Physiol 198:237–247

Strauer B, Brehm M, Zeus T, et al (2005) Regeneration of human infarcted heart muscle by intracoronary autologous bone marrow cell transplantation in chronic coronary artery disease: the IACT Study. J Am Coll Cardiol 46:1651–1658

Strauer BE, Brehm M, Zeus T, et al (2002) Repair of infarcted myocardium by autologous intracoronary mononuclear bone marrow cell transplantation in humans. Circulation 106:1913–1918

Suzuki K, Brand NJ, Allen S, et al (2001) Overexpression of connexin 43 in skeletal myoblasts: relevance to cell transplantation to the heart. J Thorac Cardiovasc Surg 122:759–766

Takahashi M, Li T, Suzuki R, et al (2006) Cytokines produced by bone marrow cells can contribute to functional improvement of the infarcted heart by protecting cardiomyocytes from ischemic injury. Am J Physiol Heart Circ Physiol 291:H886–893

Tan P, Xue S, Manunta M, et al (2006) Effect of vectors on human endothelial cell signal transduction: implications for cardiovascular gene therapy. Arterioscler Thromb Vasc Biol 26:462–467

Taylor DA, Atkins BZ, Hungspreugs P, et al (1998) Regenerating functional myocardium: improved performance after skeletal myoblast transplantation. Nat Med 4:929–933

Thattassery E, Gheorghiade M (2004) Beta blocker therapy after acute myocardial infarction in patients with heart failure and systolic dysfunction. Heart Fail Rev 9:107–113

Thom T, Haase N, Rosamond W, et al (2006) Heart disease and stroke statistics—2006 update: a report from the American Heart Association Statistics Committee and Stroke Statistics Subcommittee. Circulation 113:e85–e151

Thompson RB, Emani SM, Davis BH, et al (2003) Comparison of intracardiac cell transplantation: autologous skeletal myoblasts versus bone marrow cells. Circulation 108 [Suppl 1]:II264–II271

Thompson RB, Parsa CJ, van den Bos EJ, et al (2004) Video-assisted thoracoscopic transplantation of myoblasts into the heart. Ann Thorac Surg 78:303–307

Toma C, Pittenger MF, Cahill KS, et al (2002) Human mesenchymal stem cells differentiate to a cardiomyocyte phenotype in the adult murine heart. Circulation 105:93–98

Tomita S, Li RK, Weisel RD, et al (1999) Autologous transplantation of bone marrow cells improves damaged heart function. Circulation 100:II247–256

Torp-Pedersen C, Poole-Wilson PA, Swedberg K, et al (2005) Effects of metoprolol and carvedilol on cause-specific mortality and morbidity in patients with chronic heart failure—COMET. Am Heart J 149:370–376

Toth M, Ades P, Tischler M, et al (2006) Immune activation is associated with reduced skeletal muscle mass and physical function in chronic heart failure. Int J Cardiol 109:179–187

Tse H, Thambar S, Kwong Y, et al (2006) Safety of catheter-based intramyocardial autologous bone marrow cells implantation for therapeutic angiogenesis. Am J Cardiol 98:60–62

Tse HF, Kwong YL, Chan JK, et al (2003) Angiogenesis in ischaemic myocardium by intramyocardial autologous bone marrow mononuclear cell implantation. Lancet 361:47–49

Udelson J, Patten R, Konstam M (2003) New concepts in post-infarction ventricular remodeling. Rev Cardiovasc Med 4:S3–12

Urbanek K, Torella D, Sheikh F, et al (2005) Myocardial regeneration by activation of multipotent cardiac stem cells in ischemic heart failure. Proc Natl Acad Sci U S A 102:8692–8697

Van Den Bos EJ, Taylor DA (2003) Cardiac transplantation of skeletal myoblasts for heart failure. Minerva Cardioangiol 51:227–243

Vandervelde S, van Luyn M, Tio R, et al (2005) Signaling factors in stem cell-mediated repair of infarcted myocardium. J Mol Cell Cardiol 39:363–376

Verfaillie CM, Schwartz R, Reyes M, Jiang Y (2003) Unexpected potential of adult stem cells. Ann N Y Acad Sci 996:231–234

Vulliet PR, Greeley M, Halloran SM, et al (2004) Intra-coronary arterial injection of mesenchymal stromal cells and microinfarction in dogs. Lancet 363:783–784

Wakitani S, Saito T, Caplan AI (1995) Myogenic cells derived from rat bone marrow mesenchymal stem cells exposed to 5-azacytidine. Muscle Nerve 18:1417–1426

Wang JS, Shum-Tim D, Galipeau J, et al (2000) Marrow stromal cells for cellular cardiomyoplasty: feasibility and potential clinical advantages. J Thorac Cardiovasc Surg 120:999–1005

Wang Y, Haider H, Ahmad N, et al (2006) Combining pharmacological mobilization with intramyocardial delivery of bone marrow cells over-expressing VEGF is more effective for cardiac repair. J Mol Cell Cardiol 40:736–745

Werner N, Kosiol S, Schiegl T, et al (2005) Circulating endothelial progenitor cells and cardiovascular outcomes. N Engl J Med 353:999–1007

Wilke N, Zenovich A, Jerosch-Herold M, et al (2001) Cardiac magnetic resonance imaging for the assessment of myocardial angiogenesis. Curr Interv Cardiol Rep 3:205–212

Wollert KC, Meyer GP, Lotz J, et al (2004) Intracoronary autologous bone-marrow cell transfer after myocardial infarction: the BOOST randomised controlled clinical trial. Lancet 364:141–148

Wyatt SB, Winters KP, Dubbert PM (2006) Overweight and obesity: prevalence, consequences, and causes of a growing public health problem. Am J Med Sci 331:166–174

Yan LL, Liu K, Daviglus ML, et al (2006) Education, 15-year risk factor progression, and coronary artery calcium in young adulthood and early middle age: the Coronary Artery Risk Development in Young Adults study. JAMA 295:1793–1800

Yancy C, Fonarow G (2004) Quality of care and outcomes in acute decompensated heart failure: the ADHERE Registry. Curr Heart Fail Rep 1:121–128

Yang EH, Barsness GW, Gersh BJ, et al (2004) Current and future treatment strategies for refractory angina. Mayo Clin Proc 79:1284–1292

Yau T, Kim C, Ng D, et al (2005) Increasing transplanted cell survival with cell-based angiogenic gene therapy. Ann Thorac Surg 80:1779–1786

Zen K, Okigaki M, Hosokawa Y, et al (2006) Myocardium-targeted delivery of endothelial progenitor cells by ultrasound-mediated microbubble destruction improves cardiac function via an angiogenic response. J Mol Cell Cardiol 40:799–809

Zenovich A, Muehling O, Panse P, et al (2001) Magnetic Resonance first-pass perfusion: overview and perspectives. Rays 26:53–60

Zhai Q, Qiu L, Li Q, et al (2004) Short-term ex vivo expansion sustains the homing-related properties of umbilical cord blood hematopoietic stem and progenitor cells. Haematologica 89:265–273

Zhang M, Methot D, Poppa V, et al (2001) Cardiomyocyte grafting for cardiac repair: graft cell death and anti-death strategies. J Mol Cell Cardiol 33:907–921

Zhou R, Acton P, Ferrari V (2006) Imaging stem cells implanted in infarcted myocardium. J Am Coll Cardiol 48:2094–2106

Zibaitis A, Greentree D, Ma F, et al (1994) Myocardial regeneration with satellite cell implantation. Transplant Proc 26:3294

Zimmet J, Hare JM (2005) Emerging role for bone marrow derived mesenchymal stem cells in myocardial regenerative therapy. Basic Res Cardiol 100:471–481

Ischemic Tissue Repair by Autologous Bone Marrow-Derived Stem Cells: Scientific Basis and Preclinical Data

A. Quraishi[1] · D. W. Losordo[2] (✉)

[1] Division of Cardiology, Caritas St. Elizabeth's Medical Center, 736 Cambridge St., Boston MA, 02135, USA

[2] Northwestern University Feinberg School of Medicine, 303 East Chicago Ave., Tarry 12-703, Chicago IL, 60611, USA
d-losordo@northwestern.edu

1	Introduction	167
2	Endothelial Progenitor Cells and Mononuclear Cells	169
3	Enhancing the Efficacy of Autologous Stem Cells	171
4	Safety Concerns Regarding Therapeutic Angiogenesis	173
5	Summary	174
	References	175

Abstract The success of therapies targeting acute myocardial ischemia and the aging of the population due to improved general medical care has resulted in an increasing population of patients with chronic myocardial ischemia and congestive heart failure who remain symptomatic despite having exhausted the currently available therapeutic options. In this chapter we review the scientific underpinnings of autologous bone marrow-derived cell therapy and the early clinical experience that has fuelled interest in this approach.

Keywords Stem cell · Bone marrow · Angiogenesis · Endothelial progenitor · Heart failure · Ischemia

1 Introduction

Despite significant therapeutic advances, heart failure associated with myocardial ischemia remains the predominant cause of morbidity and mortality in the Western world. Ischemic cardiomyopathy and myocardial infarction (MI) are typified by the irreversible loss of vasculature composed of endothelial cells and smooth muscle cells and/or cardiac muscle (cardiomyocytes), which are essential for maintaining cardiac integrity and function. As medical therapies and mechanical revascularization techniques continue to be refined and the population continues to age, there is an increasingly growing population of patients who, despite optimal medical therapy and risk-factor modification, are no longer candidates for conventional mechanical revascularization. They may suffer from severe diffuse atherosclerotic disease not amenable to surgery or angioplasty, or they may have had prior revascularization procedures making future repeat procedures technically prohibitive. These patients continue to live with symptomatic obstructive vascular disease resulting in intractable angina and congestive heart failure, resulting in a sedentary lifestyle and substantially compromising the quality of life. In addition to macrovascular occlusive disease, an emerging body of data has documented the importance of microvascular insufficiency as a significant factor contributing to the overall ischemic burden, not only in patients with chronic myocardial ischemia, but also in patients with peripheral arterial disease (PAD). These patients suffer from lifestyle-limiting intermittent claudication and critical limb ischemia without options of mechanical revascularization. Despite aggressive medical therapy and modification of atherosclerotic risk factors, these patients progress to requiring amputation. Of patients who undergo an amputation, at least 10% will require a further amputation (Campbell et al. 1994; Dawson et al. 1995; Eneroth and Persson 1992; Skinner and Cohen 1996). Accordingly, attention is now increasingly focused on developing an understanding of the mechanisms that govern the maintenance, repair, and growth of the microvasculature in ischemic tissue. This discipline is embodied in the term therapeutic angiogenesis.

The recent identification of adult and embryonic stem cells has triggered attempts to directly repopulate lost tissues by stem cell transplantation as a novel therapeutic option. Currently, various BM-derived stem/progenitor cells are being investigated to regenerate or repair myocardium depending on the type of disease (predominantly deficient tissues) and the differentiation characteristics of specific stem/progenitor cell population. Reports describing provocative and hopeful examples of myocardial regeneration with adult bone marrow-derived stem and progenitor cells have furthered enthusiasm for the use of these cells, yet many questions remain regarding their therapeutic potential and the mechanisms responsible for the observed therapeutic effects. In fact, because most of the animal studies and clinical trials have been undertaken in the setting of acute MI, there has been a certain misunderstanding and limitation to interpreting the data with any stem/progenitor cells. However, emerging evidence has shown that administration of stem/progenitor cells can improve cardiac function through various mechanisms including neovascularization in animal models of chronic myocardial ischemia. Clinical trials of stem/progenitor cells in patients with chronic ischemic heart failure have also shown increases in exercise time and reductions in anginal symptoms and have provided objective evidence of improved perfusion and LV function. Small-scale placebo-controlled randomized clinical trials of cell transplantation via thoracotomy or percutaneous approach demonstrated significant improvement of both subjective symptoms and objective measures of myocardial ischemia. Larger-scale placebo-controlled randomized trials are now ongoing. This new therapeutic modality appears to be safe and well tolerated. In this chapter, we discuss the current preclinical and clinical advances in bone marrow (BM)-derived stem or progenitor cell therapies and angiogenic growth factors for chronic ischemic myocardium and peripheral limb ischemia.

2
Endothelial Progenitor Cells and Mononuclear Cells

Angiogenesis and vasculogenesis are responsible for the development of the vascular system in embryos (Gilbert 1997). Vasculogenesis refers to the in situ formation of blood vessels from endothelial progenitor cells or angioblasts (Flamme and Risau 1992; Risau and Flamme 1995) that differentiate into endothelial cells while forming new vessels in situ and give rise to capillaries (Risau 1995). In contrast, angiogenesis involves the extension of preexisting primitive vasculature by sprouting of new capillaries through migration and proliferation of already differentiated endothelial cells. Until recently, vasculogenesis was thought to be restricted to embryonic development, whereas angiogenesis was considered responsible for neovascularization in embryo and adults. Now the existence of postnatal vasculogenesis is widely accepted with the discovery that bone marrow-derived endothelial progenitor cells (EPCs) circulate in peripheral blood (Asahara et al. 1997), home to and incorporate into foci of neovascularization in adult animals (Asahara et al. 1999), and increase in number in response to tissue ischemia (Takahashi et al. 1999).

Initially, Flk-1 and a second antigen, CD34, shared by angioblasts and hematopoietic cells were used to isolate putative angioblasts from the mononuclear cell fraction of peripheral blood (Asahara et al. 1997). Meanwhile, EPCs subsequently were isolated from human umbilical cord blood, BM-derived mononuclear cells, and $CD34^+$ or $CD133^+$ hematopoietic stem cells (Asahara et al. 1999; Murohara et al. 2000; Rafii 2000; Shi et al. 1998). These cells differentiate into endothelial cells, as shown by expression of various endothelial proteins (KDR, von Willebrand factor, endothelial nitric oxidase synthase, VE-cadherin, CD146), uptake of DiI-acetylated low-density lipoprotein and binding of lectin.

In animal models of ischemia, heterologous (derived from a separate genetic source or species), homologous (derived from the same species), and autologous (derived from an individual's own) EPCs were shown to incorporate into sites of active neovascularization in ischemic tissue. Our group and others investigated the therapeutic effect of EPCs in animal models of ischemic heart diseases. Preclinical work by Kawamoto and colleagues showed the potential of endothelial progenitor cells to induce therapeutic angiogenesis and preserve function in an animal model of myocardial ischemia (Kawamoto et al. 2001). Myocardial ischemia was induced in rats by ligating the left anterior descending artery. Within 3 h of the induction of myocardial ischemia, human EPCs were injected intravenously. After 7 days, the rats were sacrificed after the intravenous injection of fluorescence-conjugated *Bandeiraea simplicifolia* lectin. Fluorescence microscopy revealed that transplanted EPCs homed into ischemic territories and incorporated into foci of myocardial neovascularization. To determine the effect on LV function, within hours after the induction of ischemia, ten rats were randomized to receive either EPCs or culture media.

At day 28, the EPC-treated group had significantly smaller left ventricle (LV) dimensions compared to the control treated group. Further, regional wall motion was better preserved in the EPC group. The animals were then sacrificed and necropsy showed that capillary density was significantly greater in the EPC group than in the control group. Immunohistochemistry confirmed that the capillaries were positive for human-specific endothelial cells.

Kocher showed similar findings using CD34 as a marker to identify human EPCs, and further showed that mononuclear cells depleted of $CD34^+$ cells were ineffective, suggesting that the CD34 population plays an essential role in ischemic tissue repair (Kocher et al. 2001). The systemic administration of ex vivo-expanded EPCs (Kawamoto et al. 2001) or direct transplantation of purified human $CD34^+$ cells (Kawamoto et al. 2001; Kocher et al. 2001) or the mononuclear cell fraction of BM (Kamihata et al. 2001) following infarction, all resulted in incorporation of the transplanted cells into foci of myocardial neovascularization, and improved neovascularization and myocardial function. The rationale of using unselected or mononuclear BM cells is based on the premise that BM or peripheral blood cells include multiple cytokines related to neovascularization and stem/progenitor cell populations which could augment vasculogenesis in animal models of experimentally induced ischemia.

At least three experimental studies have been reported on stem cell transplantation in chronic myocardial ischemia models. Kawamoto studied the effect of intramyocardial delivery of EPCs to enhance neovascularization in chronic myocardial ischemia (Kawamoto et al. 2003). $CD31^+$ and $CD31^-$ mononuclear cells were harvested from swine. The pigs were then subjected to placement of an ameroid constrictor on the left anterior descending artery to create myocardial ischemia. The pigs were then randomly assigned to receive autologous $CD31^+$ cells, $CD31^-$ cells, or placebo injected into the ischemic myocardium as identified by NOGA (Cordis, Miami Lakes, FL) electromechanical mapping. The ischemic area identified by NOGA mapping, Rentrop grade angiographic collateral development, and echocardiographic LV ejection fraction improved significant at 4 weeks in pigs treated with $CD31^+$ cells, but not those treated with $CD31^-$ cells nor control. In a parallel study, nude rats with myocardial ischemia were treated with intramyocardial injections of human $CD34^+$ cells, $CD34^-$ cells, or control. Capillary density was significantly improved in the $CD34^+$-treated rats. These studies and others served as important starting points for human studies of EPC transplantation for therapeutic angiogenesis. Fuchs et al. (2001) transplanted freshly isolated autologous unselected bone marrow cells in an ameroid constrictor model of pig using electromechanical mapping. They reported increased collateral development and myocardial perfusion and significantly improved myocardial function. In vitro experiments revealed that these cells could secrete angiogenic/arteriogenic factors such as vascular endothelial growth factor (VEGF) and macrophage chemoattractant protein-1 (MCP-1). In a canine chronic coronary occlusion model (Hamano et al. 2001), transepicardial injection of mononuclear cells after coronary occlusion led to a 50% increase in the number of microvessels observed and a significant improvement in LV wall systolic function.

Among these promising results, Yoon et al. reported potential adverse effects following adult stem cell transplantation in a rat model of acute MI (Yoon et al. 2004). In this study, severe intramyocardial calcification was observed following direct transplantation of syngeneic unselected BM cells into the infarct myocardium (Yoon et al. 2004). Although there are a few points that distinguish this study from the others or clinical studies in that the cells used were unselected filtered BM cells and the cells were directly transplanted after inducing MI, this potential adverse effect needs to be considered for follow-up plans in clinical trials.

3
Enhancing the Efficacy of Autologous Stem Cells

Although the exact mechanisms of action of local angiogenic gene therapy has yet to be understood, previous studies have shown that part of the therapeutic action is due to the recruitment of EPCs to ischemic tissue. However, recent data indicate that patients with the most severe vascular disease may have insufficient or deficient EPCs and the poorest response to angiogenic therapy. Three experimental studies were conducted with the common goal of overcoming these deficiencies by enhancing ischemic neovascularization. Shintani et al. (2006) combined human $CD34^+$ cell implantation with local vascular endothelial growth factor 2 (phVEGF2). MI was induced in 34 immunodeficient rats by ligating the left anterior descending coronary artery. The investigators compared the therapeutic efficacy of locally injected human $CD34^+$ cells alone, VEGF gene therapy alone, $CD34^-$ cells and 50 μg empty plasmid, and $CD34^+$ cells and VEGF therapy in combination, using subtherapeutic doses of $CD34^+$ cells to simulate conditions of deficient or defective EPCs. One of the major findings was that the addition of VEGF2 to EPC cultures provided a significant and dose-dependent decrease in EPC apoptosis. Western blot analysis revealed that the expression of p-Akt, which plays a role in suppressing EPC apoptosis (Dimmeler et al. 2001; Murasawa et al. 2002), is increased in cells treated with VEGF2, indicating a survival effect of VEGF on EPCs. Rats that were treated with combination therapy 4 weeks post-MI showed improved fractional shortening, increased capillary density, reduced infarct size, and an increased number of circulating EPCs. This increase was noted as early as 1 week post-MI, suggesting that combined therapy might augment certain innate mechanisms for myocardial repair after ischemic injury, including mobilization of bone-marrow-derived EPCs. Data from a recent study indicated that, in humans, the level of circulating EPCs in the peri-infarct period predicts the occurrence of cardiovascular events and death from cardiovascular causes and may help to identify patients at increased cardiovascular risk (Werner et al. 2005).

In humans and other mammals, most somatic cells undergo a finite number of cell divisions, ultimately entering a nondividing stated called senescence (Harley et al. 1990; Hastie et al. 1990; Murasawa et al. 2002; Wright and Shay 1992), which may be due to the loss of telomerase activity, a regulatory molecule for cell lifespan. True stem cells and germline cells highly express the catalytic subunit of telomerase, human telomerase reverse transcriptase (hTERT) (Harrington et al. 1997; Kilian et al. 1997; Meyerson et al. 1997; Nakamura et al. 1997); therefore, they are able to divide indefinitely. Despite the regenerative potential of EPCs, they are not pluripotent, self-renewing stem cells. Murasawa et al. (2002) deduced that constitutive expression of hTERT might induce a delay in senescence and recover/enhance regenerative properties of EPCs. Constitutive expression of hTERT conserved telomerase activity and delayed senescence, and enhanced EPC regenerative properties including mitogenesis, migration, and EPC survival. They also investigated the impact hTERT-transduced EPCs (Td/TERTs) administration on therapeutic neovascularization in athymic nude mice with unilateral hindlimb ischemia. The mice transplanted with Td/TERT demonstrated enhanced perfusion, and histologic analysis also revealed increased capillary density at day 28. The enhancement of regenerative activity in EPCs may have therapeutic implications for vascular disorders that cause severe ischemic disease.

Stromal cell-derived factor-1 (SDF-1), a member of the chemokine CXC subfamily, is considered one of the key regulators of hematopoietic stem cell trafficking. Several studies have reported that $CD34^+$ cells express CXCR4, the receptor for SDF-1; therefore, SDF-1 can induce $CD34^+$ cell migration in vitro (Mohle et al. 1998), can effect $CD34^+$ cell proliferation and migration, and can induce angiogenesis (Hattori et al. 2001; Lataillade et al. 2000; Salcedo et al. 1999; Yamaguchi et al. 2003). To further investigate whether SDF-1 augmented EPC-induced vasculogenesis, Yamaguchi et al. (2003) locally injected SDF-1 into athymic ischemic hindlimb muscle of nude mice, combined with human EPC transplantation. By day 3 after treatment, local SDF-1 administration augmented EPC accumulation within the ischemic tissue. The magnitude of EPC incorporation was similar between 3 and 7 days, which suggests that the homing of exogenously administered EPCs occurs early after transplantation. Serial measurements of hindlimb perfusion were performed at days 7, 14, 21, and 28, showing profound differences in the limb perfusion 28 days after induction of ischemia. Thus, the homing effect of local SDF-1 was accompanied by physiological evidence for enhanced neovascularization, indicating that the EPCs that were attracted to the ischemic limb by SDF-1 were incorporated into developing vasculature. Histological analysis revealed an increase in capillary density, which is a direct reflection of neovascularization, in the SDF-1-treatment group. This study was the first experimental proof for therapeutic effectiveness of augmenting local accumulation of EPCs in ischemic tissue, which subsequently enhanced vasculogenesis and contributed to neovascularization.

4
Safety Concerns Regarding Therapeutic Angiogenesis

There are several theoretical safety concerns regarding the delivery of angiogenic cytokines in an effort to induce therapeutic angiogenesis for the treatment of ischemic cardiovascular disease. These concerns can be divided into acute effects and long-term effects of angiogenesis therapy.

Potential adverse acute effects may be associated with the means of delivery of the agent. Surgical and percutaneous techniques all carry risk of morbidity and mortality. Clinical trials for therapeutic angiogenesis have been associated with certain peri-procedural adverse events in patients undergoing percutaneous procedures for the delivery of angiogenic cytokines to the myocardium. Surgical delivery of angiogenic agents is generally performed in the context of a standard cardiac surgery with its attendant risk.

The administration of VEGF protein has been shown to result in significant nitric oxide-mediated hypotension (Hariawala et al. 1996; Horowitz et al. 1997), which limits the rate and dose of protein administration. This complication, however, has not been demonstrated with VEGF gene therapy in animals or humans. Experiments in transgenic mice engineered to overexpress VEGF±angiopoietin have demonstrated lethal permeability-enhancing effects of VEGF; however, the magnitude of expression achieved in these models is likely clinically irrelevant (Thurston et al. 1999). VEGF gene transfer was associated with local edema, which manifested as pedal edema in patients treated with VEGF-A gene transfer for critical limb ischemia. It must be noted, however, that edema was present before treatment, and responded well to treatment with diuretics (Baumgartner et al. 2000). A subsequent trial of VEGF-121 gene therapy in patients with claudication also demonstrated evidence for transient local edema (Rajagopalan et al. 2003).

There are three important theoretical long-term concerns of therapeutic angiogenesis. First, there is a significant concern that the induction of angiogenesis may increase the risk of neoplastic disease and malignancy. This concern arises from the fact that angiogenic cytokines were initially discovered as factors that regulate vascularization of tumors (Folkman 1971). Despite this concern, no clinical trial has revealed any evidence that therapy with angiogenic cytokines increases the risk of tumor development. In trials with control groups, the incidence of new malignancies in the control group equals or even exceeds the incidence of new malignancies in treatment groups. In nearly 100 patients treated with angiogenesis factors at a single center, the 7-year incidence of cancer has been limited to 3 patients. Of these patients, 2 developed bladder cancer and 1 patient developed liver and brain metastases from an unknown primary.

The induction or worsening of retinopathy has also been a legitimate concern in clinical trials of therapeutic angiogenesis. Patients with proliferative retinopathy have been shown to have high levels of VEGF in their ocular fluid (Aiello et al. 1994), suggesting that VEGF may stimulate retinal neovascularization. Despite this finding, clinical trials of therapeutic angiogenesis have not demonstrated an untoward effect of angiogenic cytokines on the development of retinopathy. In nearly 100 patients treated with angiogenic cytokines at a single center, no patient developed retinopathy as assessed by an independent retina specialist (Vale et al. 1998) despite 30% of the patients being diabetic. A third concern of therapeutic angiogenesis for vascular disease is that the induction of angiogenesis may stimulate the development and progression of atherosclerotic plaques that may lead to unstable vascular syndromes. These concerns stem from the extrapolation of recent studies that have shown that inhibitors of angiogenesis can inhibit plaque growth and intimal thickening in apolipoprotein E-deficient mice (Moulton et al. 1999). While inhibitors of angiogenesis may lead to regression of atherosclerotic plaque in animal models, the converse has not been proved true in clinical trials of angiogenesis. Angiogenic therapy may, in fact, accelerate reendothelialization and suppress intimal thickening (Van Belle et al. 1996, 1997). Clinical trials of angiogenesis have shown no increase in adverse cardiovascular events in patients treated with angiogenic cytokines compared to control patients.

It is important to remember that patients with end-stage vascular disease are generally elderly and are at risk for malignancy and retinopathy because of their underlying risk factors and comorbidities (age, smoking, diabetes). Therefore, candidates for trials of vascular gene therapy are likely to develop such complications as part of the natural history of their disease states and not necessarily because of therapy with angiogenic agents. Nevertheless, painstaking efforts must continually be taken to ensure that candidates for clinical trials of angiogenesis are thoroughly screened for malignant and premalignant conditions as well as retinopathy, until these risks are completely understood.

5
Summary

As the population continues to age, there will continue to be growth in the burden of ischemic cardiovascular disease. At the present time, no form of therapy has proved beyond doubt to be effective in treating this patient population that has no other options for conventional revascularization therapies. A large body of data has now been accumulated that suggests that therapeutic angiogenesis may offer hope for this patient population.

The data regarding therapeutic angiogenesis presented in this chapter show that the available methods for inducing therapeutic angiogenesis are safe and feasible. No clear safety issues have been identified with the induction of angiogenesis, and delivery strategies are becoming less invasive thereby reducing procedural risks to patients. Despite safety and feasibility of these strategies, the efficacy of therapeutic angiogenesis has not been definitively proved in the phase I and phase II studies that have been performed thus far.

Stem/progenitor cell therapy has proved to be effective to promote neovascularization in various animal models. Moreover, most of the early clinical studies have provided promising outcomes. Although the final proof with large phase III randomized placebo-controlled trials are missing, stem cell therapy in general have the capacity to lead to improved formation of new blood vessels as a treatment for myocardial ischemia.

However, there are data to suggest that cell therapy itself may also have inherent limitations. First, the ideal agent for the induction of therapeutic angiogenesis needs to be identified. VEGF, FGF, and progenitor cells have been most actively studied, but future studies are being planned on other potential angiogenic factors such as hepatocyte growth factor, sonic hedgehog, and others. Second, transplanted cells may have low survival rates, significantly impacting on their beneficial effects (Shake et al. 2002). Strategies such as genetic modification of stem cells to overexpress prosurvival genes such as Akt have been suggested (Mangi et al. 2003). Another limitation is the advanced age and cardiovascular risk factors such as hypertension, diabetes, or hypercholesterolemia that could impair angiogenic effectiveness of cells from the patient in whom autologous cells are to be injected (Hill et al. 2003; Tepper et al. 2002; Vasa et al. 2001). A combination of cell and gene therapies that overexpress angiogenic molecules or survival may enhance therapeutic potential of the cells (Iwaguro et al. 2002; Murasawa et al. 2002; Suzuki et al. 2001).

Important challenges remain in terms of design of clinical trials of therapeutic angiogenesis. It is vital that future trials have control/placebo groups, as phase I and II studies have shown the significant placebo effect in patients undergoing control therapy. Control groups are also vitally important to assess the safety of angiogenic therapy. An understanding of the occurrence of malignancies and retinopathy with angiogenic therapy can only be assessed with a knowledge of the occurrence of these adverse events in a control group. Finally, a consensus needs to be established regarding the best clinical endpoint to use to assess efficacy and the ideal time point to assess these endpoints.

References

Aiello LP, Avery RL, Arrigg PG, Keyt BA, Jampel HD, Shah ST, Pasquale LR, Theme H, Iwamoto MA, Parke JE, Nguyen MD, Aiello LM, Ferrara N, King GL (1994) Vascular endothelial growth factor in ocular fluids of patients with diabetic retinopathy and other retinal disorders. N Engl J Med 331:1480–1487

Asahara T, Murohara T, Sullivan A, Silver M, van der Zee R, Li T, Witzenbichler B, Schatteman G, Isner JM (1997) Isolation of putative progenitor endothelial cells for angiogenesis. Science 275:964–967

Asahara T, Masuda H, Takahashi T, Kalka C, Pastore C, Silver M, Kearne M, Magner M, Isner JM (1999) Bone marrow origin of endothelial progenitor cells responsible for postnatal vasculogenesis in physiological and pathological neovascularization. Circ Res 85:221–228

Baumgartner I, Rauh G, Pieczek A, Wuensch D, Magner M, Kearney M, Schainfeld R, Isner JM (2000) Lower-extremity edema associated with gene transfer of naked DNA vascular endothelial growth factor. Ann Intern Med 132:880–884

Campbell WB, Johnston JA, Kernick VF, Rutter EA (1994) Lower limb amputation: striking the balance. Ann R Coll Surg Engl 76:205–209

Dawson I, Keller BP, Brand R, Pesch-Batenburg J, Hajo van Bockel J (1995) Late outcomes of limb loss after failed infrainguinal bypass. J Vasc Surg 21:613–622

Dimmeler S, Aicher A, Vasa M, Mildner-Rihm C, Adler K, Tiemann M, Rutten H, Fichtlscherer S, Martin H, Zeiher AM (2001) HMG-CoA reductase inhibitors (statins) increase endothelial progenitor cells via the PI 3-kinase/Akt pathway. J Clin Invest 108:391–397

Eneroth M, Persson BM (1992) Amputation for occlusive arterial disease. A multicenter study of 177 amputees. Int Orthop 16:382–387

Flamme I, Risau W (1992) Induction of vasculogenesis and hematopoiesis in vitro. Development 116:435–439

Folkman J (1971) Tumor angiogenesis: therapeutic implications. N Engl J Med 285:1182–1186

Fuchs S, Baffour R, Zhou YF, Shou M, Pierre A, Tio FO, Weissman NJ, Leon MB, Epstein SE, Kornowski R (2001) Transendocardial delivery of autologous bone marrow enhances collateral perfusion and regional function in pigs with chronic experimental myocardial ischemia. J Am Coll Cardiol 37:1726–1732

Gilbert SF (1997) Developmental biology, fourth edn. Sinauer Associates, Sunderland

Hamano K, Nishida M, Hirata K, Mikamo A, Li TS, Harada M, Miura T, Matsuzaki M, Esato K (2001) Local implantation of autologous bone marrow cells for therapeutic angiogenesis in patients with ischemic heart disease: clinical trials and preliminary results. Jpn Circ J 65:845–847

Hariawala MD, Horowitz JR, Esakof D, Sheriff DD, Walter DH, Keyt B, Isner JM, Symes JF (1996) VEGF improves myocardial blood flow but produces EDRF-mediated hypotension in porcine hearts. J Surg Res 63:77–82

Harley CB, Futcher AB, Greider CW (1990) Telomeres shorten during ageing of human fibroblasts. Nature 345:458–460

Harrington L, Zhou W, McPhail T, Oulton R, Yeung DS, Mar V, Bass MB, Robinson MO (1997) Human telomerase contains evolutionarily conserved catalytic and structural subunits. Genes Dev 11:3109–3115

Hastie ND, Dempster M, Dunlop MG, Thompson AM, Green DK, Allshire RC (1990) Telomere reduction in human colorectal carcinoma and with ageing. Nature 346:866–868

Hattori K, Heissig B, Tashiro K, Honjo T, Tateno M, Shieh JH, Hackett NR, Quitoriano MS, Crystal RG, Rafii S, Moore MA (2001) Plasma elevation of stromal cell-derived factor-1 induces mobilization of mature and immature hematopoietic progenitor and stem cells. Blood 97:3354–3360

Hill JM, Zalos G, Halcox JP, Schenke WH, Waclawiw MA, Quyyumi AA, Finkel T (2003) Circulating endothelial progenitor cells, vascular function, and cardiovascular risk. N Engl J Med 348:593–600

Horowitz JR, Rivard A, van der Zee R, Hariawala MD, Sheriff DD, Esakof DD, Chaudhry M, Symes JF, Isner JM (1997) Vascular endothelial growth factor/vascular permeability factor produces nitric oxide-dependent hypotension. Arterioscler Thromb Vasc Biol 17:2793–2800

Iwaguro H, Yamaguchi J, Kalka C, Murasawa S, Masuda H, Hayashi S, Silver M, Li T, Isner JM, Asahara T (2002) Endothelial progenitor cell vascular endothelial growth factor gene transfer for vascular regeneration. Circulation 105:732–738

Kamihata H, Matsubara H, Nishiue T, Fujiyama S, Tsutsumi Y, Ozono R, Masaki H, Mori Y, Iba O, Tateishi E, Kosaki A, Shintani S, Murohara T, Imaizumi T, Iwasaka T (2001) Implantation of bone marrow mononuclear cells into ischemic myocardium enhances collateral perfusion and regional function via side supply of angioblasts, angiogenic ligands, and cytokines. Circulation 104:1046–1052

Kawamoto A, Gwon HC, Iwaguro H, Yamaguchi JI, Uchida S, Masuda H, Silver M, Ma H, Kearney M, Isner JM, Asahara T (2001) Therapeutic potential of ex vivo expanded endothelial progenitor cells for myocardial ischemia. Circulation 103:634–637

Kawamoto A, Tkebuchava T, Yamaguchi J, Nishimura H, Yoon YS, Milliken C, Uchida S, Masuo O, Iwaguro H, Ma H, Hanley A, Silver M, Kearney M, Losordo DW, Isner JM, Asahara T (2003) Intramyocardial transplantation of autologous endothelial progenitor cells for therapeutic neovascularization of myocardial ischemia. Circulation 107:461–468

Kilian A, Bowtell DD, Abud HE, Hime GR, Venter DJ, Keese PK, Duncan EL, Reddel RR, Jefferson RA (1997) Isolation of a candidate human telomerase catalytic subunit gene, which reveals complex splicing patterns in different cell types. Hum Mol Genet 6:2011–2019

Kocher AA, Schuster MD, Szabolcs MJ, Takuma S, Burkhoff D, Wang J, Homma S, Edwards NM, Itescu S (2001) Neovascularization of ischemic myocardium by human bone-marrow-derived angioblasts prevents cardiomyocyte apoptosis, reduces remodeling and improves cardiac function. Nat Med 7:430–436

Lataillade JJ, Clay D, Dupuy C, Rigal S, Jasmin C, Bourin P, Le Bousse-Kerdiles MC (2000) Chemokine SDF-1 enhances circulating CD34(+) cell proliferation in synergy with cytokines: possible role in progenitor survival. Blood 95:756–768

Mangi AA, Noiseux N, Kong D, He H, Rezvani M, Ingwall JS, Dzau VJ (2003) Mesenchymal stem cells modified with Akt prevent remodeling and restore performance of infarcted hearts. Nat Med 9:1195–1201

Meyerson M, Counter CM, Eaton EN, Ellisen LW, Steiner P, Caddle SD, Ziaugra L, Beijersbergen RL, Davidoff MJ, Liu Q, Bacchetti S, Haber DA, Weinberg RA (1997) hEST2, the putative human telomerase catalytic subunit gene, is up-regulated in tumor cells and during immortalization. Cell 90:785–795

Mohle R, Bautz F, Rafii S, Moore MA, Brugger W, Kanz L (1998) The chemokine receptor CXCR-4 is expressed on CD34+ hematopoietic progenitors and leukemic cells and mediates transendothelial migration induced by stromal cell-derived factor-1. Blood 91:4523–4530

Moulton KS, Heller E, Konerding MA, Flynn E, Palinski W, Folkman J (1999) Angiogenesis inhibitors endostatin or TNP-470 reduce intimal neovascularization and plaque growth in apolipoprotein E-deficient mice. Circulation 99:1726–1732

Murasawa S, Llevadot J, Silver M, Isner JM, Losordo DW, Asahara T (2002) Constitutive human telomerase reverse transcriptase expression enhances regenerative properties of endothelial progenitor cells. Circulation 106:1133–1139

Murohara T, Ikeda H, Duan J, Shintani S, Sasaki KI, Eguchi H, Onitsuka I, Matsui K, Imaizumi T (2000) Transplanted cord blood-derived endothelial precursor cells augment postnatal neovascularization. J Clin Invest 105:1527–1536

Nakamura TM, Morin GB, Chapman KB, Weinrich SL, Andrews WH, Lingner J, Harley CB, Cech TR (1997) Telomerase catalytic subunit homologs from fission yeast and human. Science 277:955–959

Rafii S (2000) Circulating endothelial precursors: mystery, reality, and promise. J Clin Invest 105:17–19

Rajagopalan S, Mohler ER 3rd, Lederman RJ, Mendelsohn FO, Saucedo JF, Goldman CK, Blebea J, Macko J, Kessler PD, Rasmussen HS, Annex BH (2003) Regional angiogenesis with vascular endothelial growth factor in peripheral arterial disease: a phase II randomized, double-blind, controlled study of adenoviral delivery of vascular endothelial growth factor 121 in patients with disabling intermittent claudication. Circulation 108:1933–1938

Risau W (1995) Differentiation of endothelium. FASEB J 9:926–933

Risau W, Flamme I (1995) Vasculogenesis. Annu Rev Cell Dev Biol 11:73–91

Salcedo R, Wasserman K, Young HA, Grimm MC, Howard OM, Anver MR, Kleinman HK, Murphy WJ, Oppenheim JJ (1999) Vascular endothelial growth factor and basic fibroblast growth factor induce expression of CXCR4 on human endothelial cells: in vivo neovascularization induced by stromal-derived factor-1alpha. Am J Pathol 154:1125–1135

Shake JG, Gruber PJ, Baumgartner WA, Senechal G, Meyers J, Redmond JM, Pittenger MF, Martin BJ (2002) Mesenchymal stem cell implantation in a swine myocardial infarct model: engraftment and functional effects. Ann Thorac Surg 73:1919–1925; discussion 1926

Shi Q, Rafii S, Wu MH, Wijelath ES, Yu C, Ishida A, Fujita Y, Kothari S, Mohle R, Sauvage LR, Moore MA, Storb RF, Hammond WP (1998) Evidence for circulating bone marrow-derived endothelial cells. Blood 92:362–367

Shintani S, Kusano K, Ii M, Iwakura A, Heyd L, Curry C, Wecker A, Gavin M, Ma H, Kearney M, Silver M, Thorne T, Murohara T, Losordo DW (2006) Synergistic effect of combined intramyocardial CD34+ cells and VEGF2 gene therapy after MI. Nat Clin Pract Cardiovasc Med 3 [Suppl 1]:S123–S128

Skinner JA, Cohen AT (1996) Amputation for premature peripheral atherosclerosis: do young patients do better? Lancet 348:1396

Suzuki K, Murtuza B, Smolenski RT, Sammut IA, Suzuki N, Kaneda Y, Yacoub MH (2001) Cell transplantation for the treatment of acute myocardial infarction using vascular endothelial growth factor-expressing skeletal myoblasts. Circulation 104:I207–I212

Takahashi T, Kalka C, Masuda H, Chen D, Silver M, Kearney M, Magner M, Isner JM, Asahara T (1999) Ischemia- and cytokine-induced mobilization of bone marrow-derived endothelial progenitor cells for neovascularization. Nat Med 5:434–438

Tepper OM, Galiano RD, Capla JM, Kalka C, Gagne PJ, Jacobowitz GR, Levine JP, Gurtner GC (2002) Human endothelial progenitor cells from type II diabetics exhibit impaired proliferation, adhesion, and incorporation into vascular structures. Circulation 106:2781–2786

Thurston G, Suri C, Smith K, McClain J, Sato TN, Yancopoulos GD, McDonald DM (1999) Leakage-resistant blood vessels in mice transgenically overexpressing angiopoietin-1. Science 286:2511–2514

Vale PR, Rauh G, Wuensch DI, Pieczek A, Schainfeld RM (1998) Influence of vascular endothelial growth factor on diabetic retinopathy. Circulation 17:I-353

Van Belle E, Tio F, Couffinhal T, Maillard L, Passeri J, Isner JM (1996) Stent endothelialization: time course, impact of local catheter delivery, feasibility of recombinant protein administration, and response to cytokine expedition. Circulation 94:I-259

Van Belle E, Tio FO, Chen D, Maillard L, Kearney M, Isner JM (1997) Passivation of metallic stents following arterial gene transfer of phVEGF165 inhibits thrombus formation and intimal thickening. J Am Coll Cardiol 29:1371–1379

Vasa M, Fichtlscherer S, Aicher A, Adler K, Urbich C, Martin H, Zeiher AM, Dimmeler S (2001) Number and migratory activity of circulating endothelial progenitor cells inversely correlate with risk factors for coronary artery disease. Circ Res 89:E1–E7

Werner N, Kosiol S, Schiegl T, Ahlers P, Walenta K, Link A, Bohm M, Nickenig G (2005) Circulating endothelial progenitor cells and cardiovascular outcomes. N Engl J Med 353:999–1007

Wright WE, Shay JW (1992) The two-stage mechanism controlling cellular senescence and immortalization. Exp Gerontol 27:383–389

Yamaguchi J, Kusano KF, Masuo O, Kawamoto A, Silver M, Murasawa S, Bosch-Marce M, Masuda H, Losordo DW, Isner JM, Asahara T (2003) Stromal cell-derived factor-1 effects on ex vivo expanded endothelial progenitor cell recruitment for ischemic neovascularization. Circulation 107:1322–1328

Yoon YS, Park JS, Tkebuchava T, Luedeman C, Losordo DW (2004) Unexpected severe calcification after transplantation of bone marrow cells in acute myocardial infarction. Circulation 109:3154–3157

Cell Therapy and Gene Therapy Using Endothelial Progenitor Cells for Vascular Regeneration

T. Asahara

Department of Regenerative Medicine Science, Tokai University School of Medicine, 736 Cambridge Street, Brighton MA, 02135-2997, USA
asa777@aol.com

1	Introduction	182
2	Postnatal Neovascularization	182
3	Profiles of EPCs in Adults	183
3.1	The Isolation of EPCs in Adults	183
3.2	Diverse Identifications of Human EPCs and Their Precursors	184
4	Cell Therapy Using EPCs	185
4.1	The Potent of EPC Transplantation	185
4.2	Future Strategy of EPC Therapy	186
5	Gene-Modified EPC Therapy	187
5.1	Gene Modification of Stem/Progenitor Cells	187
5.2	Gene Modification of EPCs	189
6	Other Potential of EPCs	191
7	Conclusion	191
	References	191

Abstract The isolation of endothelial progenitor cells (EPCs) derived from adult bone marrow (BM) was an epoch-making event for the recognition of "neovessel formation" occurring as physiological and pathological responses in adults. The finding that EPCs home to sites of neovascularization and differentiate into endothelial cells (ECs) in situ is consistent with "vasculogenesis," a critical paradigm well described for embryonic neovascularization, but proposed recently in adults, in which a reservoir of stem or progenitor cells contributes to vascular organogenesis. EPCs have also been considered as therapeutic agents to supply the potent origin of neovascularization under pathological conditions. Considering the regenerative implications, gene modification of stem cells has advantages over conventional gene therapy. Ex vivo gene transfection of stem cells may avoid administration of vectors and vehicles into the recipient organism. Stem cells isolated from adults may exhibit age-related, genetic, or acquired disease-related impairment of their regenerative ability. Transcriptional or enzymatic gene modification may constitute an effective means to maintain, enhance, or inhibit EPCs' capacity to proliferate or differentiate. This chapter provides an update of EPC biology as well as EPCs' potential use for therapeutic regeneration.

Keywords Endothelial progenitor cells · Gene therapy · Therapeutic vasculogenesis · Vascular regeneration

1
Introduction

Tissue regeneration by somatic stem/progenitor cells has been recognized as a maintenance or recovery system of many organs in adults. The isolation and investigation of these somatic stem/progenitor cells have led to descriptions of how these cells contribute to postnatal organogenesis. On the basis of the regenerative potency, these stem/progenitor cells are expected as a key in strategies aimed toward therapeutic applications for damaged organs.

Recently, endothelial progenitor cells (EPCs) have been isolated from adult peripheral blood (PB). EPCs are considered to share common stem/progenitor cells with hematopoietic stem cells and have been shown to derive from bone marrow (BM) and to incorporate into foci of physiological or pathological neovascularization. The finding that EPCs home to sites of neovascularization and differentiate into endothelial cells (ECs) in situ is consistent with "vasculogenesis," a critical paradigm well described for embryonic neovascularization—but recently proposed in adults—in which a reservoir of stem/progenitor cells contributes to postnatal vascular organogenesis. The discovery of EPCs has therefore drastically changed our understanding of adult blood vessel formation. The following issue provides the update of EPC biology as well as highlights their potential utility for therapeutic vascular regeneration.

2
Postnatal Neovascularization

Through the discovery of EPCs in PB (Asahara et al. 1997; Shi et al. 1998), our understanding of postnatal neovascularization has been expanded from angiogenesis to angio/vasculogenesis. As previously described (Folkman and Shing 1992), postnatal neovascularization was originally recognized to be constituted by the mechanism of "angiogenesis," which is neovessel formation, operated by in situ proliferation and migration of preexisting endothelial cells. However, the isolation of EPCs resulted in the addition of the new mechanism, "vasculogenesis," which is de novo vessel formation by in situ incorporation, differentiation, migration, and/or proliferation of BM-derived EPCs (Asahara et al. 1999a). Furthermore, tissue-specific stem/progenitor cells with the potency to differentiate into myocytes or ECs were isolated in skeletal muscle tissue of murine hindlimb, although the origin remains to be clarified (Tamaki et al. 2002). This finding suggests that the origin of EPCs may not be limited to BM; tissue-specific stem/progenitor cells possibly provide "in situ EPCs" as a source of EPCs other than BM.

In the events of minor-scale neovessel formation, i.e., slight wounds or burns, "in situ preexisting ECs" causing postnatal angiogenesis may replicate and replace the existing cell population sufficiently, as ECs exhibit an ability at preserves their proliferative activity. Neovascularization through differentiated ECs, however, is limited in terms of cellular lifespan (Hayflick limit) and their inability to incorporate into remote target sites. In the case of large-scale tissue repair—such as is seen with patients who experience acute vascular insult secondary to burns, coronary artery bypass grafting (CABG), or acute myocardial infarction (Gill et al. 2001; Shintani et al. 2001), or in physiological cyclic organogenesis of endometrium (Asahara et al. 1999a)—BM-derived or in situ EPC kinetics are activated under the influence of appropriate cytokines, hormones, and growth factors through the autocrine, paracrine, and endocrine systems. Thus, the contemporary view of tissue regeneration is that neighboring differentiated ECs are relied upon for vascular regeneration during a minor insult, whereas tissue-specific or BM-derived stem/progenitor cells bearing EPCs/ECs are important when an emergent vascular regenerative process is required.

3
Profiles of EPCs in Adults

3.1
The Isolation of EPCs in Adults

In the embryo, evidence suggests that hematopoietic stem cells (HSCs) and EPCs (Pardanaud et al. 1987; Risau et al. 1988) are derived from a common precursor (hemangioblast) (Flamme and Risau 1992; Weiss and Orkin 1996). During embryonic development, multiple blood islands initially fuse to form a yolk sac capillary network (Risau and Flamme 1995), which provides the foundation for an arteriovenous vascular system that eventually forms following the onset of blood circulation (Risau et al. 1988). The integral relationship between the cells which circulate in the vascular system (the blood cells) and those principally responsible for the vessels themselves (ECs) is suggested by their spatial orientation within the blood islands; those cells destined to generate hematopoietic cells are situated in the center of the blood island (HSCs) versus EPCs or angioblasts, which are located at the periphery of the blood islands. In addition to this arrangement, HSCs and EPCs share certain common antigens, including CD34, KDR, Tie-2, CD117, and Sca-1 (Choi et al. 1998).

The existence of HSCs in the PB and BM, and the demonstration of sustained hematopoietic reconstitution with HSC transplantation, led to an idea that a closely related cell type, namely EPCs, may also exist in adult tissues. Recently, EPCs were successfully isolated from circulating mononuclear cells (MNCs) using KDR, CD34, and CD133 antigens shared by both embryonic EPCs and HSCs (Asahara et al. 1997; Peichev et al. 2000; Yin et al. 1997). In vitro, these cells differentiate into endothelial lineage cells, and in animal models of ischemia, heterologous, homologous, and autologous EPCs have been shown to incorporate into the foci of neovasculature, contributing to neovascularization. Recently, similar studies with EPCs isolated from human cord blood have demonstrated their analogous differentiation into ECs in vitro and in vivo (Crisa et al. 1999; Kalka et al. 2000; Murohara et al. 2000; Nieda et al. 1997).

These findings have raised important questions regarding fundamental concepts of blood vessel growth and development in adult subjects. Does the differentiation of EPCs in situ (vasculogenesis) play an important role in adult neovascularization, and would impairments in this process lead to clinical diseases? There is now a strong body of evidence suggesting that vasculogenesis does in fact make a significant contribution to postnatal neovascularization. Recent studies with animal BM transplantation (BMT) models in which BM (donor)-derived EPCs could be distinguished have shown that the contribution of EPCs to neovessel formation may range from 5% to 25% in response to granulation tissue formation (Crosby et al. 2000) or growth factor-induced neovascularization (Murayama et al. 2002). Also, in tumor neovascularization, the range is approximately 35%–45% higher than the former events (Reyes et al. 2002). The degree of EPC contribution to postnatal neovascularization is predicted to depend on each neovascularizing event or disease.

3.2
Diverse Identifications of Human EPCs and Their Precursors

Since the initial EPC report (Asahara et al. 1997; Shi et al. 1998), a number of groups have set out to better define this cell population. Because EPCs and HSCs share many surface markers, and no simple definition of EPCs exists, various methods of EPC isolation have been reported (Asahara et al. 1997; Boyer et al. 2000; Fernandez Pujol et al. 2000; Gehling et al. 2000; Gunsilius et al. 2000; Harraz et al. 2001; Kalka et al. 2000; Kang et al. 2001; Lin et al. 2000; Murohara et al. 2000; Nieda et al. 1997; Peichev et al. 2000; Quirici et al. 2001; Schatteman et al. 2000; Shi et al. 1998). The term EPC may therefore encompass a group of cells that exists in a variety of stages ranging from hemangioblasts to fully differentiated ECs. Although the true differentiation lineage of EPCs and their putative precursors remains to be determined, there is overwhelming evidence that a population of human EPCs exists in vivo.

Lin et al. cultivated peripheral MNCs from patients receiving gender-mismatched BMT and studied their growth in vitro. In this study they identified a population of BM (donor)-derived ECs with high proliferative potential (late outgrowth); these BM cells likely represent EPCs (Lin et al. 2000). Gunsilius et al. investigated a chronic myelogenous leukemia model and disclosed that BM-derived EPCs contribute to postnatal neovascularization in human (Gunsilius et al. 2000). Interestingly, in the report, BM-derived EPCs could be detected even in the wall of quiescent vessels without neovascularization events. The finding suggests that BM-derived EPCs may be related even to the turnover of ECs consisting of quiescent vessels.

Reyes et al. have isolated multipotent adult progenitor cells (MAPCs) from BM MNCs, differentiated them into EPCs, and proposed MAPCs as an origin of EPCs (Reyes et al. 2002). These studies therefore provide evidence to support the presence of BM-derived EPCs that take part in neovascularization. Also, as described above, the existence of namely "in situ EPCs" as derived from tissue-specific stem/progenitor cells in murine skeletal muscle remains to be investigated even in the other tissues (Tamaki et al. 2002).

4
Cell Therapy Using EPCs

4.1
The Potent of EPC Transplantation

The regenerative potential of stem/progenitor cells is currently under intense investigation. In vitro, stem/progenitor cells possess the capability of self-renewal and differentiation into organ-specific cell types. When placed in vivo, these cells are then provided with the proper milieu that allows them to reconstitute organ systems. The novel strategy of EPC transplantation (cell therapy) may therefore supplement the classic paradigm of angiogenesis developed by Folkman and colleagues. Our studies indicated that cell therapy with culture-expanded EPCs can successfully promote neovascularization of ischemic tissues, even when administered as "sole therapy," i.e., in the absence of angiogenic growth factors. Such a "supply-side" version of therapeutic neovascularization in which the substrate (EPCs/ECs) rather than ligand (growth factor) comprises the therapeutic agent, was first demonstrated by intravenously transplanting human EPCs to immunodeficient mice with hindlimb ischemia (Kalka et al. 2000). These findings provided novel evidence that exogenously administered EPCs rescue impaired neovascularization in an animal model of critical limb ischemia. Not only did the heterologous cell transplantation improve neovascularization and blood flow recovery, but also led to important biological outcomes—notably, the reduction of limb necrosis and auto-amputation by 50% in comparison with controls. A similar strategy applied to a model of myocardial ischemia in the nude rat demonstrated that transplanted human EPCs localize to areas of myocardial neovascularization, differentiate into mature ECs, and enhance neovascularization. These findings were associated with preserved left ventricular (LV) function and diminished myocardial fibrosis (Kawamoto et al. 2001). Murohara et al. reported similar findings in which human cord blood-derived EPCs also augmented neovascularization in hindlimb ischemic model of nude rats, followed by in situ transplantation (Murohara et al. 2000).

Other researchers have more recently explored the therapeutic potential of freshly isolated human $CD34^+$ cells (EPC-enriched fraction). Schatteman et al.

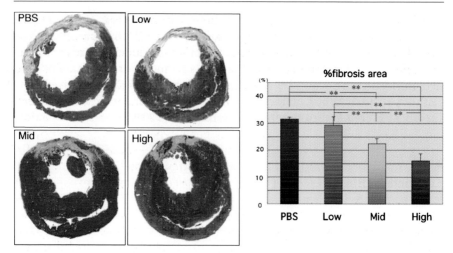

Fig. 1 Roadmap of EPC research. EPC biology research starts from the identification of endothelial progenitor cells in peripheral blood published by Asahara et al., inspired by evolving knowledge in developmental biology, hematology, and vascular biology. The development of this field is contributing to medicine through plentiful translational research throughout the world

conducted local injection of freshly isolated human $CD34^+$ cells into diabetic nude mice with hindlimb ischemia, and showed an increase in the restoration of limb flow (Schatteman et al. 2000). Similarly Kocher et al. attempted intravenous infusion of freshly isolated human $CD34^+$ cells into nude rats with myocardial ischemia, and found preservation of LV function associated with inhibition of cardiomyocyte apoptosis (Kocher et al. 2001). Thus, two approaches of EPC preparation (i.e., both cultured and freshly isolated human EPCs) may provide therapeutic benefit in vascular diseases, but as described in Sect. 5.2, will likely require further optimization techniques to acquire the ideal quality and quantity of EPCs for EPC therapy.

Very recently, Iwasaki et al. have demonstrated a $CD34^+$ cell dose-dependent contribution to recovery of LV function and neovascularization in ischemia tissues in models of myocardial ischemia (Iwasaki et al. 2006; Fig. 1). Furthermore, $CD34^+$ cells in higher dose groups differentiated into not only vasculogenic (endothelial and mural) lineage but also myocardial lineage cells. Clinical trials using mobilized $CD34^+$ may be effective in terms of vasculogenesis and myocardiogenesis.

4.2
Future Strategy of EPC Therapy

Ex vivo expansion of EPCs cultured from PB-MNCs of healthy human volunteers typically yields 5.0×10^6 cells per 100 ml of blood on day 7. Our animal studies (Kalka et al. 2000) suggest that heterologous transplantation requires systemic injection of 0.5×10^4 to approximately 2.0×10^4 human EPCs/g body weight of the recipient animal to achieve satisfactory reperfusion of an ischemic hindlimb. Rough extrapolation of these data to humans suggests that a blood volume of as much as 12 l may be necessary to obtain adequate numbers of EPCs to treat critical limb ischemia in patients. Therefore, the fundamental scarcity of EPCs in the circulation, combined with their possible functional impairment associated with a variety of phenotypes in clinical patients, such as aging, diabetes, hypercholesterolemia, and homocyst(e)inemia (vide infra), constitute major limitations of primary EPC transplantation.

Considering autologous EPC therapy, certain technical improvements may help to overcome the primary scarcity of a viable and functional EPC population. These should include: (1) local delivery of EPCs; (2) adjunctive strategies (e.g., growth factor supplements) to promote BM-derived EPC mobilization (Takahashi et al. 1999; Asahara et al. 1999b); (3) enrichment procedures, i.e., leukapheresis or BM aspiration, or (4) enhancement of EPC function by gene transduction (gene modified EPC therapy, vide infra); and (5) culture-expansion of EPCs from self-renewable primitive stem cells in BM or other tissues. Alternatively, unless the quality and quantity of autologous EPCs to satisfy the effectiveness of EPC therapy may be acquired by the technical improvements above, allogenic EPCs derived from umbilical cord blood or culture-expanded from human embryonic stem cells (Murohara et al. 2000; Levenberg et al. 2002) may be available as the sources supplying EPCs.

5
Gene-Modified EPC Therapy

5.1
Gene Modification of Stem/Progenitor Cells

Considering stem cells' regenerative ability, the gene modification of stem/progenitor cells exerts several advantages over conventional gene therapy. Ex vivo gene transduction of stem/progenitor cells may avoid the direct administration of vectors and vehicles into the recipient organism. Here are possible target mechanisms of gene-modified stem/progenitor cells for medicine demonstrated in Fig. 2.

1. Phenotype modification of stem/progenitor cell. Stem/progenitor cell target

    Stem/progenitor cells isolated from adults may exhibit a variety of impairments in regenerative ability, these being related to aging and genetic or acquired diseases. Certain properties of stem/progenitor cells can be functionally recovered by gene modifications. Transcriptional or enzymatic

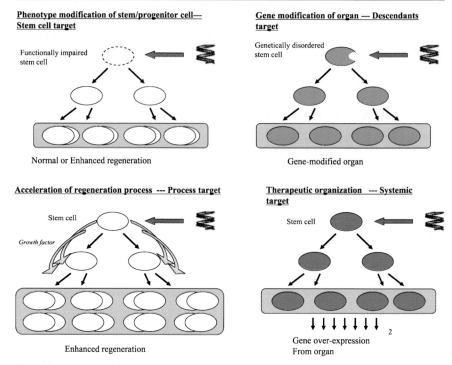

Fig. 2 The strategies of stem/progenitor cell gene modification

gene modification may constitute an effective means to maintain, enhance, or inhibit stem/progenitor cell bioactivities to proliferate or differentiate.

2. Phenotype modification of organ. Progeny target

    Genes introduced into stem cells are inherited by their progeny through differentiation cascades. Specifically, expression of the inserted gene may persist through the life of the organ system that is reconstructed by the gene-modified stem cells. Genetically disordered organs, especially in the case of hematopoietic pathology, could be replaced by this strategy following bone marrow transplantation (Cowan et al. 1999; Dunbar et al. 1998; Karlsson 1991; Reisner and Segall 1995; Wong-Staal et al. 1998).

3. Acceleration of regeneration process. Process target

    Regeneration of injured tissues is in part or primarily achieved via stem cell expansion and differentiation. Examples include endothelial progenitor cells for neovascularization, neural stem cells for neurogenesis, and HSCs and mesenchymal stem cells (MSCs) for bone marrow reconstruction. Natural reparatory processes are often too impaired from unexpectedly severe injury, basic diseases, or aging to accomplish regeneration. Gene modification of stem cells to supplement mitogens (Iwaguro et al. 2002) or to deliver inhibitory factors for negative control might accelerate retarded regenerative processes following stem cell proliferation, incorporation, and gene expression in situ.

4. Therapeutic organization. Systemic target

    To supply target molecules for therapeutic use, transplantation of genetically modified stem cells might generate expressional tissues or pseudo-organs. The established tissues or organs will continuously express a certain number of molecules locally or generally for the short- or long-term by means of vector selection and conditioning. This strategy could represent an alternative approach for therapeutic delivery in lieu of standard drug distribution and uptake.

5.2
Gene Modification of EPCs

EPC transplantation constitutes a novel therapeutic strategy that could provide a robust source of viable ECs to supplement the contribution of ECs resident in the adult vasculature that migrate, proliferate, and remodel in response to angiogenic cues, according to the classic paradigm of angiogenesis developed by Folkman and Shing (1992). Just as classical angiogenesis may be impaired in certain pathologic phenotypes, however, aging, diabetes, hypercholesterolemia, and hyperhomocysteinemia may likewise impair EPC function, including mobilization from the bone marrow and incorporation into neovascular foci. Gene transfer of EPCs during ex vivo expansion constitutes a potential means of addressing such putative liabilities in EPC function. Moreover, phenotypic modulation of EPCs ex vivo may also reduce the number of EPCs required for optimal transplantation post-ex vivo expansion, and thus serve to address a practical limitation of EPC transplantation, namely the volume of blood required to extract an optimal number of EPCs for autologous transplantation.

Iwaguro et al. have determined the impact of vascular endothelial growth factor (VEGF) gene transfer on certain properties of EPCs in vitro and the consequences of VEGF EPC transfer on neovascularization in vivo (Iwaguro et al. 2002; Fig. 3). In vitro, VEGF-gene transfer can be augmented by adenovirus vector for EPC-proliferative activity and enhanced incorporation of EPCs into endothelial cell monolayers. In vivo, transplantation of VEGF gene-transduced EPCs improved neovascularization and reduced limb necrosis by 63.7%. VEGF EPC gene transfer permitted a dose reduction of transplanted EPCs that was 30 times less than that required in previous experiments to achieve similar rates of limb salvage. These findings present one option to address the limited number of EPCs that can be isolated from peripheral blood prior to ex vivo expansion and subsequent autologous readministration for augmentation of neovascularization.

Enhanced Neovascularization by EPCs modified by VEGF gene

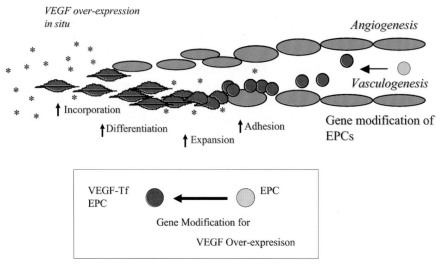

Fig. 3 Enhanced neovascularization by EPCs modified by the VEGF gene. The delivery of VEGF gene-transduced EPCs presents stimulated proliferation, migration, differentiation, and incorporation capability to the ischemic area for vasculogenesis

A strategy that may alleviate potential EPC dysfunction in ischemic disorders is considered reasonable, given the findings that EPC function and mobilization may be impaired in certain disease states.

The regulatory molecule for cell lifespan, telomerase was modified by human telomerase reverse transcriptase (hTERT) gene transfer to investigate its effect on the regenerative properties of endothelial progenitor cells (EPCs) in neovascularization (Murasawa et al. 2002). Along with high telomerase activity in hTERT-transduced EPCs, mitogenic capacity exceeded that of the control group. VEGF-induced cell migration in EPCs was also markedly enhanced by hTERT overexpression. hTERT overexpression has rescued EPCs from starvation-induced cell apoptosis, an outcome that was further enhanced in response to VEGF. The colony appearance of totally differentiated EC was detected prior to day 30 only by hTERT overexpression, whereas no EC colonies could be detected in the control group. In vivo transplantation of heterologous EPCs demonstrated that hTERT-expressing EPCs dramatically improved postnatal neovascularization in terms of limb salvage by fourfold in comparison with the control group's limb perfusion as measured by laser Doppler and capillary density. These findings provide the novel evidence that telomerase activity contributes to EPC angiogenic properties; mitogenic activity, migratory activity, and cell survival. This enhanced regenerative activity of EPC by hTERT transfer will provide a novel therapeutic strategy for postnatal neovascularization in severe ischemic disease patients.

6
Other Potential of EPCs

EPCs have recently been applied to the field of tissue engineering as a means of improving biocompatibility of vascular grafts. Artificial grafts first seeded with autologous $CD34^+$ cells from canine BM and then implanted into the aortae were found to have increased surface endothelialization and vascularization compared with controls (Bhattacharya et al. 2000). Similarly, when cultured autologous ovine EPCs were seeded onto carotid interposition grafts, the EPC-seeded grafts achieved physiological motility and remained patent for 130 days vs 15 days in nonseeded grafts (Kaushal et al. 2001). Alternatively, as previously reported, the cell sheets of cultured cardiomyocytes may be effective for the improvement of cardiac function in the damaged hearts, i.e., ischemic heart disease or cardiomyopathy (Shimizu et al. 2002a, b). The cell sheets consisting of cardiomyocytes with EPCs expectedly inducing neovessels may be attractive, as blood supply is essential to maintain the homeostasis of implanted cardiomyocytes in such cell sheets.

7
Conclusion

As the concept of BM-derived EPCs in adults and postnatal vasculogenesis are further established, clinical applications of EPCs in regenerative medicine are likely to follow. To acquire the more optimized quality and quantity of EPCs, several issues remain to be addressed, such as the development of more efficient methods of EPC purification and expansion, along with the methods of administration and senescence in EPCs. Alternatively, in case of the impossible utility of autologous BM-derived EPCs in patients with impaired BM function, appreciable EPCs isolated from umbilical cord blood or differentiated from tissue-specific stem/progenitor or embryonic stem cells need to be optimized for EPC therapy. However, the unlimited potential of EPCs along with the emerging concepts of autologous cell therapy with gene modification suggests that these treatments may soon reach clinical fruition.

References

Asahara T, Murohara T, Sullivan A, Silver M, van der Zee R, Li T, Witzenbichler B, Schatteman G, Isner JM (1997) Isolation of putative progenitor endothelial cells for angiogenesis. Science 275:964–967

Asahara T, Masuda H, Takahashi T, Kalka C, Pastore C, Silver M, Kearney M, Magner M, Isner JM (1999a) Bone marrow origin of endothelial progenitor cells responsible for postnatal vasculogenesis in physiological and pathological neovascularization. Circ Res 85:221–228

Asahara T, Takahashi T, Masuda H, Kalka C, Chen D, Iwaguro H, Inai Y, Silver M, Isner JM (1999b) VEGF contributes to postnatal neovascularization by mobilizing bone marrow-derived endothelial progenitor cells. EMBO J 18:3964–3972

Bhattacharya V, McSweeney P, Shi Q, Bruno B, Ishida A, Nash R, Storb R, Sauvage L, Hammond W, Wu M (2000) Enhanced endothelialization and microvessel formation in polyester graft seeded with $CD34^+$ bone marrow cells. Blood 95:581–585

Boyer M, Townsend L, Vogel L, Falk J, Reitz-Vick D, Trevor K, Villalba M, Bendick P, Glover J (2000) Isolation of endothelial cells and their progenitor cells from human peripheral blood. J Vasc Surg 31:181–189

Choi K, Kennedy M, Kazarov A, Papadimitriou J, Keller G (1998) A common precursor for hematopoietic and endothelial cells. Development 125:725–732

Cowan KH, Moscow JA, Huang H, Zujewski JA, O'Shaughnessy J, Sorrentino B, Hines K, Carter D, Schneider E, Cusack G, Noone M, Dunbar C, Steinberg S, Wilson W, Goldspiel B, Read EJ, Leitman SF, McDonagh K, Chow C, Abati A, Chiang Y, Chang YN, Gottesman MM, Pastan I (1999) Paclitaxel chemotherapy after autologous stem-cell transplantation and engraftment of hematopoietic cells transduced with a retrovirus containing the multidrug resistance complementary DNA (MDR1) in metastatic breast cancer patients. Clin Cancer Res 5:1619–1628

Crisa L, Cirulli V, Smith K, Ellisman M, Torbett B, Salomon D (1999) Human cord blood progenitors sustain thymic T-cell development and a novel form of angiogenesis. Blood 94:3928–3940

Crosby JR, Kaminski WE, Schatteman G, Martin PJ, Raines EW, Seifert RA, Bowen-Pope DF (2000) Endothelial cells of hematopoietic origin make a significant contribution to adult blood vessel formation. Circ Res 87:728–730

Dunbar CE, Kohn DB, Schiffmann R, Barton NW, Nolta JA, Esplin JA, Pensiero M, Long Z, Lockey C, Emmons RV, Csik S, Leitman S, Krebs CB, Carter C, Brady RO, Karlsson S (1998) Retroviral transfer of the glucocerebrosidase gene into $CD34^+$ cells from patients with Gaucher disease: in vivo detection of transduced cells without myeloablation. Hum Gene Ther 9:2629–2640

Fernandez Pujol B, Lucibello FC, Gehling UM, Lindemann K, Weidner N, Zuzarte ML, Adamkiewicz J, Elsasser HP, Muller R, Havemann K (2000) Endothelial-like cells derived from human CD14 positive monocytes. Differentiation 65:287–300

Flamme I, Risau W (1992) Induction of vasculogenesis and hematopoiesis in vitro. Development 116:435–439

Folkman J, Shing Y (1992) Angiogenesis. J Biol Chem 267:10931–10934

Gehling UM, Ergun S, Schumacher U, Wagener C, Pantel K, Otte M, Schuch G, Schafhausen P, Mende T, Kilic N, Kluge K, Schafer B, Hossfeld DK, Fiedler W (2000) In vitro differentiation of endothelial cells from AC133-positive progenitor cells. Blood 95:3106–3112

Gill M, Dias S, Hattori K, Rivera ML, Hicklin D, Witte L, Girardi L, Yurt R, Himel H, Rafii S (2001) Vascular trauma induces rapid but transient mobilization of $VEGFR2^+AC133^+$ endothelial precursor cells. Circ Res 88:167–174

Gunsilius E, Duba HC, Petzer AL, Kahler CM, Grunewald K, Stockha G, Gabl C, Dirnhofer S, Clausen J, Gastl G (2000) Evidence from a leukaemia model for maintenance of vascular endothelium by bone-marrow-derived endothelial cells. Lancet 355:1688–1691

Harraz M, Jiao C, Hanlon HD, Hartley RS, Schatteman GC (2001) $CD34^-$ blood-derived human endothelial cell progenitors. Stem Cells 19:304–312

Iwaguro H, Yamaguchi J, Kalka C, Murasawa S, Masuda H, Hayashi S, Silver M, Li T, Isner JM, Asahara T (2002) Endothelial progenitor cell vascular endothelial growth factor gene transfer for vascular regeneration. Circulation 105:732–738

Iwasaki H, Kawamoto A, Ishikawa M, Oyamada A, Nakamori S, Nishimura H, Sadamoto K, Horii M, Matsumoto T, Murasawa S, Shibata T, Suehiro S, Asahara T (2006) Dose-dependent contribution of CD34-positive cell transplantation to concurrent vasculogenesis and cardiomyogenesis for functional regenerative recovery after myocardial infarction. Circulation 113:1311–1325

Kalka C, Masuda H, Takahashi T, Kalka-Moll WM, Silver M, Kearney M, Li T, Isner JM, Asahara T (2000) Transplantation of ex vivo expanded endothelial progenitor cells for therapeutic neovascularization. Proc Natl Acad Sci USA 97:3422–3427

Kang HJ, Kim SC, Kim YJ, Kim CW, Kim JG, Ahn HS, Park SI, Jung MH, Choi BC, Kimm K (2001) Short-term phytohaemagglutinin-activated mononuclear cells induce endothelial progenitor cells from cord blood $CD34^+$ cells. Br J Haematol 113:962–969

Karlsson S (1991) Treatment of genetic defects in hematopoietic cell function by gene transfer. Blood 78:2481–2492

Kaushal S, Amiel GE, Guleserian KJ, Shapira OM, Perry T, Sutherland FW, Rabkin E, Moran AM, Schoen FJ, Atala A, Soker S, Bischoff J, Mayer JE Jr (2001) Functional small-diameter neovessels created using endothelial progenitor cells expanded ex vivo. Nat Med 7:1035–1040

Kawamoto A, Gwon HC, Iwaguro H, Yamaguchi JI, Uchida S, Masuda H, Silver M, Ma H, Kearney M, Isner JM, Asahara T (2001) Therapeutic potential of ex vivo expanded endothelial progenitor cells for myocardial ischemia. Circulation 103:634–637

Kocher AA, Schuster MD, Szabolcs MJ, Takuma S, Burkoff D, Wang J, Homma S, Edwards NM, Itescu S (2001) Neovascularization of ischemic myocardium by human bone marrow-derived angioblasts prevents cardiomyocyte apoptosis, reduces remodeling and improves cardiac function. Nat Med 7:430–436

Levenberg S, Golub JS, Amit M, Itskovitz-Eldor J, Langer R (2002) Endothelial cells derived from human embryonic stem cells. Proc Natl Acad Sci U S A 99:4391–4396

Lin Y, Weisdorf D, Solovey A, Hebbel R (2000) Origins of circulating endothelial cells and endothelial outgrowth from blood. J Clin Invest 105:71–77

Murasawa S, Llevadot J, Silver M, Isner JM, Losordo DW, Asahara T (2002) Constitutive human telomerase reverse transcriptase expression enhances regenerative properties of endothelial progenitor cells. Circulation 106:1133–1139

Murayama T, Tepper OM, Silver M, Ma H, Losordo D, Isner JM, Asahara T, Kalka C (2002) Determination of bone marrow-derived endothelial progenitor cell significance in angiogenic growth factor-induced neovascularization in vivo. Exp Hematol 30:967

Murohara T, Ikeda H, Duan J, Shintani S, Sasaki K, Eguchi H, Onitsuka I, Matsui K, Imaizumi T (2000) Transplanted cord blood-derived endothelial precursor cells augment postnatal neovascularization. J Clin Invest 105:1527–1536

Nieda M, Nicol A, Denning-Kendall P, Sweetenham J, Bradley B, Hows J (1997) Endothelial cell precursors are normal components of human umbilical cord blood. Br J Haematol 98:775–777

Pardanaud L, Altmann C, Kitos P, Dieterlen-Lievre F, Buck CA (1987) Vasculogenesis in the early quail blastodisc as studied with a monoclonal antibody recognizing endothelial cells. Development 100:339–349

Peichev M, Naiyer AJ, Pereira D, Ahu Z, Lane WJ, Williams M, Oz MC, Hicklin DJ, Witte L, Moore MAS (2000) Expression of VEGFR-2 and AC133 by circulating human $CD34^+$ cells identifies a population of functional endothelial precursors. Blood 95:952–958

Quirici N, Soligo D, Caneva L, Servida F, Bossolasco P, Deliliers GL (2001) Differentiation and expansion of endothelial cells from human bone marrow $CD133^+$ cells. Br J Haematol 115:186–194

Reisner Y, Segall H (1995) Hematopoietic stem cell transplantation for cancer therapy. Curr Opin Immunol 7:687–693

Reyes M, Dudek A, Jahagirdar B, Koodie L, Marker PH, Verfaillie CM (2002) Origin of endothelial progenitors in human postnatal bone marrow. J Clin Invest 109:337–346

Risau W, Flamme I (1995) Vasculogenesis. Annu Rev Cell Dev Biol 11:73–91

Risau W, Sariola H, Zerwes HG, Sasse J, Ekblom P, Kemler R, Doetschman T (1988) Vasculogenesis and angiogenesis in embryonic stem cell-derived embryoid bodies. Development 102:471–478

Schatteman GC, Hanlon HD, Jiao C, Dodds SG, Christy BA (2000) Blood-derived angioblasts accelerate blood-flow restoration in diabetic mice. J Clin Invest 106:571–578

Shi Q, Rafii S, Wu MH, Wijelath ES, Yu C, Ishida A, Fujita Y, Kothari S, Mohle R, Sauvage LR, Moore MA, Storb RF, Hammond WP (1998) Evidence for circulating bone marrow-derived endothelial cells. Blood 92:362–367

Shimizu T, Yamato M, Akutsu T, Shibata T, Isoi Y, Kikuchi A, Umezu M, Okano T (2002a) Electrically communicating three-dimensional cardiac tissue mimic fabricated by layered cultured cardiomyocyte sheets. J Biomed Mater Res 60:110–117

Shimizu T, Yamato M, Isoi Y, Akutsu T, Setomaru T, Abe K, Kikuchi A, Umezu M, Okano T (2002b) Fabrication of pulsatile cardiac tissue grafts using a novel 3-dimensional cell sheet manipulation technique and temperature-responsive cell culture surfaces. Circ Res 90:e40

Shintani S, Murohara T, Ikeda H, Ueno T, Honma T, Katoh A, Sasaki K, Shimada T, Oike Y, Imaizumi T (2001) Mobilization of endothelial progenitor cells in patients with acute myocardial infarction. Circulation 103:2776–2779

Takahashi T, Kalka C, Masuda H, Chen D, Silver M, Kearney M, Magner M, Isner J, Asahara T (1999) Ischemia- and cytokine-induced mobilization of bone marrow-derived endothelial progenitor cells for neovascularization. Nat Med 4:434–438

Tamaki T, Akatsuka A, Ando K, Nakamura Y, Matsuzawa H, Hotta T, Roy RR, Edgerton VR (2002) Identification of myogenic-endothelial progenitor cells in the interstitial spaces of skeletal muscle. J Cell Biol 157:571–577

Weiss M, Orkin SH (1996) In vitro differentiation of murine embryonic stem cells: new approaches to old problems. J Clin Invest 97:591–595

Wong-Staal F, Poeschla EM, Looney DJ (1998) A controlled phase 1 clinical trial to evaluate the safety and effects in HIV-1 infected humans of autologous lymphocytes transduced with a ribozyme that cleaves HIV-1 RNA. Hum Gene Ther 9:2407–2425

Yin A, Miraglia S, Zanjani E, Almeida-Porada G, Ogawa M, Leary A, Olweus J, Kearney J, Buck D (1997) AC133, a novel marker for human hematopoietic stem and progenitor cells. Blood 90:5002–5012

Mesenchymal Stem Cells for Cardiac Regenerative Therapy

K. H. Schuleri · A. J. Boyle · J. M. Hare (✉)

Miller School of Medicine, University of Miami and Johns Hopkins Medical Institutions, Stem Cell Institute and Cardiology Division, 1120 NW 14th Street, Miami FL, 33136, USA
jhare@med.miami.edu

1	Introduction		196
2	Mesenchymal Stem Cells		197
	2.1	Definition of MSCs	198
	2.2	Secreted Factors from MSCs	199
	2.3	Immunology of MSCs	199
3	MSCs for Gene Therapy, Transplantation and Tissue Engineering		200
	3.1	Genetic Modification and MSCs	201
	3.2	MSC Transplantation and Routes of Application	201
4	Mechanism of Repair		203
	4.1	Cell Tracking and Evaluation of MSC Therapy	206
		4.1.1 Invasive Imaging of MSC Therapy	206
		4.1.2 Non-invasive Imaging of MSC Therapy	207
5	Clinical Trials and MSCs		209
	5.1	Clinical Studies Using MSCs for Cardiac Applications	210
	5.2	Translational Research and Future Directions	212
6	Conclusion		212
	References		212

Abstract Until recently, the concept of treating the injured or failing heart by generating new functional myocardium was considered physiologically impossible. Major scientific strides in the past few years have challenged the concept that the heart is a post-mitotic organ, leading to the hypothesis that cardiac regeneration could be therapeutically achieved. Bone marrow-derived adult stem cells were among the first cell populations that were used to test this hypothesis. Animal studies and early clinical experience support the concept that therapeutically delivered mesenchymal stem cells (MSCs) safely improve heart function after an acute myocardial infarction (MI). MSCs produce a variety of cardio-protective signalling molecules, and have the ability to differentiate into both myocyte and vascular lineages. Additionally, MSCs are attractive as a cellular vehicle for gene delivery, cell transplantation or for tissue engineering because they offer several practical advantages. They can be obtained in relatively large numbers through standard clinical procedures, and they are easily expanded in culture. The multi-lineage potential of MSC, in combination with their immunoprivileged status, make MSCs a promising source for cell therapy in cardiac diseases. Here we provide an overview of biological characteristics of MSCs, experimental animal studies and early clinical trials with MSCs. In addition, we discuss the routes of cell delivery, cell tracking experiments and current knowledge of the mechanistic underpinnings of their action.

Keywords Mesenchymal stem cells · Acute myocardial infarction · Heart failure · Cell transplantation

1
Introduction

The mammalian heart develops early in embryogenesis and is the first organ that becomes functional. A primary heart tube is formed from the mesoderm, which shows peristaltic contraction at 3 weeks of gestation in the human embryo (Sissman 1970). In embryonic and early post-natal life, the increase in cardiac mass reflects the well-controlled balance between the addition of new myocytes and the death of unnecessary cells. Postpartum cardiac growth occurs due to an increase in the number and volume of myocytes, which leads to the development of the adult heart phenotype (Anversa et al. 1986).

The traditional paradigm that adult cardiac myocytes are terminally differentiated and unable to divide has recently been challenged. The re-entry of cardiac myocytes into the cell cycle has been demonstrated by many investigators. However, the outcome of this cell cycle re-entry remains unknown. There is little conclusive evidence of true cytokinesis of adult cardiomyocytes and some investigators have shown that cell cycle re-entry in adult cardiomyocytes results in apoptosis of those cells (Agah et al. 1997; Liu and Kitsis 1996; von Harsdorf et al. 1999). Thus, the division of cardiac myocytes as an endogenous mechanism to repair myocardial damage, or as a potential therapeutic option, remains controversial.

A major discovery from the laboratory of Piero Anversa provided a pivotal mechanism by which cardiac myocytes could regenerate. In this study, c-Kit$^+$ precursor cells were identified in rodent myocardium that had the capacity to differentiate into all major cardiac lineages (Beltrami et al. 2003). This observation offered a new perspective on cardiac homeostasis, which indicates that the steady-state complement of cellular constituents represents a balance between cell loss and cell appearance. Throughout life, there is loss and concurrent replacement of adult cells. Cardiac failure can occur when the rate of myocyte apoptosis and necrosis exceeds the capacity of the intrinsic stem cell pool to replenish the lost myocytes (Leri et al. 2005).

Heart failure remains the leading cause of death and hospitalisation in industrialised countries and is an emerging public health concern in the developing world despite advances in medical strategies aimed at post-infarct remodelling and the development of aggressive reperfusion strategies. In the United States alone, 5.2 million people are living with congestive heart failure (CHF) (American Heart Association 2006). Following a diagnosis of CHF, 1 in 5 patients will be dead within 12 months (American Heart Association 2006). The estimated cost for treating heart failure patients in 2007 is U.S. $33.2 billion. Therefore any new treatment modality that benefits heart failure patients has the potential to result in dramatic improvements in health outcomes as well as substantial cost saving for the community. The utilisation of mesenchymal stem cell therapy for cardiac repair may realise this potential.

2
Mesenchymal Stem Cells

McCulloch and Till (1960) identified a bone marrow stem cell capable of reconstituting haematopoiesis in mice. The clonality of these haematopoietic stem cells was subsequently demonstrated (Becker et al. 1963). This groundbreaking discovery led to the concept of adult stem cells.

In addition to haematopoietic stem cells, the marrow was shown to contain additional cells with the capacity for self-renewal and differentiation. Residing in the stromal compartment of bone marrow, these cells were termed mesenchymal stem cells (MSCs) and first described by Friedenstein at the Gamaleya Institute in Moscow (Friedenstein 1961). To characterise these cells, bone marrow was harvested from guinea-pigs and cultured in nutrient medium (see Fig. 1; Friedenstein et al. 1970). Adherent cultured cells re-implanted in ectopic sites developed the histological and morphological appearance of bone and cartilage, supporting their osteoprogenitor potential.

Subsequently, the culture conditions necessary for the expansion and differentiation of these cells were further refined. The original work of MSCs focussed on osteo-chondral differentiation and their potential to repair car-

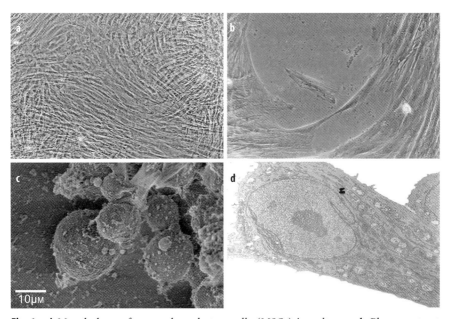

Fig. 1a–d Morphology of mesenchymal stem cells (MSCs) in culture. **a,b** Phase-contrast images of MSC in culture. MSCs show fibroblast-like features and growth patterns in culture. **c** Scanning electron microscope (SEM) image showing newly seeded MSCs attaching to the surface of a cell culture dish. **d** Transmission electron microscope (TEM) image of an MSC demonstrating the ultra-structural characteristics of a stem cell

tilage and bone. Human MSCs were first isolated by Arnold Caplan at Case Western Reserve University from small bone marrow samples (Caplan 1991; Haynesworth et al. 1992), demonstrating the clinical feasibility of harvesting and expanding these cells ex vivo. Caplan suggested that the actual MSC is an adherent fibroblastic cell (see Fig. 1) isolated by Percoll density centrifugation. The disadvantage of this definition is that this isolation method is prone to contamination with haematopoietic stem cells (HSCs), which leads to cellular heterogeneity in the final cell preparation. However, the most current cell separation and culture techniques result in high levels of viable MSCs for research and clinical use.

2.1
Definition of MSCs

There is no consensus on the definition of MSCs. Originally, MSCs were thought to correspond to the marrow cells that self-renew and give rise to mesenchymal tissues (Minguell et al. 2001). However, the marrow stromal cell population that contains the MSC pool can also differentiate into tissues other than those that originate in the embryonic mesoderm. Therefore, the term mesenchymal for these stem cell has been questioned by some investigators (Vogel et al. 2003; Woodbury et al. 2000).

MSCs are currently operationally defined by their ability to adhere to the surface of cell culture dishes (see Fig. 1) and the absence of haematopoietic markers. MSCs represent only 0.001–0.01% of the total population of nucleated cells in the bone marrow (Pittenger et al. 1999). However, they can be easily expanded in culture, and retain their growth and multilineage potential over several passages, although they lack immortality (Muraglia et al. 2000; Pittenger et al. 1999).

Individual MSC isolates have slightly different cell characteristics depending on the exact isolation method and culturing conditions used. To date, three different subpopulations of MSCs are described:

1. The smallest, most rapidly dividing cells have been labelled as recycling stem (RS) cells and are considered the most primitive form of MSCs (Prockop et al. 2001). They have a different multi-lineage differentiation potential. By definition, RS cells lack the expression of haematopoietic cell surface markers, but in contrast to other MSC populations, RS cells do not express c-kit (CD117).

2. Another cell subpopulation is the so-called multipotent adult progenitor cells (MAPCs). These cells, unlike other MSC populations, are immortal in culture and show in vitro capacity to differentiate in all three germ layer cell types (Jiang et al. 2002a, b).

3. Human bone marrow-derived multipotent stem cells (hBMSCs) are the latest subpopulation described to date. Like MAPCs, they can give rise to all three

germ layers and share the ability with MSCs to engraft and differentiate to multiple lineages in a rat model of acute infarction (Yoon et al. 2005).

In humans, MSCs are found in bone marrow, muscle, skin and adipose tissue, and are known to differentiate into muscle, fibroblasts, bone, tendon, ligament and adipose tissue (Caplan 1991). For cardiovascular research MSCs have been harvested from bone marrow and adipose tissue, and successfully expanded in culture.

2.2
Secreted Factors from MSCs

Early experiments focusing on cartilaginous and osseous tissue showed that cultured MSCs participated in new tissue formation, rather than inducing endogenous cells to differentiate by paracrine signalling. However, from the early days MSCs were studied for their ability to function as a stromal network and support haematopoietic stem cells. MSCs indeed produce an array of growth factors and cytokines, including macrophage, granulocyte and granulocyte-macrophage colony stimulating factors (M-, G-, and GM-CSF), stem cell factor (SCF-1), leukaemia inhibitory factor (LIF), stromal cell-derived factor (SDF-1), Flt-3 ligand, and interleukin-1, -6, -7, -8, -11, -14 and-15 (Haynesworth et al. 1996; Majumdar et al. 1998). Most of these cytokines are known to be important signalling molecules in modulating acute and chronic immune responses.

More recently, Kinnaird et al. analysed conditioned medium from cultured MSCs and added vascular endothelial growth factor (VEGF), basic fibroblast growth factor (bFGF), placental growth factor (PIGF) and monocyte chemo-attractant protein (MCP-1) to the list of secreted cytokines. Notably, these cytokines are involved in vasculogenesis and angiogenesis (Kinnaird et al. 2004b). Subsequent experiments demonstrated that injection of cell-free conditioned medium from MSC cultures is able to enhance angiogenesis, suggesting that MSCs exert their effects, at least in part, by paracrine actions. Beneficial effects on blood flow and limb function have been demonstrated in a hindlimb ischaemia model (Kinnaird et al. 2004a). Improved neovascularisation and cardiac function have been shown in mice following MI after intra-myocardial application of this cell-free medium (Gnecchi et al. 2005). It therefore appears that MSCs are capable of releasing a variety cytokines and may act via the release of these soluble factors to enhance angiogenesis.

2.3
Immunology of MSCs

One of the most exciting aspects of MSC biology is their apparent immuno-privilege. While expressing major histocompatibility complex I (MHC I), MSCs have low levels of MHC II, suggesting they may be able to avoid the initiation of an allogeneic immune response. The unique immunologic characteristics of MSCs were first demonstrated in sheep, where human MSCs were shown to evade rejection, persist long-term at the transplantation sites, and exhibit multipotential, site-specific differentiation (Liechty et al. 2000). Experiments in non-human primates have evaluated the homing, immunologic reaction, and long-term cell fate of autologous and allogeneic MSCs. These studies demonstrated that allogenic MSCs were not rejected, showed similar effects to the autologous MSCs and could be detected in the recipient 9 months after transplantation (Bartholomew et al. 2002; Devine et al. 2001, 2003; see also Fig. 3 below).

In vitro experiments elucidated the unique immunologic properties of MSCs. After interferon-γ incubation, the expression of MHC II, a classical stimulator for the T cell response, is upregulated; however, the lack of co-stimulatory molecules B7–1 (CD80), B7–2 (CD86) and CD40 prevents T cell proliferation (Bartholomew et al. 2002; Tse et al. 2003). In fact, allogeneic MSCs inhibit T cell interaction in a local fashion, but T cell viability is not affected after transplantation (Di et al. 2002).

Aggarwal and Pittenger investigated the lack of T cell response to allogeneic MSC transplantation by exposing MSCs to isolated subpopulations of immune cells, such as T_{Helper} -1(T_H1), T_{Helper} -2 (T_H2), T regulatory (T_R) and natural killer (NK) cells. Their results indicate that MSCs interact with these immune cells and alter their cytokine profile. Pro-inflammatory T_H1 cells and NK cells decreased the secretion of interferon-γ. Anti-inflammatory T_H2 cells increased the immuno-protective secretion of interleukin (IL)-4, and furthermore MSCs caused an increase in the number of T_R cells and the associated secretion of anti-inflammatory IL-10 (Aggarwal and Pittenger 2005).

Overall, these experiments indicate that allogeneic MSCs can modify immune response in a dose-dependent manner and can reduce the inflammatory response when present in appropriate amounts. Currently, phase III clinical trials are evaluating the optimal MSC dose and frequency of administration to prevent or treat graft-vs-host disease during allogeneic HSC transplantation.

3
MSCs for Gene Therapy, Transplantation and Tissue Engineering

For several reasons, MSCs are attractive as a cellular vehicle for gene delivery, cell transplantation and tissue engineering applications. They can be obtained in relatively large quantities through standard clinical procedures, such as bone marrow harvest or liposuction. MSC are easily expanded in culture to numbers that make clinical application practical. Their unique immune-privileged status may allow them to be used as an allogeneic pre-prepared "off-the-shelf" product; therefore they can be administered immediately when required, rather than waiting weeks or months for adequate numbers of cells to be achieved by cell culture.

3.1
Genetic Modification and MSCs

MSCs are capable of long-term transgene expression (Zhang et al. 2002). Genetically modified MSCs were first successfully used to deliver pacemaker genes into the heart in an attempt to develop biological pacemakers (Potapova et al. 2004; Valiunas et al. 2004). These studies demonstrated that MSCs can be used to deliver a variety of genes and influence the function of syncytial tissues by expressing gap junction proteins.

Recently, genetic modification of MSCs has been used to overcome the problem of limited viability and engraftment of the transplanted cells in the myocardium. Indeed, genetic modification can improve survival, metabolic characteristics, proliferative capacity or differentiation of MSCs. In addition, cells can be engineered to be a vehicle for gene therapy whose secreted product can exert paracrine or endocrine actions that may results in further therapeutic benefits. This approach was first conceptualised by Mangi and colleagues. MSCs over-expressing the anti-apoptotic gene *Akt-1* became more resistant to apoptosis in vitro and in vivo, and showed a dramatic improvement in cardiac function in a rodent MI model, with almost complete replacement of the infarct scar with functioning myocardium (Mangi et al. 2003).

Subsequently, Tang et al. demonstrated that MSCs transduced with haemoxygenase, an enzyme preventing oxidative damage, showed both better engraftment and enhanced cell survival following intra-myocardial delivery relative to non-transduced MSCs (Tang et al. 2005). In a recent study, myocardial transplantation of VEGF gene-transfected MSCs demonstrated a better improvement in myocardial perfusion and in restoration of heart function than either cellular or gene therapy alone (Yang et al. 2006).

Though the function of MSCs might be further enhanced by genetic manipulation, this approach is currently restricted to basic and translational research because of the limited clinical experience with gene therapy and genetically modified cell products in the setting. However, these approaches will help to further elucidate the biology of stem cell function and improve our understanding of the mechanisms of repair. It is possible that in the future, genetically modified MSCs may be of value for clinical therapy.

3.2
MSC Transplantation and Routes of Application

In animal studies and human trials MSCs have been delivered using four routes: (1) intra-coronary infusion, (2) intra-venous infusions, (3) direct injection into the myocardium via percutaneous endomyocardial approaches, and (4) surgical epicardial administration. Each approach offers unique advantages and disadvantages.

The advantage of intra-coronary infusion is that cells can travel directly into myocardial regions in which nutrient blood flow and oxygen supply are abundant, providing a favourable environment for stem cell survival and engraftment. However, the timing of delivery is crucial. In the setting of an acute MI with damaged endothelium, cells may traverse the endothelial barrier more easily and cross the vessel wall from the vascular lumen into the tissue where they are needed. If this proves to be the case, early delivery of MSCs may result in the optimal repair of tissue damage. However, the acute inflammatory process and the local oedema following MI might limit the engraftment and survival of transplanted cells. If this proves to be the case, then application of MSC therapy at later time points may result in more effective repair. Thus, the most appropriate time point for cell transplantation after an acute MI remains to be determined.

Recent in vitro and in vivo experiments confirm that MSCs can easily traverse endothelial barriers, and that the majority of cells migrate within 30 min (Schmidt et al. 2006). Conversely, homing of intra-arterially injected MSCs requires migration out of the vessel into the surrounding tissue, so that inadequately perfused areas of the myocardium could be targeted less efficiently. Mitigating this, however, is the demonstration that MSCs home specifically to the site of an acute ischaemic injury in animal models after systemic delivery (Barbash et al. 2003; Nagaya et al. 2004; Saito et al. 2002). Similar homing-to-injury results have been obtained from a clinically relevant porcine model of an acutely reperfused myocardial infarct (Price et al. 2005).

The underlying mechanism for cell homing relates to ischaemia and hypoxia-induced release of cytokines and the induced expression of adhesion molecules into the injured myocardium. SDF1 and its cognate receptor CXCR4 are well known to be necessary for stem cell homing to the bone marrow (Peled et al. 1999). In ischaemically damaged myocardium, SDF-1 expression is upregulated and remains elevated for days (Abbott et al. 2004; Ma et al. 2005). Studies have shown that MSCs express the CXCR4 receptor and migrate towards a SDF-1 gradient in vitro and across endothelium in vivo (Peled et al. 1999).

Matrix metalloproteinases, especially MMP-9 which is highly activated in the post-infarcted myocardium (Wilson et al. 2003), are also implicated in this process. MMP-9 induces the soluble kit ligand, which facilitates stem cell mobilisation from the bone marrow (Heissig et al. 2002). The notion of a stem cell homing signal from the ischaemic-injured myocardium provides the rationale for using intravenous infusions in patients with acute myocardial infarction.

Safety issues have been raised about the intra-coronary delivery route for MSCs. The fear is that cells administered this way may result in blockage of the coronary arteries and further damage to the myocardium. One study using autologous mesenchymal stem cells from dog bone marrow injected into normal coronary arteries demonstrated micro-infarction in the region supplied by that artery, suggesting that these cells aggregate in vivo leading to arterial or arteriolar obstruction (Vulliet et al. 2004). Nevertheless, intra-coronary mesenchymal stem cell therapy in one human trial resulted in improved cardiac function and was safe (Chen et al. 2004b).

Although intra-coronary infusion of bone marrow-derived stem cells is currently the preferred method of cell delivery in clinical trials, and it seems to be an attractive option for MSC application in patients with acute myocardial infarction as well, intra-myocardial injections are an emerging option especially in patients with old infarcts or chronic heart failure, where the homing signal released from acutely damaged tissue are likely to be less intense.

Open heart surgery offers the advantage that cells can be placed under direct vision. Especially in patients with ischaemic cardiomyopathy and chronic infarct scars, this can be a valuable approach to deliver cells directly to the areas where they are needed the most. Because of its invasive nature, this kind of procedure is likely to be performed only in patients with existing clinical indications for cardiac surgery. Taken together, intra-myocardial cell delivery during coronary bypass grafting and catheter-based delivery systems appear to be safe and feasible.

Recently, Miyahara et al. reported a novel tissue engineering approach of MSC grafting into the myocardium (Miyahara et al. 2006). MSCs were cultured from adipose tissue, and grown as a monolayer on cell sheets. After myocardial transplantation, the engrafted MSC sheets gradually grew to form a thick stratum that included newly formed vessels, undifferentiated cells and a few cardiomyocytes. Compared to fibroblast cell sheets, the monolayered MSCs reversed wall thinning in the scar area and improved cardiac function in rats with myocardial infarction.

The ideal method of MSC transplantation still remains to be determined, but it is already apparent that MSCs can be safely administered via several different modes of delivery to treat cardiovascular diseases. Cell delivery may have to be specifically tailored to different disease states and conditions.

4
Mechanism of Repair

The equilibrium between myocyte death and formation of new myocytes suggests that the cardiac stem cell pool receives signals from cells undergoing apoptosis or necrosis, which induce resident progenitor cells to grow and functionally replace lost myocytes. Although myocardial infarct scars do not regenerate spontaneously, the endogenous repair mechanisms of the mammalian heart may be manipulated and exploited to achieve a *restitutio ad integrum* after ischaemic damage. Whether the stem cell reserve in the adult heart suffices to reconstitute acute and massive loss of myocardium needs to be resolved. Experimental animal studies have demonstrated the ability of cardiac stem cells to translocate to the site of injury (Beltrami et al. 2001; Oh et al. 2003). Therefore, we have to distinguish between the regenerative capacity

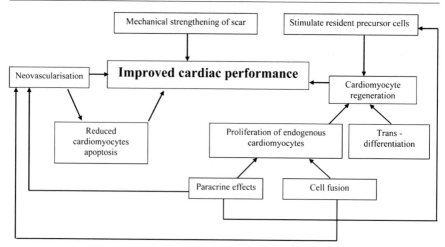

Fig. 2 Mechanisms of cardiac repair. (Modified from Boyle et al. 2006, with permission)

of the damaged and the non-damaged myocardium. Anversa and co-workers showed that myocyte formation is enhanced in patients with acute and chronic heart failure (Beltrami et al. 2001; Urbanek et al. 2005b).

This theory lays the foundation on which stem cell therapy is built. A number of different mechanisms operating alone or in concert are proposed to underlie successful cardiac cell therapy (see Fig. 2):

1. Restitution of the depleted stem cell pool at the site of injury for cardiac myogenesis
2. Induction of vasculo- and angiogenesis for adequate oxygen and blood supply, which may enhance myocyte survival and limit apoptosis
3. Enhancement of the signalling from the injury site to attract more endogenous cardiac stem cells
4. Mechanical strengthening of the infarct scar, preventing on-going left ventricular (LV) remodelling and allowing endogenous repair mechanisms to improve LV function
5. Modulation the inflammatory response to acute MI, thereby limiting the amount of initial tissue damage
6. Stimulation of endogenous cardiomyocytes to re-enter the cell cycle and proliferate
7. Direct trans-differentiation into cardiomyocytes, endothelial cells and vascular smooth muscle cells

This list of putative mechanisms is by no means exhaustive. Indeed, any number of these proposed mechanisms, or another as-yet-unrecognised means of repair, may be acting in concert to achieve cardiac regeneration.

Human and animal data suggest that bone marrow-derived cells play a role in the repair process after injury, but the precise means by which the repair process occur remains incompletely defined. Whether circulating and resident progenitor cells repair the damaged heart by trans-differentiation into new cardiac muscle, or by any of the other mechanisms listed above continues to be debated (Mouquet et al. 2005; Urbanek et al. 2005a, b).

MSCs have the ability to trans-differentiate into cardiac myocytes. While in vitro observations after MSC treatment with DNA demethylating agent 5-azacytidine support differentiation (Makino et al. 1999), in vivo studies demonstrate incomplete differentiation with MSCs harbouring myocyte-specific proteins but lacking full cardiomyocyte phenotypes (Amado et al. 2005; Shake et al. 2002). MSCs are reported to produce cardiac connexin 43 (Valiunas et al. 2004) and therefore could in theory electromechanically couple to host myocardium.

We have demonstrated the presence of proliferating cardiomyocytes in the scar region after MSC transplantation in the setting of an acute MI (Amado et al. 2005; see Fig. 4). Furthermore, our recent data suggest growth of new tissue replacing areas of scar tissue with viable and functional myocardium. However, whether this occurs due to MSC differentiation into cardiomyocytes, proliferation of endogenous cardiomyocytes or their precursors, or other currently unappreciated mechanisms remains to be determined (see Fig. 2). Indeed, it has been shown that exogenous signals introduced following MI can cause proliferation and differentiation of endogenous cardiac stem cells, resulting in myocardial regeneration and improved LV function (Limana et al. 2005). MSC transplantation may act in a paracrine or cell-to-cell signalling fashion to effect cardiac repair via cytokine-induced enhancement of endogenous cardiac stem cell function.

Other actions of MSC therapies include immunomodulatory effects on inflammatory cells as discussed, which might be beneficial in the setting of an acute MI, and signalling to modulate direct scar healing and remodelling. There is strong evidence that MSCs participate in the regulation of collagen formation. Xu et al. have reported that MSCs influence the composition of extracellular matrix at the scar, altering the expression of MMP-1 and tissue inhibitor of matrix metalloproteinases (TIMPS), among other components (Xu et al. 2005a, b). Others have demonstrated improved ventricular compliance after MSC therapy and suggested that this, in concert with reduced scar formation, may contribute to the improvements in LV function (Berry et al. 2006).

Furthermore, MSCs are also reported to participate in vasculogenesis and angiogenesis by paracrine signalling and incorporation into newly formed vessels (Silva et al. 2005). Kinnaird et al. first reported a beneficial effect of MSCs on collateral remodelling and perfusion in a mouse model of hindlimb ischaemia (Kinnaird et al. 2004a). With regard to the heart, there are reports in small and large animal models (Silva et al. 2005; Tang et al. 2004; Tomita et al. 2002; Zhang et al. 2004) that implanted MSCs differentiate into endothelial cells and enhance vascular density.

4.1
Cell Tracking and Evaluation of MSC Therapy

A variety of methods to label and track MSCs have been used in order to follow the engraftment, distribution and bioactivity in the recipients. However, each cell-labelling method has its limitation, and caution is necessary in interpreting and combining results from different experimental studies.

Both invasive and non-invasive cellular imaging requires labelling of the target cells. The techniques used so far include:

1. Dye labelling of membrane or surface proteins markers such as DiI (1,1″-dioleyl-3,3,3″,3″-tetramethylindocarbocyanine methanesulphonate)
2. Labelling of DNA with intercalating fluors such as DAPI (4′,6-diamidino-2-phenylindole) or the false nucleotide BrdU (bromodeoxyuridine)
3. Gene tagging of cells by transfection such as the reporter gene β-galactosidase (lacZ) or green fluorescent protein (GFP)
4. In situ hybridisation for Y-chromosome in female recipients of male cells
5. PCR for a transgene or Y-chromosome
6. Iron labelling for magnetic resonance imaging (MRI)
7. Radiolabelling with ^{111}indium for γ-imaging (SPECT)
8. Iridium labelling of MSCs for detection by neutron activation analysis (NAA)
9. PET imaging of transfected cells carrying an emitting reporter gene

4.1.1
Invasive Imaging of MSC Therapy

Identification, localisation and evaluation of the fate of transplanted MSCs are mandatory to understand cell function and elucidate the mechanisms of repair of MSC therapy. This can be best achieved by evaluation of MSC distribution in tissue specimens.

To date, gender mismatch-labelling seems to be the best approach for cell tracking, because the native properties of cells are used without affecting cell function. This approach is particularly interesting for the evaluation of MSC therapy, where allogeneic cell delivery is possible without immunosuppression. The advantages of male-to-female transplantation are that the Y-chromosome is present in virtually every intact donor cell, and that current paint probes make it relatively easy to image the Y-chromosome in situ. Quaini et al. used this technique to demonstrate the presence of male cells after orthotopic transplantation of female hearts into male patients (Quaini et al. 2002). However, the Y-chromosome can only be visualised in tissue sections that sample the

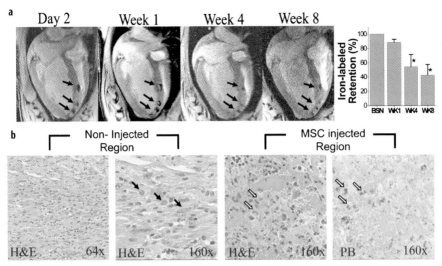

Fig. 3a,b Engraftment of allogeneic porcine mesenchymal stem cells assessed with magnetic resonance imaging. **a** Pseudo-long axis image obtained after MSC injection. Magnetically iron-labelled (Feridex) MSCs appear as hypo-enhanced areas (*arrows*) visible 2 days, 1 week, 4 weeks and 8 weeks after injection. Bar graph (*right*) illustrating Feridex retention over 8 weeks. **b** Histologic evaluation of post-infarct myocardium. H&E stains (*left*) obtained from non-injected infarcted area, demonstrating an abundance of inflammatory cells (*closed arrows*) surrounding the necrotic region (160× magnification). H&E and Prussian blue stains (PB) of MSC injection sites *(right)* demonstrating the absence of inflammatory cells. *Open arrows* indicate iron-labelled MSCs. (Reprinted from Amado et al. 2005, with permission)

nuclei, and many of the transplanted cells will not be detected. This is a problem with all nuclear stains, and it can cause a problem when the abundance of transplanted cells is low to begin with, or diminishes over time (see Fig. 3).

In preclinical studies, MSCs have been either labelled with iron oxide (Amado et al. 2005; Kraitchman et al. 2005; see Fig. 3), DiI/DAPI (Amado et al. 2005; Silva et al. 2005), BrdU (Tomita et al. 2002), or have been transfected with lacZ (Toma et al. 2002). Immuno double labelling for proteins expressed early in cardiac myocyte development has been used to determine the fate of transplanted cells. However, each technique has its own strengths and limitations, requiring caution with data interpretation. Further experiments and novel approaches, such as promoter-driven expression of labelled cardiomyocytes and endothelial markers, are needed to ascertain with certainty the fate of cells after transplantation.

4.1.2
Non-invasive Imaging of MSC Therapy

Non-invasive imaging is playing an important role in assessing the myocardial response to MSC therapy. The correlation of tissue specimen, allowing cellu-

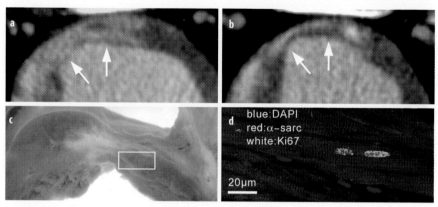

Fig. 4a–d Cardiac regeneration after intramyocardial MSC injection. MDCT scan of the same MSC-treated animal at 5 days (**a**) and 8 weeks (**b**) following MSC injection. There is a time-dependent reduction in the thickness of the infarct scar and a corresponding increase in the thickness of the sub-endocardial rim of tissue at 8 weeks. **c** Representative pathology specimen of a MSC-treated animal showing a patchy scar pattern, and a thick endomyocardial rim. **d** Ki67-positive myocytes (white nuclear stain) in the endomyocardial rim, demonstrating the regenerative capacity of the heart (*red*: α-sarcomeric actin). (Parts a and b reprinted from Amado et al. 2006, with permission)

lar and molecular characterisation, with non-invasive imaging of myocardial function, viability, perfusion and cell tracking is the basis to elucidate the effects of cell therapy.

MRI and contrast-enhanced multi-detector computed tomography (MDCT) have been used to serially monitor the regional and global function and myocardial structure in MSC-treated hearts, demonstrating in vivo that the reappearance of myocardial tissue is accompanied by time-dependent restoration of contractile function, consistent with a regenerative process (see Fig. 4; Amado et al. 2006).

Most current MRI in vivo-labelling strategies for MSCs have applied iron oxide particles—so-called ultra-small superparamagnetic iron oxide (SPIO) compounds. The compound is internalised into the living cell because surface labels may become detached and transferred to other cells. Another advantage of SPIO labelling is that when the cells are lysed the iron oxide is only taken up by phagocytic cells and ultimately recycled in the iron pool, thereby preventing damage to tissue at the injection side. On the other hand, this is also the biggest problem of SPIO labelling because iron particles from degraded cells are incorporated by adjacent phagocytic cells. Thus, hypo-intense artefacts in images at later time points after cell delivery may represent cells other than the originally labelled ones. Therefore strict quantification of cell numbers based on the change in volume of the hypo-intense signal is difficult to determine.

In a hybrid approach, Kraitchman et al. added the radiotracer [111]indium-oxime as a cell surface label to the internalised SPIO label, thereby providing two independent cell labels for in vivo imaging (Kraitchman et al. 2005). ^{111}Indium-oxime can be used to obtain semi-quantitative information about MSC abundance in different tissues. After intravenous infusion in a dog model of myocardial infarction, MSCs were seen first to deposit in the lung at 24 h. Subsequently, the MSCs migrated to the infarcted areas of the heart and persisted for at least a week.

Furthermore, it is possible to label human MSCs with ^{111}indium-oxime without affecting the viability and differentiation capacity of the MSCs (Bindslev et al. 2006). This could be a tool for the control of stem cell delivery and dose response in clinical cardiovascular trials.

5
Clinical Trials and MSCs

Mesenchymal stem cells have been used in a number of phase I and II clinical trials to treat several different conditions (see Table 1). These clinical trials demonstrate the safety and feasibility of both autologous and allogeneic MSC therapy in a variety of clinical settings and set the scene for future large-scale trials of MSC therapy for cardiac repair.

In an early study of MSC therapy, Koc and colleagues (2000) used autologous culture-expanded MSCs as an adjunct to peripheral blood progenitor cell transplant in 28 breast cancer patients. They demonstrated the feasibility of ex

Table 1 Clinical studies using mesenchymal stem cells for non-cardiac applications

Study	Number treated	Autologous/ allogeneic	Indication	Randomised
Lazarus et al. 2005	46	Allogeneic	Allogeneic BM transplant	–
Koc et al. 2002	11	Allogeneic	Hurler's syndrome or MLD	–
Koc et al. 2000	28	Autologous	PBPC transplant for breast cancer	–
Bang et al. 2005	5	Autologous	Stroke	–
Garcia-Olmo et al. 2005	5	Autologous	Crohn's disease fistula	–
Horwitz et al. 2002	6	Allogeneic	Osteogenesis Imperfecta	–
Le et al. 2004	1	Allogeneic	GVHD	–

BM, bone marrow; GVHD, graft-vs-host disease; MI, myocardial infarction; MLD, metachromatic leukodystrophy; PBPC, peripheral blood progenitor cell vivo culture to expand MSC numbers and the safety of intravenous infusion of these expanded autologous MSCs. Subsequently, the same researchers have used allogeneic MSCs as adjunctive therapy for various diseases (Koc et al. 2002; Lazarus et al. 2005), and have demonstrated the safety and feasibility of this allogeneic cell therapy approach.

5.1
Clinical Studies Using MSCs for Cardiac Applications

There are currently two published studies utilising MSC therapy for cardiac repair, as well as two studies registered with clinicaltrials.gov (see Table 2). In the first published trial, Chen and colleagues (2004a, b) randomised 69 patients after primary angioplasty within 12 h of the onset of MI to receive intra-coronary autologous MSCs or saline 18 days following MI. Bone marrow was harvested from all patients on day 8 following the infarct, and MSCs were purified from the marrow by the Percoll density gradient centrifugation method. No difference in arrhythmias or mortality was reported. This demonstrated the safety and feasibility of intra-coronary MSC infusion in post-MI patients, as well as a reduction in the size of the perfusion defect and an improvement in overall LV function. Katritsis and co-workers (2005) also tested intra-coronary MSCs after MI. They also showed safety of intra-coronary autologous MSCs, and their results suggested that MSC therapy may lead to a reduction in infarct scar size compared to a non-randomised untreated control group.

Just as the clinical trials using MSC therapy for non-cardiac applications advanced from autologous to allogeneic therapy, so the clinical trials of MSC therapy for cardiac repair are evolving to utilise allogeneic cells. The first clinical phase I trial using allogeneic human MSC therapy following MI (see Table 2) has finished enrolling patients. The results, however, have not been published to date. This transition from autologous to allogeneic cell therapy may represent the most exciting breakthrough in cell therapy to date, for two reasons. First, it obviates the need for painful bone-marrow aspirate on post-infarction cardiac patients, who are generally receiving both anti-platelet agents and anti-coagulants and are therefore at risk for haemorrhagic complications from this invasive procedure. Second, it allows for administration of an "off-the-shelf" product to MI patients. This has several distinct advantages, as there is no need for ex vivo expansion of cells from patients and thus the attendant risk of contamination is prevented. Additionally, the cells can be administered immediately without the need for expansion over several weeks to achieve adequate numbers to effect clinically meaningful results. This delay in treatment may reduce the clinical efficacy of the cellular transplantation. Finally, there is also the ability to administer stem cells from young healthy donors, which may have more therapeutic potential than those from older adults. On-going and future clinical trials are investigating both autologous and allogeneic MSC therapy for cardiac repair and the results are eagerly anticipated.

Table 2 Clinical studies using mesenchymal stem cells for cardiac applications

Study author/ sponsor	Number treated	Trial design	Condition treated	Cells and delivery	Primary outcome	Randomised	Region	Status
Chen et al. 2004a,b	34	Phase I/II	MI	Intracoronary autologous MSC	Cardiac death	+	China	Published
Katritsis et al. 2005	11	Phase I	MI	Intracoronary autologous MSC	Feasibility and safety	−	Greece	Published
Rigshospitalet, Denmark	40	Phase I/II	Myocardial ischaemia	Intramyocardial autologous MSC	Perfusion	−	Denmark	Registered[a], not yet recruiting
Osiris Therapeutics	53	Phase I	MI	Intravenous MSC	Safety	+	USA	Registered[a], finished recruiting

[a] Registered with clinicaltrials.gov;
MI, myocardial infarction; MSC, mesenchymal stem cell

5.2 Translational Research and Future Directions

At present, there is an abundance of safety and feasibility data supporting the advancement of MSC therapy into larger phase II and III clinical trials (Boyle et al. 2006). However, there remain many questions that can only be answered by rigorously controlled clinical trials. In particular, how many cells are needed for myocardial regeneration in cardiac patients and what kind of cell preparations are optimal to obtain a clinically meaningful result? Also, the preferred delivery method for MSC therapy in each type of cardiac disease must be tested in clinical trials. It is likely that different diseases, and even different stages of the same disease process, such as acute vs chronic myocardial infarct scaring, will require different delivery and dosing approaches. In fact, multiple treatments at different time-points may even be required. Only clinical trials conducted in parallel with basic research of MSC biology and pre-clinical translation studies will elucidate the understanding of cellular repair mechanisms and guide this emerging therapy into a clinical reality.

6 Conclusion

Although at this time there are no data to support MSC therapy as standard clinical practice for cardiac applications, there is a wealth of preclinical and early clinical data showing safety, feasibility and early efficacy of MSC therapy. MSC cell therapies for cardiac diseases are appropriately poised to enter randomised placebo-controlled clinical trials.

References

Abbott JD, Huang Y, Liu D, Hickey R, Krause DS, Giordano FJ (2004) Stromal cell-derived factor-1alpha plays a critical role in stem cell recruitment to the heart after myocardial infarction but is not sufficient to induce homing in the absence of injury. Circulation 110:3300–3305

Agah R, Kirshenbaum LA, Abdellatif M, Truong LD, Chakraborty S, Michael LH, Schneider MD (1997) Adenoviral delivery of E2F-1 directs cell cycle reentry and p53-independent apoptosis in postmitotic adult myocardium in vivo. J Clin Invest 100:2722–2728

Aggarwal S, Pittenger MF (2005) Human mesenchymal stem cells modulate allogeneic immune cell responses. Blood 105:1815–1822

Amado LC, Saliaris AP, Schuleri KH, St JM, Xie JS, Cattaneo S, Durand DJ, Fitton T, Kuang JQ, Stewart G, Lehrke S, Baumgartner WW, Martin BJ, Heldman AW, Hare JM (2005) Cardiac repair with intramyocardial injection of allogeneic mesenchymal stem cells after myocardial infarction. Proc Natl Acad Sci U S A 102:11474–11479

Amado LC, Schuleri KH, Saliaris AP, Boyle AJ, Helm R, Oskouei B, Centola M, Eneboe V, Young R, Lima JA, Lardo AC, Heldman AW, Hare JM (2006) Multi-modality non-invasive imaging demonstrates in-vivo cardiac regeneration following mesenchymal stem cell therapy. J Am Coll Cardiol 48:2116–2124

American Heart Association (2006) Heart Disease and Stroke Statistics—2007 Update. A report from the American Heart Association Statistics Committee and Stroke Statistics Subcommittee (2006) http://circ.ahajournals.org/cgi/content/short/CIRCULATION AHA.106.179918. Circulation. Published online ahead of print, 28 December 2006. Cited 16 January 2007

Anversa P, Ricci R, Olivetti G (1986) Quantitative structural analysis of the myocardium during physiologic growth and induced cardiac hypertrophy: a review. J Am Coll Cardiol 7:1140–1149

Bang OY, Lee JS, Lee PH, Lee G (2005) Autologous mesenchymal stem cell transplantation in stroke patients. Ann Neurol 57:874–882

Barbash IM, Chouraqui P, Baron J, Feinberg MS, Etzion S, Tessone A, Miller L, Guetta E, Zipori D, Kedes LH, Kloner RA, Leor J (2003) Systemic delivery of bone marrow-derived mesenchymal stem cells to the infarcted myocardium: feasibility, cell migration, and body distribution. Circulation 108:863–868

Bartholomew A, Sturgeon C, Siatskas M, Ferrer K, McIntosh K, Patil S, Hardy W, Devine S, Ucker D, Deans R, Moseley A, Hoffman R (2002) Mesenchymal stem cells suppress lymphocyte proliferation in vitro and prolong skin graft survival in vivo. Exp Hematol 30:42–48

Becker AJ, McCulloch EA, Till JE (1963) Cytological demonstration of the clonal nature of spleen colonies derived from transplanted mouse marrow cells. Nature 197:452–454

Beltrami AP, Urbanek K, Kajstura J, Yan SM, Finato N, Bussani R, Nadal-Ginard B, Silvestri F, Leri A, Beltrami CA, Anversa P (2001) Evidence that human cardiac myocytes divide after myocardial infarction. N Engl J Med 344:1750–1757

Beltrami AP, Barlucchi L, Torella D, Baker M, Limana F, Chimenti S, Kasahara H, Rota M, Musso E, Urbanek K, Leri A, Kajstura J, Nadal-Ginard B, Anversa P (2003) Adult cardiac stem cells are multipotent and support myocardial regeneration. Cell 114:763–776

Berry MF, Engler AJ, Woo YJ, Pirolli TJ, Bish LT, Jayasankar V, Morine KJ, Gardner TJ, Discher DE, Sweeney HL (2006) Mesenchymal stem cell injection after myocardial infarction improves myocardial compliance. Am J Physiol Heart Circ Physiol 290:H2196–H2203

Bindslev L, Haack-Sorensen M, Bisgaard K, Kragh L, Mortensen S, Hesse B, Kjaer A, Kastrup J (2006) Labelling of human mesenchymal stem cells with indium-111 for SPECT imaging: effect on cell proliferation and differentiation. Eur J Nucl Med Mol Imaging 33:1171–1177

Boyle AJ, Schulman SP, Hare JM, Oettgen P (2006) Controversies in cardiovascular medicine: ready for the next step. Circulation 114:339–352

Caplan AI (1991) Mesenchymal stem cells. J Orthop Res 9:641–650

Chen SL, Fang WW, Qian J, Ye F, Liu YH, Shan SJ, Zhang JJ, Lin S, Liao LM, Zhao RC (2004a) Improvement of cardiac function after transplantation of autologous bone marrow mesenchymal stem cells in patients with acute myocardial infarction. Chin Med J (Engl) 117:1443–1448

Chen SL, Fang WW, Ye F, Liu YH, Qian J, Shan SJ, Zhang JJ, Chunhua RZ, Liao LM, Lin S, Sun JP (2004b) Effect on left ventricular function of intracoronary transplantation of autologous bone marrow mesenchymal stem cell in patients with acute myocardial infarction. Am J Cardiol 94:92–95

Devine SM, Bartholomew AM, Mahmud N, Nelson M, Patil S, Hardy W, Sturgeon C, Hewett T, Chung T, Stock W, Sher D, Weissman S, Ferrer K, Mosca J, Deans R, Moseley A, Hoffman R (2001) Mesenchymal stem cells are capable of homing to the bone marrow of non-human primates following systemic infusion. Exp Hematol 29:244–255

Devine SM, Cobbs C, Jennings M, Bartholomew A, Hoffman R (2003) Mesenchymal stem cells distribute to a wide range of tissues following systemic infusion into nonhuman primates. Blood 101:2999–3001

Di NM, Carlo-Stella C, Magni M, Milanesi M, Longoni PD, Matteucci P, Grisanti S, Gianni AM (2002) Human bone marrow stromal cells suppress T-lymphocyte proliferation induced by cellular or nonspecific mitogenic stimuli. Blood 99:3838–3843

Friedenstein AJ (1961) Osteogenetic activity of transplanted transitional epithelium. Acta Anat (Basel) 45:31–59

Friedenstein AJ, Chailakhjan RK, Lalykina KS (1970) The development of fibroblast colonies in monolayer cultures of guinea-pig bone marrow and spleen cells. Cell Tissue Kinet 3:393–403

Garcia-Olmo D, Garcia-Arranz M, Herreros D, Pascual I, Peiro C, Rodriguez-Montes JA (2005) A phase I clinical trial of the treatment of Crohn's fistula by adipose mesenchymal stem cell transplantation. Dis Colon Rectum 48:1416–1423

Gnecchi M, He H, Liang OD, Melo LG, Morello F, Mu H, Noiseux N, Zhang L, Pratt RE, Ingwall JS, Dzau VJ (2005) Paracrine action accounts for marked protection of ischemic heart by Akt-modified mesenchymal stem cells. Nat Med 11:367–368

Haynesworth SE, Goshima J, Goldberg VM, Caplan AI (1992) Characterization of cells with osteogenic potential from human marrow. Bone 13:81–88

Haynesworth SE, Baber MA, Caplan AI (1996) Cytokine expression by human marrow-derived mesenchymal progenitor cells in vitro: effects of dexamethasone and IL-1 alpha. J Cell Physiol 166:585–592

Heissig B, Hattori K, Dias S, Friedrich M, Ferris B, Hackett NR, Crystal RG, Besmer P, Lyden D, Moore MA, Werb Z, Rafii S (2002) Recruitment of stem and progenitor cells from the bone marrow niche requires MMP-9 mediated release of kit-ligand. Cell 109:625–637

Horwitz EM, Gordon PL, Koo WK, Marx JC, Neel MD, McNall RY, Muul L, Hofmann T (2002) Isolated allogeneic bone marrow-derived mesenchymal cells engraft and stimulate growth in children with osteogenesis imperfecta: implications for cell therapy of bone. Proc Natl Acad Sci U S A 99:8932–8937

Jiang Y, Jahagirdar BN, Reinhardt RL, Schwartz RE, Keene CD, Ortiz-Gonzalez XR, Reyes M, Lenvik T, Lund T, Blackstad M, Du J, Aldrich S, Lisberg A, Low WC, Largaespada DA, Verfaillie CM (2002a) Pluripotency of mesenchymal stem cells derived from adult marrow. Nature 418:41–49

Jiang Y, Vaessen B, Lenvik T, Blackstad M, Reyes M, Verfaillie CM (2002b) Multipotent progenitor cells can be isolated from postnatal murine bone marrow, muscle, and brain. Exp Hematol 30:896–904

Katritsis DG, Sotiropoulou PA, Karvouni E, Karabinos I, Korovesis S, Perez SA, Voridis EM, Papamichail M (2005) Transcoronary transplantation of autologous mesenchymal stem cells and endothelial progenitors into infarcted human myocardium. Catheter Cardiovasc Interv 65:321–329

Kinnaird T, Stabile E, Burnett MS, Lee CW, Barr S, Fuchs S, Epstein SE (2004a) Marrow-derived stromal cells express genes encoding a broad spectrum of arteriogenic cytokines and promote in vitro and in vivo arteriogenesis through paracrine mechanisms. Circ Res 94:678–685

Kinnaird T, Stabile E, Burnett MS, Shou M, Lee CW, Barr S, Fuchs S, Epstein SE (2004b) Local delivery of marrow-derived stromal cells augments collateral perfusion through paracrine mechanisms. Circulation 109:1543–1549

Koc ON, Gerson SL, Cooper BW, Dyhouse SM, Haynesworth SE, Caplan AI, Lazarus HM (2000) Rapid hematopoietic recovery after coinfusion of autologous-blood stem cells and culture-expanded marrow mesenchymal stem cells in advanced breast cancer patients receiving high-dose chemotherapy. J Clin Oncol 18:307–316

Koc ON, Day J, Nieder M, Gerson SL, Lazarus HM, Krivit W (2002) Allogeneic mesenchymal stem cell infusion for treatment of metachromatic leukodystrophy (MLD) and Hurler syndrome (MPS-IH). Bone Marrow Transplant 30:215–222

Kraitchman DL, Tatsumi M, Gilson WD, Ishimori T, Kedziorek D, Walczak P, Segars WP, Chen HH, Fritzges D, Izbudak I, Young RG, Marcelino M, Pittenger MF, Solaiyappan M, Boston RC, Tsui BM, Wahl RL, Bulte JW (2005) Dynamic imaging of allogeneic mesenchymal stem cells trafficking to myocardial infarction. Circulation 112:1451–1461

Lazarus HM, Koc ON, Devine SM, Curtin P, Maziarz RT, Holland HK, Shpall EJ, McCarthy P, Atkinson K, Cooper BW, Gerson SL, Laughlin MJ, Loberiza FR Jr, Moseley AB, Bacigalupo A (2005) Cotransplantation of HLA-identical sibling culture-expanded mesenchymal stem cells and hematopoietic stem cells in hematologic malignancy patients. Biol Blood Marrow Transplant 11:389–398

Le BK, Rasmusson I, Sundberg B, Gotherstrom C, Hassan M, Uzunel M, Ringden O (2004) Treatment of severe acute graft-versus-host disease with third party haploidentical mesenchymal stem cells. Lancet 363:1439–1441

Leri A, Kajstura J, Anversa P (2005) Cardiac stem cells and mechanisms of myocardial regeneration. Physiol Rev 85:1373–1416

Liechty KW, MacKenzie TC, Shaaban AF, Radu A, Moseley AM, Deans R, Marshak DR, Flake AW (2000) Human mesenchymal stem cells engraft and demonstrate site-specific differentiation after in utero transplantation in sheep. Nat Med 6:1282–1286

Limana F, Germani A, Zacheo A, Kajstura J, Di CA, Borsellino G, Leoni O, Palumbo R, Battistini L, Rastaldo R, Muller S, Pompilio G, Anversa P, Bianchi ME, Capogrossi MC (2005) Exogenous high-mobility group box 1 protein induces myocardial regeneration after infarction via enhanced cardiac C-kit+ cell proliferation and differentiation. Circ Res 97:e73–e83

Liu Y, Kitsis RN (1996) Induction of DNA synthesis and apoptosis in cardiac myocytes by E1A oncoprotein. J Cell Biol 133:325–334

Ma J, Ge J, Zhang S, Sun A, Shen J, Chen L, Wang K, Zou Y (2005) Time course of myocardial stromal cell-derived factor 1 expression and beneficial effects of intravenously administered bone marrow stem cells in rats with experimental myocardial infarction. Basic Res Cardiol 100:217–223

Majumdar MK, Thiede MA, Mosca JD, Moorman M, Gerson SL (1998) Phenotypic and functional comparison of cultures of marrow-derived mesenchymal stem cells (MSCs) and stromal cells. J Cell Physiol 176:57–66

Makino S, Fukuda K, Miyoshi S, Konishi F, Kodama H, Pan J, Sano M, Takahashi T, Hori S, Abe H, Hata J, Umezawa A, Ogawa S (1999) Cardiomyocytes can be generated from marrow stromal cells in vitro. J Clin Invest 103:697–705

Mangi AA, Noiseux N, Kong D, He H, Rezvani M, Ingwall JS, Dzau VJ (2003) Mesenchymal stem cells modified with Akt prevent remodeling and restore performance of infarcted hearts. Nat Med 9:1195–1201

McCulloch EA, Till JE (1960) The radiation sensitivity of normal mouse bone marrow cells, determined by quantitative marrow transplantation into irradiated mice. Radiat Res 13:115–125

Minguell JJ, Erices A, Conget P (2001) Mesenchymal stem cells. Exp Biol Med (Maywood) 226:507–520

Miyahara Y, Nagaya N, Kataoka M, Yanagawa B, Tanaka K, Hao H, Ishino K, Ishida H, Shimizu T, Kangawa K, Sano S, Okano T, Kitamura S, Mori H (2006) Monolayered mesenchymal stem cells repair scarred myocardium after myocardial infarction. Nat Med 12:459–465

Mouquet F, Pfister O, Jain M, Oikonomopoulos A, Ngoy S, Summer R, Fine A, Liao R (2005) Restoration of cardiac progenitor cells after myocardial infarction by self-proliferation and selective homing of bone marrow-derived stem cells. Circ Res 97:1090–1092

Muraglia A, Cancedda R, Quarto R (2000) Clonal mesenchymal progenitors from human bone marrow differentiate in vitro according to a hierarchical model. J Cell Sci 113:1161–1166

Nagaya N, Fujii T, Iwase T, Ohgushi H, Itoh T, Uematsu M, Yamagishi M, Mori H, Kangawa K, Kitamura S (2004) Intravenous administration of mesenchymal stem cells improves cardiac function in rats with acute myocardial infarction through angiogenesis and myogenesis. Am J Physiol Heart Circ Physiol 287:H2670–H2676

Oh H, Bradfute SB, Gallardo TD, Nakamura T, Gaussin V, Mishina Y, Pocius J, Michael LH, Behringer RR, Garry DJ, Entman ML, Schneider MD (2003) Cardiac progenitor cells from adult myocardium: homing, differentiation, and fusion after infarction. Proc Natl Acad Sci U S A 100:12313–12318

Peled A, Petit I, Kollet O, Magid M, Ponomaryov T, Byk T, Nagler A, Ben-Hur H, Many A, Shultz L, Lider O, Alon R, Zipori D, Lapidot T (1999) Dependence of human stem cell engraftment and repopulation of NOD/SCID mice on CXCR4. Science 283:845–848

Pittenger MF, Mackay AM, Beck SC, Jaiswal RK, Douglas R, Mosca JD, Moorman MA, Simonetti DW, Craig S, Marshak DR (1999) Multilineage potential of adult human mesenchymal stem cells. Science 284:143–147

Potapova I, Plotnikov A, Lu Z, Danilo P Jr, Valiunas V, Qu J, Doronin S, Zuckerman J, Shlapakova IN, Gao J, Pan Z, Herron AJ, Robinson RB, Brink PR, Rosen MR, Cohen IS (2004) Human mesenchymal stem cells as a gene delivery system to create cardiac pacemakers. Circ Res 94:952–959

Price MJ, Chou CC, Frantzen M, Miyamoto T, Kar S, Lee S, Shah PK, Martin BJ, Lill M, Forrester JS, Chen PS, Makkar RR (2005) Intravenous mesenchymal stem cell therapy early after reperfused acute myocardial infarction improves left ventricular function and alters electrophysiologic properties. Int J Cardiol Oct 21

Prockop DJ, Sekiya I, Colter DC (2001) Isolation and characterization of rapidly self-renewing stem cells from cultures of human marrow stromal cells. Cytotherapy 3:393–396

Quaini F, Urbanek K, Beltrami AP, Finato N, Beltrami CA, Nadal-Ginard B, Kajstura J, Leri A, Anversa P (2002) Chimerism of the transplanted heart. N Engl J Med 346:5–15

Saito T, Kuang JQ, Bittira B, Al-Khaldi A, Chiu RC (2002) Xenotransplant cardiac chimera: immune tolerance of adult stem cells. Ann Thorac Surg 74:19–24

Schmidt A, Ladage D, Steingen C, Brixius K, Schinkothe T, Klinz FJ, Schwinger RH, Mehlhorn U, Bloch W (2006) Mesenchymal stem cells transmigrate over the endothelial barrier. Eur J Cell Biol July 3

Shake JG, Gruber PJ, Baumgartner WA, Senechal G, Meyers J, Redmond JM, Pittenger MF, Martin BJ (2002) Mesenchymal stem cell implantation in a swine myocardial infarct model: engraftment and functional effects. Ann Thorac Surg 73:1919–1925

Silva GV, Litovsky S, Assad JA, Sousa AL, Martin BJ, Vela D, Coulter SC, Lin J, Ober J, Vaughn WK, Branco RV, Oliveira EM, He R, Geng YJ, Willerson JT, Perin EC (2005) Mesenchymal stem cells differentiate into an endothelial phenotype, enhance vascular density, and improve heart function in a canine chronic ischemia model. Circulation 111:150–156

Sissman NJ (1970) Developmental landmarks in cardiac morphogenesis: comparative chronology. Am J Cardiol 25:141–148

Tang YL, Zhao Q, Zhang YC, Cheng L, Liu M, Shi J, Yang YZ, Pan C, Ge J, Phillips MI (2004) Autologous mesenchymal stem cell transplantation induce VEGF and neovascularization in ischemic myocardium. Regul Pept 117:3–10

Tang YL, Tang Y, Zhang YC, Qian K, Shen L, Phillips MI (2005) Improved graft mesenchymal stem cell survival in ischemic heart with a hypoxia-regulated heme oxygenase-1 vector. J Am Coll Cardiol 46:1339–1350

Toma C, Pittenger MF, Cahill KS, Byrne BJ, Kessler PD (2002) Human mesenchymal stem cells differentiate to a cardiomyocyte phenotype in the adult murine heart. Circulation 105:93–98

Tomita S, Mickle DA, Weisel RD, Jia ZQ, Tumiati LC, Allidina Y, Liu P, Li RK (2002) Improved heart function with myogenesis and angiogenesis after autologous porcine bone marrow stromal cell transplantation. J Thorac Cardiovasc Surg 123:1132–1140

Tse WT, Pendleton JD, Beyer WM, Egalka MC, Guinan EC (2003) Suppression of allogeneic T-cell proliferation by human marrow stromal cells: implications in transplantation. Transplantation 75:389–397

Urbanek K, Rota M, Cascapera S, Bearzi C, Nascimbene A, De AA, Hosoda T, Chimenti S, Baker M, Limana F, Nurzynska D, Torella D, Rotatori F, Rastaldo R, Musso E, Quaini F, Leri A, Kajstura J, Anversa P (2005a) Cardiac stem cells possess growth factor-receptor systems that after activation regenerate the infarcted myocardium, improving ventricular function and long-term survival. Circ Res 97:663–673

Urbanek K, Torella D, Sheikh F, De AA, Nurzynska D, Silvestri F, Beltrami CA, Bussani R, Beltrami AP, Quaini F, Bolli R, Leri A, Kajstura J, Anversa P (2005b) Myocardial regeneration by activation of multipotent cardiac stem cells in ischemic heart failure. Proc Natl Acad Sci U S A 102:8692–8697

Valiunas V, Doronin S, Valiuniene L, Potapova I, Zuckerman J, Walcott B, Robinson RB, Rosen MR, Brink PR, Cohen IS (2004) Human mesenchymal stem cells make cardiac connexins and form functional gap junctions. J Physiol 555:617–626

Vogel W, Grunebach F, Messam CA, Kanz L, Brugger W, Buhring HJ (2003) Heterogeneity among human bone marrow-derived mesenchymal stem cells and neural progenitor cells. Haematologica 88:126–133

von Harsdorf, Hauck L, Mehrhof F, Wegenka U, Cardoso MC, Dietz R (1999) E2F-1 overexpression in cardiomyocytes induces downregulation of p21CIP1 and p27KIP1 and release of active cyclin-dependent kinases in the presence of insulin-like growth factor I. Circ Res 85:128–136

Vulliet PR, Greeley M, Halloran SM, MacDonald KA, Kittleson MD (2004) Intra-coronary arterial injection of mesenchymal stromal cells and microinfarction in dogs. Lancet 363:783–784

Wilson EM, Moainie SL, Baskin JM, Lowry AS, Deschamps AM, Mukherjee R, Guy TS, St John-Sutton MG, Gorman JH III, Edmunds LH Jr, Gorman RC, Spinale FG (2003) Region- and type-specific induction of matrix metalloproteinases in post-myocardial infarction remodeling. Circulation 107:2857–2863

Woodbury D, Schwarz EJ, Prockop DJ, Black IB (2000) Adult rat and human bone marrow stromal cells differentiate into neurons. J Neurosci Res 61:364–370

Xu X, Xu Z, Xu Y, Cui G (2005a) Effects of mesenchymal stem cell transplantation on extracellular matrix after myocardial infarction in rats. Coron Artery Dis 16:245–255

Xu X, Xu Z, Xu Y, Cui G (2005b) Selective down-regulation of extracellular matrix gene expression by bone marrow derived stem cell transplantation into infarcted myocardium. Circ J 69:1275–1283

Yang J, Zhou W, Zheng W, Ma Y, Lin L, Tang T, Liu J, Yu J, Zhou X, Hu J (2006) Effects of myocardial transplantation of marrow mesenchymal stem cells transfected with vascular endothelial growth factor for the improvement of heart function and angiogenesis after myocardial infarction. Cardiology 107:17–29

Yoon YS, Wecker A, Heyd L, Park JS, Tkebuchava T, Kusano K, Hanley A, Scadova H, Qin G, Cha DH, Johnson KL, Aikawa R, Asahara T, Losordo DW (2005) Clonally expanded novel multipotent stem cells from human bone marrow regenerate myocardium after myocardial infarction. J Clin Invest 115:326–338

Zhang S, Guo J, Zhang P, Liu Y, Jia Z, Ma K, Li W, Li L, Zhou C (2004) Long-term effects of bone marrow mononuclear cell transplantation on left ventricular function and remodeling in rats. Life Sci 74:2853–2864

Zhang XY, La Russa VF, Bao L, Kolls J, Schwarzenberger P, Reiser J (2002) Lentiviral vectors for sustained transgene expression in human bone marrow-derived stromal cells. Mol Ther 5:555–565

Autotransplantation of Bone Marrow-Derived Stem Cells as a Therapy for Neurodegenerative Diseases

I. Kan · E. Melamed · D. Offen (✉)

Laboratory of Neurosciences, Felsenstein Medical Research Center, Rabin Medical Center, Beilinson Campus Tel Aviv University, Sackler School of Medicine, 49100 Petah-Tikva, Israel
doffen@post.tau.ac.il

1	Introduction	220
2	Selective Pattern of Neurodegeneration	222
3	Bone Marrow Stem Cells: Cell Candidates for Autologous Transplantation	222
3.1	Hematopoietic Stem Cells	222
3.2	Mesenchymal Stem Cells	223
4	Autologous Transplantation: Mechanisms of Action	224
4.1	Neuroregeneration: Cell Replacement	224
4.1.1	The Concept of Cell Replacement	224
4.1.2	Mesenchymal Stem Cell Transplantation	225
4.2	Trophic Factor Delivery	227
4.2.1	The Concept of Trophic Factor Delivery	228
4.2.2	Mesenchymal Stem Cell Transplantation	230
4.3	Immunomodulation	231
4.3.1	The Concept of Immunomodulation	231
4.3.2	Mesenchymal Stem Cell Transplantation	233
5	Conclusions	233
	References	234

Abstract Neurodegenerative diseases are characterized by a progressive degeneration of selective neural populations. This selective hallmark pathology and the lack of effective treatment modalities make these diseases appropriate candidates for cell therapy. Bone marrow-derived mesenchymal stem cells (MSCs) are self-renewing precursors that reside in the bone marrow and may further be exploited for autologous transplantation. Autologous transplantation of MSCs entirely circumvents the problem of immune rejection, does not cause the formation of teratomas, and raises very few ethical or political concerns. More than a few studies showed that transplantation of MSCs resulted in clinical improvement. However, the exact mechanisms responsible for the beneficial outcome have yet to be defined. Possible rationalizations include cell replacement, trophic factors delivery, and immunomodulation. Cell replacement theory is based on the idea that replacement of degenerated neural cells with alternative functioning cells induces long-lasting clinical improvement. It is reasoned that the transplanted cells survive, integrate into the endogenous neural network, and lead to functional improvement. Trophic factor delivery presents a more practical short-term approach. According to this approach, MSC effectiveness may be credited to the production of neurotrophic factors that support neuronal cell survival, induce endogenous cell proliferation, and promote nerve fiber regeneration at sites of injury. The third potential mechanism of action is supported by the recent reports claiming that neuroinflammatory mechanisms play an important role in the pathogenesis of neurodegenerative disorders. Thus, inhibiting chronic inflammatory stress might explain the beneficial effects induced by MSC transplantation. Here, we assemble evidence that supports each theory and review the latest studies that have placed MSC transplantation into the spotlight of biomedical research.

Keywords Neurodegenerative disease · Autologous transplantation · Adult stem cells · Mesenchymal stem cells

1
Introduction

Many of the neurodegenerative disorders are attributed to the degeneration of specific neurons with subsequent functional loss. The susceptibility of specific neuronal populations has yet to be solved. The same uncertainty applies to the underling causes of progressive neurodegeneration. Nevertheless, the sporadic occurrence of these disorders counters the theory of programmed cell degeneration and supports the existence of multifactorial stimulating factors. Accumulating evidence implies that the physiological process of aging has an important role in the occurrence of such diseases (Mattson and Magnus 2006). During the process of aging the occurrence of neurodegenerative disorder is determined by genetic and environmental factors that counteract or facilitate fundamental mechanisms of aging. Therapeutic modalities of neurodegenerative diseases are limited, and although several treatments have been shown to modify the course of the disease, no treatments have proved to halt the degeneration.

Mounting evidence now suggests that cell therapy may be of functional benefit in many neurodegenerative diseases. Each condition has specific requirements for the phenotype, developmental stage, and number of cells required. The ideal cells for universal application in cell therapy would possess several key properties. First, they have to be highly proliferative, allowing optimal exploitation of donor material. Second, the cells must have wide differentiation potential, allowing differentiation into appropriate neural and glial cell types. Importantly, both proliferation and differentiation would be controllable. Stem cells are feasible candidates for cell therapy of neurodegenerative diseases. Different therapeutic strategies based on stem cells have been developed and studied. Several strategies support cell replacement of the damaged tissue while others rely on therapeutic effects induced by cell transplantation.

The most primitive of all stem cell populations, offering the most potential, are embryonic stem cells (ESCs) obtained from the inner cell mass of developing blastocysts. ESCs are pluripotent stem cells that can give rise to cells of the three germ layers found in the implanted embryo, fetus, or developed organism, but not to embryonic components of the trophoblast and placenta. Application of the embryonic pluripotent stem cells to clinical studies have been impeded due to: potential immune rejection in allogeneic transplantation (Vogel 2002), formation of teratomas (Reubinoff et al. 2000), lack of their availability, and serious ethical and political issues (Perin et al. 2003). In view of the latter, autologous cell sources may prove to be more beneficial and acceptable as a therapeutic tool in the future.

Adult stem cells remain in an undifferentiated, or unspecialized, state in different tissues of the adult organism and may be exploited in autologous transplantation. They possess the ability to self-renew, and can differentiate into at least one mature, specialized cell type. In the traditional developmental paradigm, adult stem cells are able to differentiate only to the tissue in which they reside. Recent data challenge the committed fate of the adult stem cells and present evidence for their plasticity. Thus, adult stem cell therapy may offer an accessible, therapeutic tool for damaged tissue replacement and tissue engineering that is free of ethical debate. Stem cell therapy in general and adult stem cell exploitation in particular are predominantly relevant in tissues and organs that have little capacity for self-repair. One such organ is the brain.

Traditionally, the mammalian central nervous system (CNS) was considered to be a nonrenewable tissue, but this principle has been challenged in the past decade. Studies have demonstrated that neural stem cells (NSCs) exist not only in the developing mammalian nervous system but also in the adult nervous system of all mammalian organisms, including humans (Gage 2000; Rakic 2002). NSCs are capable of undergoing expansion and differentiation into neurons, astrocytes, and oligodendrocytes in vitro (Reynolds and Weiss 1992; Gage et al. 1995) and after transplantation in vivo (Svendsen et al. 1997). These studies imply that NSCs may further be utilized for treatment of severe brain disorders. However, because of the inaccessibility of NSC sources deep in the brain, stimulation of endogenous neurogenesis may provide a more applicable treatment modality. During the past few years several groups studied the ongoing process of neurogenesis in the intact and diseased adult nervous system. Zhao et al. (2003) have provided evidence for the continuous turnover of dopaminergic cells in the adult substantia nigra pars compacta (SNpc). Moreover, several studies showed a selective increase in the production of dopaminergic neurons in the adult olfactory bulb in response to dopaminergic deficiency (Yamada et al. 2004; Winner et al. 2006). A recent report of Tande et al. (2006), however, was unable to obtain evidence for the existence of neurogenesis in the striatum of normal or the 1-methyl-4-phenyl-1,2,3,6-tetrahydropyridine (MPTP)-induced monkey model of Parkinson's disease (PD). The process of augmented neurogenesis has also been observed in animal models of Alzheimer's disease (AD) (Jin et al. 2004a) and in patients with AD (Jin et al. 2004b). Chi et al. (2006) showed that motor neuron degeneration promotes neural progenitor cell proliferation, migration, and neuronal differentiation in the spinal cords of amyotrophic lateral sclerosis (ALS) mice. However, the amount of neuronal cell replacement by endogenous stem cells without additional manipulation is minimal and does not allow significant functional recovery of the damaged tissue. Nevertheless, stimulation with molecules that govern proliferation and differentiation might dramatically change the therapeutic spectrum of neurodegenerative diseases (Lie et al. 2004).

2
Selective Pattern of Neurodegeneration

Selective vulnerability is most readily appreciated in the context of neurodegenerative disorders such as PD, AD, and ALS. Each disease is characterized by its own, unique pattern of degeneration.

In PD, whether sporadic or inherited, dopaminergic neurons of the SNpc (A9) progressively degenerate. Interestingly, other dopaminergic populations are relatively spared, including the adjacent ventral tegmental area (VTA or A10) neurons (Chung et al. 2005).

In AD the earliest and the most consistent degeneration occurs in the forebrain cholinergic projection system particularly in a structure called nucleus basalis of Meynert (Whitehouse et al. 1981, 1982). West et al. (1994) showed that although neuronal and synaptic loss occurs diffusely across the brain, neurons in layer II of the entorhinal cortex and hippocampal CA1 neurons are particularly vulnerable.

ALS (or Lou Gehrig's disease) involves the loss of upper and lower motor neurons. However, in ALS motor neurons do not display universal vulnerability. Some motor nuclei (III, IV, VI, and Onuf's nucleus) remain relatively intact during terminal stages of the disease, while others (V, VII, XII, and most of the spinal nuclei) usually degenerate (Laslo et al. 2000).

Understanding the basis of the selective vulnerability that characterizes many neurodegenerative diseases might generate a more refined therapeutic approaches in which transplanted cells more closely reflect a lost neuronal subtype.

3
Bone Marrow Stem Cells: Cell Candidates for Autologous Transplantation

Autologous cell transplantation is a promising strategy for treatment of several CNS pathologies and offers the prospect of permanent cure. Deriving cells from an adult patient's own tissues entirely circumvents the problem of immune rejection. In addition, adult stem cells do not form teratomas, and their application raises very few ethical or political issues associated with the use of human embryos.

Postnatal bone marrow (BM) is a readily accessible source of adult stem cells for autologous transplantation and contains two major stem cell populations—hematopoietic stem cells (HSCs) and mesenchymal stem cells (MSCs).

3.1
Hematopoietic Stem Cells

Human HSCs—characterized as being $CD34^+$, $c\text{-}Kit^+$, $Thy\text{-}1^{low}$, $CD10^-$, $CD14^-$, $CD15^-$, $CD16^-$, $CD19^-$, and $CD20^-$ (Shizuru et al. 2005)—differentiate into progenitor cells and mature blood cells of all hematopoietic lineages. Krause et al. (2001) demonstrated that a single HSCs was not only able to repopulate the hematopoietic system in irradiated mice, but differentiated into lung epithelium, skin, liver, and the gastrointestinal tract. In addition, several studies have reported that HSCs may give rise to neurons in vitro and in vivo (Koshizuka et al. 2004; Sigurjonsson et al. 2005; Reali et al. 2006). These studies suggest that transdifferentiation of HSCs into neurons is a rare event in the intact adult brain, but can be induced by a particular microenvironment present in the injured tissues. Other experiments dispute these findings and claim that HSCs and their progeny maintain lineage fidelity in the brain and do not adopt neural cell fates with any measurable frequency (Massengale et al. 2005).

3.2
Mesenchymal Stem Cells

Over 30 years ago Friedenstein et al. (1966, 1974, 1976) defined another BM cell population, fibroblast-colony-forming cells, which adhere to cell culture plastic surfaces and can differentiate into osteoblasts, adipocytes, and chondrocytes—mesenchymal cell types, ex vivo and in vivo. In later experiments, fibroblast-colony forming-cells were termed MSCs or BM stromal cells (BMSC) (Prockop 1997; Pittenger et al. 1999). Subsequently, most of the research concentrated on the BMSC-created microenvironment, which is essential for lineage commitment and differentiation of HSCs and has regulatory roles in hematopoiesis (Cherry et al. 1994; Moreau et al. 1993). MSCs represent a very small fraction of BM cells, only 0.001%–0.01% of the total population of nucleated cells in the BM. Although millions of cells could be harvested after several passages of the cells in culture (Pittenger et al. 1999), prolonged culture might change MSC characteristics, thus reducing their therapeutic potential. Previous studies revealed that during culture expansion MSCs undergo an aging process in which their early progenitor properties, proliferation, and homing capability are gradually lost (Banfi et al. 2000; DiGirolamo et al. 1999; Rombouts and Ploemacher 2003). This negative effect of in vitro expansion has been correlated with the rate of their telomere loss. Current culture protocols that involve population expansion stimulate rapid aging of MSCs, with large telomere shortening and little remaining proliferative capacity. Surprisingly, some MSC cultures were able to maintain telomere length for over 40 population doublings after an initial telomere shortening. Telomere length preservation may point toward a selection of more primitive MSCs, a small population that may not always present in all samples due to the small volume of the BM primary sample (Baxter et al. 2004). Several studies report the existence of rare BM-derived multipotent adult progenitor cells (MAPCs) that can be cultured for more than 70 doublings and have long telomeres that do not shorten during culture (Jiang et al. 2002; Reyes and Verfaillie 2001). MAPCs, selected from the BM mononuclear cells by depletion of CD45 and glycophorin-A (Gly-A) cells, can be identified in the adherent cultured MSCs only after the cells undergo approximately 30 or more population doublings. The MAPC population is characterized by rapid replication and can differentiate into multiple mesenchymal cell types as well as hematopoietic lines, representing a primitive progenitor cell. To date, no research has determined whether MAPCs constitute a small rare subpopulation of MSCs or a new cell population developed under unique in vitro cell culture conditions. Several researchers have tried to characterize MSCs. Minguell et al. (2001) reviewed these studies and concluded that the antigenic phenotype of MSCs is not unique, but borrows features of mesenchymal, endothelial, epithelial, and muscle cells, and does not include the typical hematopoietic antigens CD45, CD34, and CD14. In addition, the constitutive or stimulated production of growth factors, interleukins, chemokines, matrix molecules, and the expression of their receptors provide evidence that MSCs contribute to the formation and function of a stromal microenvironment, responsible for the inductive/regulatory signals for MSCs, hematopoietic progenitor cells, and other nonmesenchymal stromal cells present in the BM.

In this chapter we will assess the reported benefits and mechanisms of action of MSC transplantation for treatment of neurodegenerative diseases concentrating on AD, PD and ALS.

4
Autologous Transplantation: Mechanisms of Action

4.1
Neuroregeneration: Cell Replacement

Cell replacement is a single application of cell therapy for the treatment of neurodegenerative diseases. It aims to replace the degenerated neural cells with alternative functioning cells. It is reasoned that these cells will survive, integrate into the endogenous neural network, and lead to significant clinical improvement.

4.1.1
The Concept of Cell Replacement

The clinical trials with cell therapy in PD patients are based on the idea that restoration of striatal dopaminergic transmission by grafted dopaminergic neurons will induce long-lasting clinical improvement in PD symptoms. In support, open-label clinical transplantation trials of embryonic dopaminergic neurons show that intrastriatal transplantation can give substantial symptomatic relief in advanced PD patients and provide the proof-of-principle for the cell replacement strategy in PD (Freed et al. 1992; Lindvall et al. 1990).

In these studies clinical improvement correlated with graft survival and host reinnervation. Two recent NIH-sponsored placebo-controlled trials, however, have given disappointing results, which included the development of graft-induced dyskinesias (Freed et al. 2001; Olanow et al. 2003). These studies stressed that issues such as graft standardization, slowly developing inflammatory responses, and dyskinesias need to be addressed before the application of fetal dopaminergic tissues for treatment of PD. In addition, ethical issues associated with the use of tissue from aborted fetuses make it necessary to develop alternative sources of cells for transplantation.

In AD, long projections and the expression of nerve growth factor receptors led some researchers to declare basal forebrain cholinergic neurons irreplaceable (Sugaya 2003). Although there are no reports at the moment of clinical use of embryonic transplants in AD patients, several works have demonstrated that cholinergic-rich cells of fetal origin can improve the performance in animal models of cholinergic depletion (Gage and Bjorklund 1986; Hodges et al. 1991a, b; Muir et al. 1992; Grigoryan et al. 2000).

In ALS, if stem cells could be used to generate large projection neurons that would appropriately connect to their targets upon transplantation, there is a prospect for replacement and a possible cure. Wichterle et al. (2002) reported that motor neurons derived from mouse embryonic stem cells could populate the embryonic spinal cord, extend axons, and form synapses with target muscles. A later study by Deshpande et al. (2006) administered ESC-derived motor neurons along with phosphodiesterase type 4 inhibitor and dibutyryl cyclic AMP (dbcAMP) to overcome myelin-mediated repulsion into paralyzed adult rats. In addition, the researchers applied a focal attractant, glial cell-derived neurotrophic factor (GDNF), within the peripheral nervous system (PNS) to direct the transplanted embryonic stem cell-derived axons toward skeletal muscle targets. This well-designed study showed that transplant-derived axons reached muscle, formed neuromuscular junctions, were physiologically active and mediated partial recovery from paralysis. In animal models of ALS, several studies showed that transplantation of cells derived from the human teratocarcinoma cell line or human umbilical cord blood resulted in beneficial effects. However, neuroprotection was suggested to be the main cause of the observed benefits rather than the direct motor neuron replacement (Ende et al. 2000; Garbuzova-Davis et al. 2002, 2003).

4.1.2
Mesenchymal Stem Cell Transplantation

Several recent studies report the generation of dopaminergic neurons from BM-derived MSCs. Woodbury et al. (2002) reported that rat MSCs might be induced to differentiate into neuron-like cells expressing neurotransmission-related genes. Moreover, a small portion of the differentiated cells expressed tyrosine hydroxylase (TH), a rate-limiting enzyme in the synthesis of dopamine.

Hermann et al. (2004) showed that human MSCs could be converted into a clonogenic neural stem cell-like population growing in neurosphere-like structures. In addition, following neuronal differentiation, these cells demonstrated dopamine production and potassium-dependent release. Our laboratory demonstrated that following induced neuronal differentiation, human MSCs displayed neuron-like morphology, neuronal markers, and transcription factors that characterize midbrain dopaminergic neurons and secreted dopamine in response to depolarization (Blondheim et al. 2006; I. Kan, Y. Barhum, T. Ben-Zur, T. Charlow, Y. Levy, A. Burstein, B. Bulvik, E. Melamed, and D. Offen, submitted).

Schwarz et al. (1999, 2001) utilized rat MSCs, genetically engineered by transgene to express human TH and guanosine triphosphate cyclohydrolase I (GTPCH), an enzyme required for the biosynthesis of tetrahydropterin cofactor for TH (BH4). These cells released L-3,4-dihydroxyphenylalanine (L-DOPA), a precursor of dopamine, and induced behavioral recovery following the transplantation into the rat model of PD. However, the ameliorative effect of the transplanted rat MSCs was short-lived (up to 9 days), presumably due to inactivation of the transgenes. Recently, Dezawa et al. (2004) demonstrated a highly efficient and specific induction of rat and human MSCs using gene transfer of Notch intracellular domain (NICD) and subsequent treatment with basic fibroblast growth factor (bFGF), forskolin, ciliary neurotrophic factor (CNF), brain-derived growth factor (BDNF) and nerve growth factor (NGF). Following the transfection, MSCs expressed neural stem cell markers such as microtubule-associated protein 2 (MAP-2), neurofilament M (NF-M) and β-tubulin III. Following subsequent trophic factor administration, action potentials, compatible with characteristics of functional neurons, were recorded. Further treatment of the induced neuronal cells with GDNF increased the proportion of TH-positive and dopamine-producing cells. Transplantation of these GDNF-treated cells showed functional improvement when grafted into a 6-hydroxydopamine (6-OHDA) rat model of PD. Li et al. (2001) showed in vivo differentiation of mouse MSCs, prelabeled with bromo-deoxyuridine (BrdU) and grafted into the striatum of the MPTP mouse model of PD. The transplanted mice exhibited significant motor improvement 35 days after transplantation. At least 4 weeks after the transplantation BrdU-reactive cells were revealed in the grafted mice striatum and approximately 0.8% of these cells expressed TH. These studies show that engraftment of MSCs, naive or following in vitro induction, may result in a significant clinical improvement in animal models of PD. However, whether or not cell replacement underlies the recovery mechanism has yet to be determined.

The evidence for cholinergic-induced differentiation of MSCs is limited. As mentioned, Woodbury et al. (2002) showed that MSCs were induced into neuron-like cells and expressed neurotransmitter-related genes. Precise analysis showed that a large population of MSC-derived neurons expressed choline acetyltransferase (ChAT), which catalyzes the synthesis of the excitatory transmitter acetylcholine and a smaller subpopulation expressed TH. Chen et al. (2006) demonstrated that the α-secretase-cleaved fragment of the amyloid precursor protein (sAPPα), a potent neurotrophic factor, potentiates the NGF/retinoic acid (RA)-induced transdifferentiation of MAPCs into neural progenitor cells and, more specifically, enhances their terminal differentiation into a cholinergic-like neuronal phenotype. In addition, after the intravenous transplantation of sAPPα-transfected MAPCs into transgenic PS/APP mice, sAPPα-immunopositive MAPCs were identified within the septohippocampal system and found in close proximity to the cerebral vasculature.

In ALS, Mazzini et al. (2003, 2004) have evaluated the feasibility and safety of intraspinal cord implantation of autologous MSCs in 7 patients affected by ALS. Following in vitro expansion, the cells were suspended in the autologous cerebrospinal fluid and directly injected into the surgically exposed spinal cord at the T7–T9 levels. No patients manifested severe adverse effects and the researchers noticed a significant slowing down in the linear decline of the forced vital capacity of 4 out of 7 patients 24 months after transplantation of MSCs. This effect was correlated with the number of implanted cells. To date, the specific replacement of motor neurons poses a formidable challenge that might require multisegmental delivery to sites of need, reestablishment of appropriate afferent innervations, and the long-distance extension of their axons through often degenerating nerve roots to specific loci in the distant musculature. Our poor understanding of the biology underlying these processes suggests that despite the recent progress described in generating and isolating motor neurons, cell replacement treatment of the motor neuronopathies remains a difficult goal. Clement et al. (2003) showed that in SOD1G93A chimeric mice, motor neuron degeneration requires damage from mutant SOD1 acting in nonneuronal cells. Wild-type nonneuronal (glial) cells could delay degeneration and extend survival of mutant-expressing motor neurons. This theory implies that replacing the degenerated neurons might be insufficient and the administration of supportive nonneuronal cells might be required. These nonneuronal cells might protect the grafted cells from the progressive damage and might even prolong the survival of endogenous degenerating neural cells.

4.2
Trophic Factor Delivery

Using stem cells to generate new neurons and replace those lost in neurodegenerative diseases would be a major breakthrough. A more practical short-term approach may be to use stem cells to protect neurons dying in these diseases.

4.2.1
The Concept of Trophic Factor Delivery

Trophic factors are believed to have the capability of providing neuroprotective or restorative effects in PD. Collier and Sortwell (1999) summarized the effects of 29 molecules that promote survival and growth of dopaminergic neurons. These molecules are either naturally occurring factors or synthetic molecules that access endogenous receptors. The authors concluded that the single greatest challenge to neurotrophic therapies is the development of techniques for targeted delivery of these potential materials. Among these factors GDNF, a powerful neuroprotective agent, became the subject of extensive studies. Its neuroprotective properties coupled to its reduced levels in the basal ganglia in PD (Chauhan et al. 2001) made it the first trophic factor to be tested in PD patients. The research of Nutt et al. (2003) revealed that the drug was not efficacious when given via an intracerebroventricular catheter. Researchers speculated that GDNF did not reach the target tissues—namely putamen and substantia nigra—and therefore did not improve Parkinsonism. In later trials GDNF was administered directly into the site of greatest dopamine loss in PD—the posterior putamen. In a small open-label study GDNF delivery showed clear positive clinical effects, with evidence of dopaminergic-fiber sprouting at the site of GDNF delivery, using both fluorine-18-labeled-dopa positron emission tomography (PET) scanning and postmortem analysis of a single case (Love et al. 2005; Gill et al. 2003). By contrast, Lang and colleagues' (2006) double-blind, placebo-controlled trial showed no beneficial clinical effect. In this study, 34 patients with moderately advanced PD were randomly assigned intraputaminal placebo or GDNF, in equal numbers, with the primary endpoint being the unified PD rating scale (UPDRS) "off" motor score at 6 months. No significant difference was seen in this or any other measures, apart from ^{18}F-dopa scanning of the posterior putamen, but not of the whole striatum. Additionally, device-related adverse events (infection and catheter misplacement) and anti-GDNF antibodies were reported in 3 patients. However, the change in the outcome of open-label and double-blind, placebo-controlled studies should not discourage further research but set a challenge to researchers to improve and refine the GDNF delivery mode. Other neurotrophic factors that have been extensively studied in PD are neurturin (NTN), BDNF, and sonic hedgehog (SHH). Several studies showed the protective effect of NTN, a trophic factor of the GDNF family of ligands, on dopaminergic neurons in substantia nigra (Fjord-Larsen et al. 2005; Li et al. 2003; Oiwa et al. 2002). These neuroprotective effects have led to the initiation of a phase I clinical study of NTN delivery via an adeno-associated viral vector delivery system. SHH has also displayed neuroprotective effects protecting nigral dopaminergic neurons from 6-OHDA-induced toxicity following adenoviral delivery (Dass et al. 2005; Hurtado-Lorenzo et al. 2004). BDNF, however, failed to protect dopaminergic neurons from 6-OHDA-induced death but did block the amphetamine-induced, ipsiversive, turning-behavior caused by the partial unilateral lesion of the nigrostriatal pathway (Klein et al. 1999). Researchers suggested that exogenous BDNF exerts a modulatory influence on the remaining dopaminergic or other types of neurons, perhaps leading to an enhanced release of dopamine in the striatum.

In AD, the influence of NGF, which specifically targets basal forebrain cholinergic neurons, nociceptive dorsal root ganglion neurons, and some third-order sympathetic neurons (Levi-Montalcini 1987) has been the subject of extensive research. Various studies showed that NGF increases the synthesis of ChAT and prevents basal forebrain cholinergic neurons atrophy caused by experimental injury or associated with physiological aging, suggesting that NGF might reduce cholinergic cell loss in AD (Hefti et al. 1984; Fischer et al. 1987; Koliatsos et al. 1993; Tuszynski et al. 1991; Markowska et al. 1994, 1996). However, development of an effective NGF delivery mode to AD patients encounters several challenges. First, NGF does not cross the blood–brain barrier when administered peripherally. Second, adverse side effects were revealed following direct infusion of NGF into the brain ventricular system (Tuszynski 2002). Consequently, several alternative methods have been proposed for NGF delivery. Tuszynski et al. (2005) showed that autologous fibroblasts, obtained from small skin biopsies, genetically modified to produce and secrete human NGF, ameliorated cognitive deficits of AD patients after stereotaxic injections. Capsoni et al. (2002) proposed another way of NGF administration—intranasal delivery. By using the intranasal route of administration, this group showed that NGF could rescue all the histological hallmarks characterizing the AD-like neurodegeneration in AD11 mice.

In ALS, several growth factors have been proposed and tested. In a transgenic mouse model of ALS, intrathecal infusion of insulin-like growth factor (IGF)-1 improved motor performance, delayed the onset of clinical disease, and extended survival (Nagano et al. 2005b). In order to evaluate the potential benefits of intrathecal IGF-1 administration in ALS patients a double-blind clinical trial was performed. However, only a modest beneficial effect was recorded (Nagano et al. 2005a). Beck et al. (2005) explored the effects of intrathecal administered BDNF on autonomic functions in patients with ALS. Unfortunately, the group concluded that autonomic nervous system function deteriorated along with motor performance independently from treatment with BDNF. In another study, human neural progenitor cells (hNPC) were isolated from the cortex, expanded in culture, and modified using lentivirus to secrete GDNF. These cells were transplanted into the lumbar spinal cord of a rat model of ALS. The cells survived up to 11 weeks following transplantation, integrated into both gray and white matter, and secreted GDNF within the region of cell survival, but not outside this area. However, no positive clinical effects were recorded (Klein et al. 2005). Additional clinical trials in ALS are being carried by a Chinese neurosurgeon Huang Hongyun. The procedure includes isolation of olfactory ensheathing cells (OECs) from the glomerular layer of olfactory bulbs inside the noses of aborted fetuses, and in vitro expansion and injection into the atrophied area of the frontal lobes. The group claimed that this procedure could stabilize ALS in about half the patients treated. The beneficial effects were assigned to the secretion of growth factors, which encourage the regeneration of nerve axons, rather than to the cell replacement (Watts 2005). To date no double-blind, controlled study has been performed that can verify these results. As a final point, the safety issues of these procedures need to be addressed.

4.2.2
Mesenchymal Stem Cell Transplantation

Several recent studies suggest that MSC effectiveness may be credited to production of neurotrophic factors that support neuronal cell survival, induce endogenous cell proliferation, and promote nerve fiber regeneration at sites of injury (Li et al. 2002; Mahmood et al. 2004). Indeed, Crigler et al. (2006) demonstrated that MSCs express a variety of neuro-regulatory proteins in addition to BDNF and β-NGF, which promote survival and induce neurite formation in neuroblastoma cells and primary nerves from the lumbar spine. In addition, Arnhold et al. (2006) showed that naive MSCs express BDNF, NGF, and GDNF when cultivated in standard medium and elevate the expression levels after exposure to neural precursor selection medium or to differentiation medium. Synthesis and release of growth factors by the grafted cells or indirect stimulation of neurotrophic release from the host tissue may be in part accountable for the functional recovery induced by MSC transplantation. Moreover, MSCs may be well suited as vehicles for treating neurological deficits due to their propensity to home to sites of tissue injury (Ji et al. 2004), which may target neuroprotective agents to brain or spinal cord lesions. Part of the recovery mechanism may include elevated neurogenesis. Chopp and Li (2002) showed that the presence of MSCs promoted induction and migration of new cells from a primary source within the ventricular zone and the choroid plexus into the injured brain. The authors claim that these cells may contribute to functional repair after stroke, although the relation of the induction of neurogenesis and the migration of these cells to the restoration of function has not been directly tested. A recent study of Rivera et al. (2006) revealed that the interactions between adult MSCs and NSCs in vitro, mediated by soluble factors, induce oligodendrogenic fate decision in NSCs at the expense of astrogenesis. Thus, trophic factor delivery and/or elevated neurogenesis might underlie the beneficial effects of MSC transplantation in neurodegenerative diseases.

4.3
Immunomodulation

4.3.1
The Concept of Immunomodulation

Recent evidence clearly indicates that neuroinflammatory mechanisms may play an important role in the pathogenesis of PD (Hunot and Hirsch 2003; McGeer and McGeer 2004; Marchetti et al. 2005). Under normal conditions, astrocytes exert a fundamental protective function against oxidative stress, and the interaction between astrocytes and neurons has been variously demonstrated to exert striking neurotrophic, differentiation, and neuroprotective effects (Takeshima et al. 1994; McNaught and Jenner 1999; Marchetti et al. 2005). However, under conditions of chronic inflammatory stress, activated astrocytes and microglia may become dysfunctional and overexpress a variety of cytotoxic mediators eventually resulting in dopaminergic neuron death (Morale et al. 2006). Indeed, studies accumulated over the last two decades have clearly indicated the activation of microglia and astroglia in the nigrostriatal system of PD patients and animal models of PD (Hunot and Hirsch 2003; McGeer and McGeer 2004). Activated microglia release proinflammatory and neurotoxic factors that are thought to contribute to neuronal damage. These factors include proinflammatory cytokines such as tumor necrosis factor α (TNF-α) and interleukin 1 (IL-1), reactive nitrogen species, proteases, reactive oxygen species (ROS), eicosanoids, and excitatory amino acids (Liu and Hong 2003; Merrill and Benveniste 1996). The inflammatory process, induced by proinflammatory agents, can result in the degeneration of dopamine-containing neurons and may be an important contributor to the neuronal loss in PD. Moreover, the experimental evidence that inhibition of the inflammatory process correlates with less neuronal impairment supports the notion that inflammation plays a deleterious role in the neurodegeneration in PD. Thus, inflammatory inhibition might become a promising therapeutic intervention for PD (Gao et al. 2003).

Activated astrocytes and microglia (gliosis) have been documented in studies of AD (Giulian et al. 1995; Sasaki et al. 1997). As mentioned already, when activated, astrocytes and microglia produce several proinflammatory signal molecules, including cytokines, growth factors, complement molecules, and adhesion molecules. Of particular interest in AD are the cytokines S100β, which is mainly produced by astrocytes, and IL-10, which is mainly produced by activated microglia. A review article by Griffin (2006) showed that inflammation is a driving force in the neuropathology of AD, mediated by proinflammatory cytokines and creating a chronic and self-sustaining inflammatory interaction between activated microglia and astrocytes, stressed neurons and β-amyloid plaques. The author brings evidence that the key initiating factor appears to be overexpression of IL-1, which may be a result of any number of events, including disease, trauma, genetic polymorphisms, or simply age-related wearing away. Through various pathways, IL-1 overexpression causes neuronal death, which activates more microglia, which in turn release more IL-1 in a self-amplifying fashion. Over the years, this inflammation destroys sufficient neurons to cause the clinical signs of AD. Interestingly, nearly all the cytokines and chemokines that have been studied in AD, including IL-1β, IL-6, TNF-α, IL-8, transforming growth factor β (TGF-β), and macrophage inflammatory protein-1α (MIP-1α) seem to be upregulated in AD compared with control individuals (Akiyama et al. 2000). Epidemiological studies have documented a beneficial effect of nonsteroidal antiinflammatory drugs (NSAIDs) in AD (McGeer et al. 1996; Stewart et al. 1997; Akiyama et al. 2000; In 't Veld et al. 2001; Szekely et al. 2004). Long-term NSAID therapy delayed the onset and the progression of the disease, reduced symptomatic severity, and significantly slowed the rate of cognitive impairment (Rich et al. 1995). The putative target of NSAID actions is thought to be microglia associated with the senile plaques (Sastre et al. 2006).

The evidence for involvement of microglial cells in ALS pathology is abundant and very strong. In a review of the inflammatory processes in ALS, McGeer and McGeer (2002) presented extensive evidence that activated microglia, astrocytes, and infiltrating leukocytes are present in affected tissues in ALS. These investigators noted that overreactive innate immune defenses, where microglia play a major part, could generate a neuroinflammatory environment that is harmful to the viable host tissue. This process, known as autotoxicity, could evolve from a persistent stimulation of microglia. Accumulation of microglia-activating signals next to the injured neurons in ALS could lead to the initiation of autotoxicity. The Sargsyan et al. (2005) review brings evidence that microglia are the likely glial cell type responsible for the secretion of microglia-activating signals and for propagation of glial activation and inflammation in ALS. Activated microglia and factors secreted by these cells appear to play a direct role in ALS pathology, although it is still unclear whether they contribute as initiators, propagators, or both initiators and propagators of motor neuron injury. Further evidence of microglial involvement in ALS pathology comes from studies with pharmacological suppression of microglial activation using antibiotics, such as minocycline, which has been reported to reduce or suppress microglial activation after neuronal injury (Kriz et al. 2002; Zhu et al. 2002; van den Bosch et al. 2002; Tikka et al. 2002; Zhang et al. 2003). These results led to phase I/II studies of minocycline in ALS patients (Gordon et al. 2004). Although no difference was observed between treated and untreated groups in these studies, pivotal phase III trials are ongoing. Cyclooxygenase-2 (COX-2) is a key molecule in the inflammatory pathway in ALS (Kiaei et al. 2005). Celecoxib (Celebrex) and rofecoxib are inhibitors of COX-2. Treatment with these COX-2 inhibitors combined with creatine increased survival by up to 30% in SOD1 mutant mice (Drachman et al. 2002; Pompl et al. 2003; Klivenyi et al. 2004). However, to date none of the COX inhibitors tested has shown efficacy in human ALS patients (Moisse and Strong 2006).

It seems that the brain's immune response during the neurodegenerative disease is a double-edged sword, simultaneously beneficial and detrimental. Inflammation seems to start as a time- and site-specific defense mechanism aimed at eliminating irreversibly damaged neurons and at favoring survival of cells that retain a chance for recovery. However, at later stages, inflammation can evolve as an uncontrolled chronic reaction. Therefore, inhibiting chronic inflammatory stress and persistent microglial stimulation would be an effective therapeutic approach to slow the progression of neurodegenerative diseases.

4.3.2
Mesenchymal Stem Cell Transplantation

Several studies have reported the in vitro and in vivo immunosuppressive properties of BM-derived MSCs (Bartholomew et al. 2002; Djouad et al. 2003; Le Blanc 2003; Zappia et al. 2005; Corcione et al. 2006). It is reasoned that the transplanted MSCs face a foreign, inflammatory environment and may induce immunomodulating processes that limit local inflammation in order to enhance their survival. The underling mechanism of action may include the secretion of soluble factors that create an immunosuppressive milieu (Ryan et al. 2005). An additional mechanism may be a reduction in infiltration of blood-borne inflammatory cells. This mechanism of action has been associated with the beneficial effect of MSC transplantation in the animal model of multiple sclerosis (MS) (Zhang et al. 2005). The proposed processes may profoundly change the inflammatory environment and promote partial recovery. However, additional studies are needed to determine the precise influence of MSC transplantation on innate chronic inflammation.

5
Conclusions

Several studies have demonstrated the neuroprotective and neuroregenerative effects that were associated with functional improvements following MSC transplantation. Functional recovery may be due to the replacement of degenerated neurons, neurotrophic factors delivery, and/or immunomodulation of the ongoing inflammatory reaction. These beneficial effects and the lack of ethical issues or immune rejections involved in autologous transplantation of MSCs have drawn the attention of biomedical research to this therapeutic approach. There are, however, many unresolved questions and problems regarding the safety and the feasibility of this approach. These issues are not obstacles but challenges to overcome in order to develop a permanent cure for neurodegenerative diseases.

Acknowledgements This work was performed in partial fulfillment of the requirements for a PhD degree of Inna Kan, Sackler School of Medicine, Tel Aviv University, Israel. This work was supported, in part, by the Israel Ministry of Health, the National Parkinson's Foundation, Miami, FL, USA, and the Norma and Alan Aufzein Chair for Research in Parkinson's Disease, Tel Aviv University, Israel.

References

Akiyama H, Barger S, Barnum S, Bradt B, Bauer J, Cole GM, Cooper NR, Eikelenboom P, Emmerling M, Fiebich BL (2000) Inflammation and Alzheimer's disease. Neurobiol Aging 21:383–421

Arnhold S, Klein H, Klinz FJ, Absenger Y, Schmidt A, Schinkothe T, Brixius K, Kozlowski J, Desai B, Bloch W, Addicks K (2006) Human bone marrow stroma cells display certain neural characteristics and integrate in the subventricular compartment after injection into the liquor system. Eur J Cell Biol 85:551–565

Banfi A, Muraglia A, Dozin B, Mastrogiacomo M, Cancedda R, Quarto R (2000) Proliferation kinetics and differentiation potential of ex vivo expanded human bone marrow stromal cells: implications for their use in cell therapy. Exp Hematol 28:707–715

Bartholomew A, Sturgeon C, Siatskas M, Ferrer K, McIntosh K, Patil S, Hardy W, Devine S, Ucker D, Deans R (2002) Mesenchymal stem cells suppress lymphocyte proliferation in vitro and prolong skin graft survival in vivo. Exp Hematol 30:42–48

Baxter MA, Wynn RF, Jowitt SN, Wraith JE, Fairbairn LJ, Bellantuono I (2004) Study of telomere length reveals rapid aging of human marrow stromal cells following in vitro expansion. Stem Cells 22:675–682

Beck M, Flachenecker P, Magnus T, Giess R, Reiners K, Toyka KV, Naumann M (2005) Autonomic dysfunction in ALS: a preliminary study on the effects of intrathecal BDNF. Amyotroph Lateral Scler Other Motor Neuron Disord 6:100–103

Blondheim NR, Levy YS, Ben-Zur T, Burshtein A, Cherlow T, Kan I, Barzilai R, Bahat-Stromza M, Barhum Y, Bulvik S, Melamed E, Offen D (2006) Human mesenchymal stem cells express neural genes, suggesting a neural predisposition. Stem Cells Dev 15:141–164

Capsoni S, Giannotta S, Cattaneo A (2002) Nerve growth factor and galantamine ameliorate early signs of neurodegeneration in anti-nerve growth factor mice. Proc Natl Acad Sci U S A 99:12432–12437

Chauhan NB, Siegel GJ, Lee JM (2001) Depletion of glial cell line-derived neurotrophic factor in substantia nigra neurons of Parkinson's disease brain. J Chem Neuroanat 21:277–288

Chen CW, Boiteau RM, Lai WF, Barger SW, Cataldo AM (2006) sAPPalpha enhances the transdifferentiation of adult bone marrow progenitor cells to neuronal phenotypes. Curr Alzheimer Res 3:63–70

Cherry B, Yasumizu R, Toki J, Asou H, Nishino T, Komatsu Y, Ikehara S (1994) Production of hematopoietic stem cell-chemotactic factor by bone marrow stromal cells. Blood 83:964–971

Chi L, Ke Y, Luo C, Li B, Gozal D, Kalyanaraman B, Liu R (2006) Motor neuron degeneration promotes neural progenitor cell proliferation, migration, and neurogenesis in the spinal cords of amyotrophic lateral sclerosis mice. Stem Cells 24:34–43

Chopp M, Li Y (2002) Treatment of neural injury with marrow stromal cells. Lancet Neurol 1:92–100

Chung CY, Seo H, Sonntag KC, Brooks A, Lin L, Isacson O (2005) Cell type-specific gene expression of midbrain dopaminergic neurons reveals molecules involved in their vulnerability and protection. Hum Mol Genet 14:1709–1725

Clement AM, Nguyen MD, Roberts EA, Garcia ML, Boillee S, Rule M, McMahon AP, Doucette W, Siwek D, Ferrante RJ, Brown RH Jr, Julien JP, Goldstein LSB, Cleveland DW (2003) Wild-type nonneuronal cells extend survival of SOD1 mutant motor neurons in ALS mice. Science 302:113–117

Collier TJ, Sortwell CE (1999) Therapeutic potential of nerve growth factors in Parkinson's disease. Drugs Aging 14:261–287

Corcione A, Benvenuto F, Ferretti E, Giunti D, Cappiello V, Cazzanti F, Risso M, Gualandi F, Mancardi GL, Pistoia V, Uccelli A (2006) Human mesenchymal stem cells modulate B-cell functions. Blood 107:367–372

Crigler L, Robey RC, Asawachaicharn A, Gaupp D, Phinney DG (2006) Human mesenchymal stem cell subpopulations express a variety of neuro-regulatory molecules and promote neuronal cell survival and neuritogenesis. Exp Neurol 198:54–64

Dass B, Iravani MM, Huang C, Barsoum J, Engber TM, Galdes A, Jenner P (2005) Sonic hedgehog delivered by an adeno-associated virus protects dopaminergic neurones against 6-OHDA toxicity in the rat. J Neural Transm 112:763–778

Deshpande DM, Kim YS, Martinez T, Carmen J, Dike S, Shats I, Rubin LL, Drummond J, Krishnan C, Hoke A, Maragakis N, Shefner J, Rothstein JD, Kerr DA (2006) Recovery from paralysis in adult rats using embryonic stem cells. Ann Neurol 60:32–44

Dezawa M, Kanno H, Hoshino M, Cho H, Matsumoto N, Itokazu Y, Tajima N, Yamada H, Sawada H, Ishikawa H, Mimura T, Kitada M, Suzuki Y, Ide C (2004) Specific induction of neuronal cells from bone marrow stromal cells and application for autologous transplantation. J Clin Invest 113:1701–1710

DiGirolamo CM, Stokes D, Colter D, Phinney DG, Class R, Prockop DJ (1999) Propagation and senescence of human marrow stromal cells in culture: a simple colony-forming assay identifies samples with the greatest potential to propagate and differentiate. Br J Haematol 107:275–281

Djouad F, Plence P, Bony C, Tropel P, Apparailly F, Sany J, Noel D, Jorgensen C (2003) Immunosuppressive effect of mesenchymal stem cells favors tumor growth in allogeneic animals. Blood 102:3837–3844

Drachman DB, Frank K, Dykes-Hoberg M, Teismann P, Almer G, Przedborski S, Rothstein JD (2002) Cyclooxygenase 2 inhibition protects motor neurons and prolongs survival in a transgenic mouse model of ALS. Ann Neurol 52:771–778

Ende N, Weinstein F, Chen R, Ende M (2000) Human umbilical cord blood effect on sod mice (amyotrophic lateral sclerosis). Life Sci 67:53–59

Fischer W, Wictorin K, Bjorklund A, Williams LR, Varon S, Gage FH (1987) Amelioration of cholinergic neuron atrophy and spatial memory impairment in aged rats by nerve growth factor. Nature 329:65–68

Fjord-Larsen L, Johansen JL, Kusk P, Tornoe J, Gronborg M, Rosenblad C, Wahlberg LU (2005) Efficient in vivo protection of nigral dopaminergic neurons by lentiviral gene transfer of a modified Neurturin construct. Exp Neurol 195:49–60

Freed CR, Breeze RE, Rosenberg NL, Schneck SA, Kriek E, Qi JX, Lone T, Zhang YB, Snyder JA, Wells TH (1992) Survival of implanted fetal dopamine cells and neurologic improvement 12 to 46 months after transplantation for Parkinson's disease. N Engl J Med 327:1549–1555

Freed CR, Greene PE, Breeze RE, Tsai WY, DuMouchel W, Kao R, Dillon S, Winfield H, Culver S, Trojanowski JQ, Eidelberg D, Fahn S (2001) Transplantation of embryonic dopamine neurons for severe Parkinson's disease. N Engl J Med 344:710–719

Friedenstein AJ, Piatetzky-Shapiro II, Petrakova KV (1966) Osteogenesis in transplants of bone marrow cells. J Embryol Exp Morphol 16:381–390

Friedenstein AJ, Deriglasova UF, Kulagina NN, Panasuk AF, Rudakowa SF, Luria EA, Ruadkow IA (1974) Precursors for fibroblasts in different populations of hematopoietic cells as detected by the in vitro colony assay method. Exp Hematol 2:83–92

Friedenstein AJ, Gorskaja U, Kalugina NN (1976) Fibroblast precursors in normal and irradiated mouse hematopoietic organs. Exp Hematol 4:267–274

Gage FH (2000) Mammalian neural stem cells. Science 287:1433–1438

Gage FH, Bjorklund A (1986) Cholinergic septal grafts into the hippocampal formation improve spatial learning and memory in aged rats by an atropine-sensitive mechanism. J Neurosci 6:2837–2847

Gage FH, Coates PW, Palmer TD, Kuhn HG, Fisher LJ, Suhonen JO, Peterson DA, Suhr ST, Ray J (1995) Survival and differentiation of adult neuronal progenitor cells transplanted to the adult brain. Proc Natl Acad Sci U S A 92:11879–11883

Gao HM, Liu B, Zhang W, Hong JS (2003) Novel anti-inflammatory therapy for Parkinson's disease. Trends Pharmacol Sci 24:395–401

Garbuzova-Davis S, Willing AE, Milliken M, Saporta S, Zigova T, Cahill DW, Sanberg PR (2002) Positive effect of transplantation of hNT neurons (NTera 2/D1 cell-line) in a model of familial amyotrophic lateral sclerosis. Exp Neurol 174:169–180

Garbuzova-Davis S, Willing AE, Zigova T, Saporta S, Justen EB, Lane JC, Hudson JE, Chen N, Davis CD, Sanberg PR (2003) Intravenous administration of human umbilical cord blood cells in a mouse model of amyotrophic lateral sclerosis: distribution, migration, and differentiation. J Hematother Stem Cell Res 12:255–270

Gill SS, Patel NK, Hotton GR, O'Sullivan K, McCarter R, Bunnage M, Brooks DJ, Svendsen CN, Heywood P (2003) Direct brain infusion of glial cell line-derived neurotrophic factor in Parkinson disease. Nat Med 9:589–595

Giulian D, Haverkamp LJ, Li J, Karshin WL, Yu J, Tom D, Li X, Kirkpatrick JB (1995) Senile plaques stimulate microglia to release a neurotoxin found in Alzheimer brain. Neurochem Int 27:119–137

Gordon PH, Moore DH, Gelinas DF, Qualls C, Meister ME, Werner J, Mendoza M, Mass J, Kushner G, Miller RG (2004) Placebo-controlled phase I/II studies of minocycline in amyotrophic lateral sclerosis. Neurology 62:1845–1847

Griffin WS (2006) Inflammation and neurodegenerative diseases. Am J Clin Nutr 83:470S–474S

Grigoryan GA, Gray JA, Rashid T, Chadwick A, Hodges H (2002) Conditionally immortal neuroepithelial stem cell grafts restore spatial learning in rats with lesions at the source of cholinergic forebrain projections cholinergic forebrain projections. Restor Neurol Neurosci 174:183–202

Hefti F, Dravid A, Hartikka J (1984) Chronic intraventricular injections of nerve growth factor elevate hippocampal choline acetyltransferase activity in adult rats with partial septo-hippocampal lesions. Brain Res 293:305–311

Hermann A, Gastl R, Liebau S, Popa MO, Fiedler J, Boehm BO, Maisel M, Lerche H, Schwarz J, Brenner R, Storch A (2004) Efficient generation of neural stem cell-like cells from adult human bone marrow stromal cells. J Cell Sci 117:4411–4422

Hodges H, Allen Y, Kershaw T, Lantos PL, Gray JA, Sinden J (1991a) Effects of cholinergic-rich neural grafts on radial maze performance of rats after excitotoxic lesions of the forebrain cholinergic projection system—I. Amelioration of cognitive deficits by transplants into cortex and hippocampus but not into basal forebrain. Neuroscience 45:587–607

Hodges H, Allen Y, Sinden J, Lantos PL, Gray JA (1991b) Effects of cholinergic-rich neural grafts on radial maze performance of rats after excitotoxic lesions of the forebrain cholinergic projection system—II. Cholinergic drugs as probes to investigate lesion-induced deficits and transplant-induced functional recovery. Neuroscience 45:609–623

Hunot SM, Hirsch EC (2003) Neuroinflammatory processes in Parkinson's disease. Ann Neurol 53:49–60

Hurtado-Lorenzo A, Millan E, Gonzalez-Nicolini V, Suwelack D, Castro MG, Lowenstein PR (2004) Differentiation and transcription factor gene therapy in experimental Parkinson's disease: sonic hedgehog and gli-1, but not Nurr-1, protect nigrostriatal cell bodies from 6-OHDA-induced neurodegeneration. Mol Ther 10:507–524

in 't Veld BA, Ruitenberg A, Hofman A, Launer LJ, van Duijn CM, Stijnen T, Breteler MMB, Stricker BHC (2001) Nonsteroidal antiinflammatory drugs and the risk of Alzheimer's disease. N Engl J Med 345:1515–1521

Ji JF, He BP, Dheen ST, Tay SSW (2004) Interactions of chemokines and chemokine receptors mediate the migration of mesenchymal stem cells to the impaired site in the brain after hypoglossal nerve injury. Stem Cells 22:415–427

Jiang Y, Jahagirdar BN, Reinhardt RL, Schwartz RE, Keene CD, Ortiz-Gonzalez XR, Reyes M, Lenvik T, Lund T, Blackstad M, Du J, Aldrich S, Lisberg A, Low WC, Largaespada DA, Verfaillie CM (2002) Pluripotency of mesenchymal stem cells derived from adult marrow. Nature 418:41–49

Jin K, Galvan V, Xie L, Mao XO, Gorostiza OF, Bredesen DE, Greenberg DA (2004a) Enhanced neurogenesis in Alzheimer's disease transgenic (PDGF-APPSw,Ind) mice. Proc Natl Acad Sci U S A 101:13363–13367

Jin K, Peel AL, Mao XO, Xie L, Cottrell BA, Henshall DC, Greenberg DA (2004b) Increased hippocampal neurogenesis in Alzheimer's disease. Proc Natl Acad Sci U S A 101:343–347

Kiaei M, Kipiani K, Petri S, Choi DK, Chen J, Calingasan NY, Beal MF (2005) Integrative role of cPLA2 with COX-2 and the effect of non-steroidal anti-inflammatory drugs in a transgenic mouse model of amyotrophic lateral sclerosis. J Neurochem 93:403–411

Klein RL, Lewis MH, Muzyczka N, Meyer EM (1999) Prevention of 6-hydroxydopamine-induced rotational behavior by BDNF somatic gene transfer. Brain Res 847:314–320

Klein SM, Behrstock S, McHugh J, Hoffmann K, Wallace K, Suzuki M, Aebischer P, Svendsen CN (2005) GDNF delivery using human neural progenitor cells in a rat model of ALS. Hum Gene Ther 16:509–521

Klivenyi P, Kiaei M, Gardian G, Calingasan NY, Beal MF (2004) Additive neuroprotective effects of creatine and cyclooxygenase 2 inhibitors in a transgenic mouse model of amyotrophic lateral sclerosis. J Neurochem 88:576–582

Koliatsos VE, Price DL, Clatterbuck RE, Markowska AL, Olton DS, Wilcox BJ (1993) Neurotrophic strategies for treating Alzheimer's disease: lessons from basic neurobiology and animal models. Ann NY Acad Sci 695:292–299

Koshizuka SM, Okada SM, Okawa AM, Koda MM, Murasawa MM, Hashimoto MM, Kamada TM, Yoshinaga KM, Murakami MM, Moriya HM, Yamazaki MM (2004) Transplanted hematopoietic stem cells from bone marrow differentiate into neural lineage cells and promote functional recovery after spinal cord injury in mice. J Neuropathol Exp Neurol 63:64–72

Krause DS, Theise ND, Collector MI, Henegariu O, Hwang S, Gardner R, Neutzel S, Sharkis SJ (2001) Multi-organ, multi-lineage engraftment by a single bone marrow-derived stem cell. Cell 105:369–377

Kriz J, Nguyen MD, Julien JP (2002) Minocycline slows disease progression in a mouse model of amyotrophic lateral sclerosis. Neurobiol Dis 10:268–278

Lang AE, Gill S, Patel NK, Lozano A, Nutt JG, Penn R, Brooks DJ, Hotton G, Moro E, Heywood P, Brodsky MA, Burchiel K, Kelly P, Dalvi A, Scott B, Stacy M, Turner D, Wooten VG, Elias WJ, Laws ER, Dhawan V, Stoessl A, Matcham JJ, Coffey RJ, Traub M (2006) Randomized controlled trial of intraputamenal glial cell line-derived neurotrophic factor infusion in Parkinson disease. Ann Neurol 59:459–466

Laslo P, Lipski J, Nicholson LFB, Miles GB, Funk GD (2000) Calcium binding proteins in motoneurons at low and high risk for degeneration in ALS. Neuroreport 11:3305–3308

Le Blanc K (2003) Immunomodulatory effects of fetal and adult mesenchymal stem cells. Cytotherapy 5:485–489

Levi-Montalcini R (1987) The nerve growth factor 35 years later. Science 237:1154–1162

Li H, He Z, Su T, Ma Y, Lu S, Dai C, Sun M (2003) Protective action of recombinant neurturin on dopaminergic neurons in substantia nigra in a rhesus monkey model of Parkinson's disease. Neurol Res 25:263–267

Li Y, Chen J, Wang L, Zhang L, Lu M, Chopp M (2001) Intracerebral transplantation of bone marrow stromal cells in a 1-methyl-4-phenyl-1,2,3,6-tetrahydropyridine mouse model of Parkinson's disease. Neurosci Lett 316:67–70

Li Y, Chen J, Chen XG, Wang L, Gautam SC, Xu YX, Katakowski M, Zhang LJ, Lu M, Janakiraman N, Chopp M (2002) Human marrow stromal cell therapy for stroke in rat: neurotrophins and functional recovery. Neurology 59:514–523

Lie DC, Song H, Colamarino SA, Ming Gl, Gage FH (2004) Neurogenesis in the adult brain: new strategies for central nervous system diseases. Annu Rev Pharmacol Toxicol 44:399–421

Lindvall O, Brundin P, Widner H, Rehncrona S, Gustavii B, Frackowiak R, Leenders KL, Sawle G, Rothwell JC, Marsden CD, et al (1990) Grafts of fetal dopamine neurons survive and improve motor function in Parkinson's disease. Science 247:574–577

Liu B, Hong JS (2003) Role of microglia in inflammation-mediated neurodegenerative diseases: mechanisms and strategies for therapeutic intervention. J Pharmacol Exp Ther 304:1–7

Love S, Plaha P, Patel NK, Hotton GR, Brooks DJ, Gill SS (2005) Glial cell line-derived neurotrophic factor induces neuronal sprouting in human brain. Nat Med 11:703–704

Mahmood A, Lu D, Chopp M (2004) Marrow stromal cell transplantation after traumatic brain injury promotes cellular proliferation within the brain. Neurosurgery 55:1185–1193

Marchetti B, Serra PA, Tirolo C, L'Episcopo F, Caniglia S, Gennuso F, Testa N, Miele E, Desole S, Barden N, Morale MC (2005) Glucocorticoid receptor-nitric oxide crosstalk and vulnerability to experimental parkinsonism: pivotal role for glia-neuron interactions. Brain Res Brain Res Rev 48:302–321

Markowska AL, Koliatsos VE, Breckler SJ, Price DL, Olton DS (1994) Human nerve growth factor improves spatial memory in aged but not in young rats. J Neurosci 14:4815–4824

Markowska AL, Price D, Koliatsos VE (1996) Selective effects of nerve growth factor on spatial recent memory as assessed by a delayed nonmatching-to-position task in the water maze. J Neurosci 16:3541–3548

Massengale M, Wagers AJ, Vogel H, Weissman IL (2005) Hematopoietic cells maintain hematopoietic fates upon entering the brain. J Exp Med 201:1579–1589

Mattson MP, Magnus T (2006) Ageing and neuronal vulnerability. Nat Rev Neurosci 7:278–294

Mazzini L, Fagioli F, Boccaletti R, Mareschi K, Oliveri G, Olivieri C, Pastore I, Marasso R, Madon E (2003) Stem cell therapy in amyotrophic lateral sclerosis: a methodological approach in humans. Amyotroph Lateral Scler Other Motor Neuron Disord 4:158–161

Mazzini L, Fagioli F, Boccaletti R (2004) Stem-cell therapy for amyotrophic lateral sclerosis. Lancet North Am Ed 364:1936–1937

McGeer PL, McGeer EG (2002) Inflammatory processes in amyotrophic lateral sclerosis. Muscle Nerve 26:459–470

McGeer PL, McGeer EG (2004) Inflammation and neurodegeneration in Parkinson's disease. Parkinsonism Relat Disord 10:S3–S7

McGeer PL, Schulzer M, McGeer EG (1996) Arthritis and anti-inflammatory agents as possible protective factors for Alzheimer's disease: a review of 17 epidemiologic studies. Neurology 47:425–432

McNaught KS, Jenner P (1999) Altered glial function causes neuronal death and increases neuronal susceptibility to 1-methyl-4-phenylpyridinium- and 6-hydroxydopamine-induced toxicity in astrocytic/ventral mesencephalic co-cultures. J Neurochem 73:2469–2476

Merrill JE, Benveniste EN (1996) Cytokines in inflammatory brain lesions: helpful and harmful. Trends Neurosci 19:331–338

Minguell JJ, Erices A, Conget P (2001) Mesenchymal stem cells. Exp Biol Med 226:507–520

Moisse K, Strong MJ (2006) Innate immunity in amyotrophic lateral sclerosis. Biochim Biophys Acta 1762:1083–1093

Morale MC, Serra PA, L'Episcopo F, Tirolo C, Caniglia S, Testa N, Gennuso F, Giaquinta G, Rocchitta G, Desole MS (2006) Estrogen, neuroinflammation and neuroprotection in Parkinson's disease: glia dictates resistance versus vulnerability to neurodegeneration. Neuroscience 138:869–878

Moreau I, Duvert V, Caux C, Galmiche MC, Charbord P, Banchereau J, Saeland S (1993) Myofibroblastic stromal cells isolated from human bone marrow induce the proliferation of both early myeloid and B-lymphoid cells. Blood 82:2396–2405

Muir JL, Dunnett SB, Robbins TW, Everitt BJ (1992) Attentional functions of the forebrain cholinergic systems: effects of intraventricular hemicholinium, physostigmine, basal forebrain lesions and intracortical grafts on a multiple-choice serial reaction time task. Exp Brain Res 89:611–622

Nagano I, Shiote M, Murakami T, Kamada H, Hamakawa Y, Matsubara E, Yokoyama M, Moritaz K, Shoji M, Abe K (2005a) Beneficial effects of intrathecal IGF-1 administration in patients with amyotrophic lateral sclerosis. Neurol Res 27:768–772

Nagano I, Ilieva H, Shiote M, Murakami T, Yokoyama M, Shoji M, Abe K (2005b) Therapeutic benefit of intrathecal injection of insulin-like growth factor-1 in a mouse model of amyotrophic lateral sclerosis. J Neurol Sci 235:61–68

Nutt JG, Burchiel KJ, Comella CL, Jankovic J, Lang AE, Laws ER Jr, Lozano AM, Penn RD, Simpson RK Jr, Stacy M, Wooten GF (2003) Randomized, double-blind trial of glial cell line-derived neurotrophic factor (GDNF) in PD. Neurology 60:69–73

Oiwa Y, Yoshimura R, Nakai K, Itakura T (2002) Dopaminergic neuroprotection and regeneration by neurturin assessed by using behavioral, biochemical and histochemical measurements in a model of progressive Parkinson's disease. Brain Res 947:271–283

Olanow CW, Goetz GC, Kordower HJ, Stoessl AJ, Sossi V, Brin FM, Shannon MK, Nauert GM, Perl PD, Godbold J, Freeman BT (2003) A double-blind controlled trial of bilateral fetal nigral transplantation in Parkinson's disease. Ann Neurol 54:403–414

Perin EC, Geng YJ, Willerson JT (2003) Adult stem cell therapy in perspective. Circulation 107:935–938

Pittenger MF, Mackay AM, Beck SC, Jaiswal RK, Douglas R, Mosca JD, Moorman MA, Simonetti DW, Craig S, Marshak DR (1999) Multilineage potential of adult human mesenchymal stem cells. Science 284:143–147

Pompl PN, Ho L, Bianchi M, McManus T, Qin W, Pasinetti GM (2003) A therapeutic role for cyclooxygenase-2 inhibitors in a transgenic mouse model of amyotrophic lateral sclerosis. FASEB J 17:725–727

Prockop DJ (1997) Marrow stromal cells as stem cells for nonhematopoietic tissues. Science 276:71–74

Rakic P (2002) Neurogenesis in adult primate neocortex: an evaluation of the evidence. Nat Rev Neurosci 3:65–71

Reali C, Scintu F, Pillai R, Cabras S, Argiolu F, Ristaldi MS, Sanna MA, Badiali M, Sogos V (2006) Differentiation of human adult CD34+ stem cells into cells with a neural phenotype: role of astrocytes. Exp Neurol 197:399–406

Reubinoff BE, Pera MF, Fong CY, Trounson A, Bongso A (2000) Embryonic stem cell lines from human blastocysts: somatic differentiation in vitro. Nat Biotechnol 18:399–404

Reyes M, Verfaillie CM (2001) Characterization of multipotent adult progenitor cells, a subpopulation of mesenchymal stem cells. Ann NY Acad Sci 938:231–235

Reynolds BA, Weiss S (1992) Generation of neurons and astrocytes from isolated cells of the adult mammalian central nervous system. Science 255:1707–1710

Rich JB, Rasmusson DX, Folstein MF, Carson KA, Kawas C, Brandt J (1995) Nonsteroidal anti-inflammatory drugs in Alzheimer's disease. Neurology 45:51–55

Rivera FJ, Couillard-Despres S, Pedre X, Ploetz S, Caioni M, Lois C, Bogdahn U, Aigner L (2006) Mesenchymal stem cells instruct oligodendrogenic fate decision on adult neural stem cells. Stem Cells 24:2209–2219

Rombouts WJ, Ploemacher RE (2003) Primary murine MSC show highly efficient homing to the bone marrow but lose homing ability following culture. Leukemia 17:160–170

Ryan J, Barry F, Murphy JM, Mahon B (2005) Mesenchymal stem cells avoid allogeneic rejection. J Inflamm 2:8

Sargsyan SA, Monk PN, Shaw PJ (2005) Microglia as potential contributors to motor neuron injury in amyotrophic lateral sclerosis. Glia 51:241–253

Sasaki A, Yamaguchi H, Ogawa A, Sugihara S, Nakazato Y (1997) Microglial activation in early stages amyloid beta protein deposition. Acta Neuropathol (Berl) 94:316–322

Sastre M, Klockgether T, Heneka MT (2006) Contribution of inflammatory processes to Alzheimer's disease: molecular mechanisms. Int J Dev Neurosci 24:167–176

Schwarz EJ, Alexander GM, Prockop DJ, Azizi SA (1999) Multipotential marrow stromal cells transduced to produce L-DOPA: engraftment in a rat model of Parkinson disease. Hum Gene Ther 10:2539–2549

Schwarz EJ, Reger RL, Alexander GM, Class R, Azizi SA, Prockop DJ (2001) Rat marrow stromal cells rapidly transduced with a self-inactivating retrovirus synthesize L-DOPA in vitro. Gene 8:1214–1223

Shizuru JA, Negrin RS, Weissman IL (2005) Hematopoietic stem and progenitor cells: clinical and preclinical regeneration of the hematolymphoid system. Annu Rev Med 56:509–538

Sigurjonsson OE, Perreault MC, Egeland T, Glover JC (2005) Adult human hematopoietic stem cells produce neurons efficiently in the regenerating chicken embryo spinal cord. Proc Natl Acad Sci U S A 102:5227–5232

Stewart WF, Kawas C, Corrada M, Metter EJ (1997) Risk of Alzheimer's disease and duration of NSAID use. Neurology 48:626–632

Sugaya K (2003) Neuroreplacement therapy and stem cell biology under disease conditions. Cell Mol Life Sci 60:1891–1902

Svendsen CN, Caldwell MA, Shen J, ter Borg MG, Rosser AE, Tyers P, Karmiol S, Dunnett SB (1997) Long-term survival of human central nervous system progenitor cells transplanted into a rat model of Parkinson's disease. Exp Neurol 148:135–146

Szekely CA, Thorne JE, Zandi PP, Ek M, Messias E, Breitner JCS, Goodman SN (2004) Nonsteroidal anti-inflammatory drugs for the prevention of Alzheimer's disease: a systematic review. Neuroepidemiology 23:159–169

Takeshima T, Johnston JM, Commissiong JW (1994) Mesencephalic type 1 astrocytes rescue dopaminergic neurons from death induced by serum deprivation. J Neurosci 14:4769–4779

Tande D, Hoglinger G, Debeir T, Freundlieb N, Hirsch EC, Francois C (2006) New striatal dopamine neurons in MPTP-treated macaques result from a phenotypic shift and not neurogenesis. Brain 129:1194–1200

Tikka TM, Vartiainen NE, Goldsteins G, Oja SS, Andersen PM, Marklund SL, Koistinaho J (2002) Minocycline prevents neurotoxicity induced by cerebrospinal fluid from patients with motor neurone disease. Brain 125:722–731

Tuszynski MH (2002) Growth-factor gene therapy for neurodegenerative disorders. Lancet Neurol 1:51–57

Tuszynski MH, Sang H, Yoshida K, Gage FH (1991) Recombinant human nerve growth factor infusions prevent cholinergic neuronal degeneration in the adult primate brain. Ann Neurol 30:625–636

Tuszynski MH, Thal L, Pay M, Salmon DP, Hoi S, Bakay R, Patel P, Blesch A, Vahlsing HL, Ho G, Tong G, Potkin SG, Fallon J, Hansen L, Mufson EJ, Kordower JH, Gall C, Conner J (2005) A phase 1 clinical trial of nerve growth factor gene therapy for Alzheimer disease. Nat Med 11:551–555

Van Den Bosch LC, Tilkin P, Lemmens G, Robberecht W (2002) Minocycline delays disease onset and mortality in a transgenic model of ALS. Neuroreport 13:1067–1070

Vogel G (2002) Stem cells not so stealthy after all. Science 297:175

Watts J (2005) Controversy in China. Lancet 365:109–110

West MJ, Coleman PD, Flood DG, Troncoso JC (1994) Differences in the pattern of hippocampal neuronal loss in normal ageing and Alzheimer's disease. Lancet 344:769–772

Whitehouse PJ, Price DL, Clark JT, Coyle JT, DeLong MR (1981) Alzheimer disease: evidence for selective loss of cholinergic neurons in the nucleus basalis. Ann Neurol 10:122–126

Whitehouse PJ, Price DL, Struble RG, Clark AW, Coyle JT, DeLong MR (1982) Alzheimer's disease and senile dementia: loss of neurons in the basal forebrain. Science 215:1237–1239

Wichterle H, Lieberam I, Porter JA, Jessell TM (2002) Directed differentiation of embryonic stem cells into motor neurons. Cell 110:385–397

Winner B, Geyer M, Couillard-Despres S, Aigner R, Bogdahn U, Aigner L, Kuhn G, Winkler J (2006) Striatal deafferentation increases dopaminergic neurogenesis in the adult olfactory bulb. Exp Neurol 197:113–121

Woodbury D, Reynolds K, Black IB (2002) Adult bone marrow stromal stem cells express germline, ectodermal, endodermal, and mesodermal genes prior to neurogenesis. J Neurosci Res 69:908–917

Yamada M, Onodera M, Mizuno Y, Mochizuki H (2004) Neurogenesis in olfactory bulb identified by retroviral labeling in normal and 1-methyl-4-phenyl-1,2,3,6-tetrahydropyridine-treated adult mice. Neuroscience 124:173–181

Zappia E, Casazza S, Pedemonte E, Benvenuto F, Bonanni I, Gerdoni E, Giunti D, Ceravolo A, Cazzanti F, Frassoni F, Mancardi G, Uccelli A (2005) Mesenchymal stem cells ameliorate experimental autoimmune encephalomyelitis inducing T-cell anergy. Blood 106:1755–1761

Zhang J, Li Y, Chen J, Cui Y, Lu M, Elias SB, Mitchell JB, Hammill L, Vanguri P, Chopp M (2005) Human bone marrow stromal cell treatment improves neurological functional recovery in EAE mice. Exp Neurol 195:16–26

Zhang W, Narayanan M, Friedlander RM (2003) Additive neuroprotective effects of minocycline with creatine in a mouse model of ALS. Ann Neurol 53:267–270

Zhao M, Momma S, Delfani K, Carlen M, Cassidy RM, Johansson CB, Brismar H, Shupliakov O, Frisen J, Janson AM (2003) Evidence for neurogenesis in the adult mammalian substantia nigra. Proc Natl Acad Sci U S A 100:7925–7930

Zhu S, Stavrovskaya IG, Drozda M, Kim BYS, Ona V, Li M, Sarang S, Liu AS, Hartley DM, Wu DC, Gullans S, Ferrante RJ, Przedborski S, Kristal BS, Friedlander RM (2002) Minocycline inhibits cytochrome c release and delays progression of amyotrophic lateral sclerosis in mice. Nature 417:74–78

Stem Cells as a Treatment for Chronic Liver Disease and Diabetes

N. Levičar[1] (✉) · I. Dimarakis[1] · C. Flores[2] · J. Tracey[1] · M. Y. Gordon[2] · N. A. Habib[1]

[1] Department of Surgical Oncology and Technology, Faculty of Medicine, Imperial College London, Hammersmith Hospital, Du Cane Road, London W12 0NN, UK
n.levicar@imperial.ac.uk

[2] Department of Haematology, Faculty of Medicine, Imperial College London, Hammersmith Hospital, Du Cane Road, London W12 0NN, UK

1	Cell Therapy	244
2	Haematopoietic Stem Cells	244
3	Injured Liver	245
3.1	Generation of Hepatocytes by Haematopoietic Stem Cells	246
3.2	Haematopoietic Stem Cell Therapy for Liver Diseases	247
3.2.1	Animal Studies	247
3.2.2	Human Studies	249
4	Diabetes	250
4.1	Haematopoietic Stem Cells for β-Cell Generation	251
4.2	Haematopoietic Stem Cells for Diabetes	254
4.2.1	Animal Studies	254
4.2.2	Human Studies	256
5	Conclusions	257
	References	257

Abstract Advances in stem cell biology and the discovery of pluripotent stem cells have made the prospect of cell therapy and tissue regeneration a clinical reality. Cell therapies hold great promise to repair, restore, replace or regenerate affected organs and may perform better than any pharmacological or mechanical device. There is an accumulating body of evidence supporting the contribution of adult stem cells, in particular those of bone marrow origin, to liver and pancreatic islet cell regeneration. In this review, we will focus on the cell therapy for the diseased liver and pancreas by adult haematopoietic stem cells, as well as their possible contribution and application to tissue regeneration. Furthermore, recent progress in the generation, culture and targeted differentiation of human haematopoietic stem cells to hepatic and pancreatic lineages will be discussed. We will also explore the possibility that stem cell technology may lead to the development of clinical modalities for human liver disease and diabetes.

Keywords Stem cell therapy · Haematopoietic stem cells · Liver disease · Diabetes

1
Cell Therapy

Cell therapy has emerged as a novel approach for the treatment of many human degenerative diseases. Cell-based regenerative strategies aim to replace, repair or enhance the biological function of damaged tissues or organs by utilising cells and bioactive molecules to trigger, enhance, support and complement the residual capacity for repair. This can be achieved by the transplantation of cells, which are typically manipulated *ex vivo*, into a target organ in sufficient numbers for them to survive long enough to restore the normal function of organs and tissues. Possible candidate cells to be used include autologous primary cells, cell lines, various stem cells including bone marrow stem cells, cord blood stem cells and embryonic stem (ES) cells (Fodor 2003).

In recent years, advances in stem cell biology, including embryonic and somatic stem cells, have made the prospect of tissue regeneration a potential clinical reality, and several studies have shown the great promise that stem cells hold for therapy (Assmus et al. 2002; Wollert et al. 2004). Despite the unquestioned totipotency of ES cells, there are numerous unanswered biological questions as to the regulation of their growth and differentiation. The safety profile of unselected ES cells for transplantation early on demonstrated dysregulated cell growth resulting in teratoma formation (Reubinoff et al. 2000). Moreover, the ethical and legal issues associated with ES cells have shifted the focus to adult stem cells, and their regenerative potential has been under intense investigation.

The main role of adult stem cells, which are present in approximately 1%–2% of the total cell population within a specific tissue, is to replenish of the tissue's functional cells in appropriate proportions and numbers in response to 'wear and tear' loss or direct organ damage (Fang et al. 2004). They are vital in the maintenance of tissue homeostasis by continuously contributing to tissue regeneration and replacing cell lost during apoptosis or direct injury (Li and Xie 2005). Adult stem and progenitor cells possess the capacity of self-renewal and differentiation into one or more mature cell types. They are able to maintain their populations within the human body through asymmetric and symmetric divisions, resulting in the creation of differentiated and undifferentiated progeny (Preston et al. 2003). These properties make them ideal candidates for stem cell-based therapies and tissue engineering.

2
Haematopoietic Stem Cells

Haematopoietic stem cells (HSC) are multipotent bone marrow cells that sustain the formation of the blood and immune system throughout life. First identified in 1961, HSC have been by far the best characterised and most studied example of adult stem cells (Till and McCulloch 1961). The bone marrow compartment is largely made up of committed progenitor cells, non-circulating stromal cells [called mesenchymal stem cells (MSC)] that have the ability to develop into mesenchymal lineages and HSC (Pittenger et al. 1999; Bianco et al. 2001). It was previously thought that adult stem cells were lineage restricted, but recent studies demonstrated that bone marrow-derived progenitors in addition to haematopoiesis also participate in regeneration of ischaemic myocardium (Orlic et al. 2001), damaged skeletal muscle (Gussoni et al. 1999) and neurogenesis (Mezey et al. 2000).

MSC are the second major population of stem cells residing in bone marrow. MSC can be isolated as a growing adherent cell population and can differentiate into osteoblasts, adipocytes and chondrocytes (Pittenger et al. 1999).

3
Injured Liver

Liver diseases impose a heavy burden on society and affect approximately 17% of the population. Cirrhosis, the end result of long-term liver damage, has long been an important cause of death in UK and showed large rises in death rates over the past 20 years (Ministry of Health 2001). The main causes of cirrhosis globally are hepatitis B and C and alcohol abuse. Changing patterns of alcohol consumption and the increasing incidence of obesity and diabetes suggest that the burden of fibrosis and cirrhosis related to alcohol and non-alcohol steatohepatitis (NASH) will continue to increase (Fallowfield and Iredale 2004).

Cirrhosis is a progressive liver disease and is marked by the gradual destruction of liver tissue over time. Persistent injuries lead to hepatic scarring (fibrosis), which, if unopposed, leads to cirrhosis and demise of liver function. End-stage liver fibrosis is cirrhosis, whereby normal liver architecture is disrupted by fibrotic bands, parenchymal nodules and vascular distortion. Portal hypertension and hepatocyte dysfunction are the end results and give rise to major systemic complications and premature death. At the cirrhotic stage, liver disease is considered irreversible and the only solution is orthotopic liver transplantation (OLT). However, the increasing shortage of donor organs restricts liver transplantation. With the widening donor-recipient gap, the increasing incidence of liver disease, life-long dependence on immunosuppression and the poor outcome in patients not supported by liver transplantation, there is obviously a demand for new strategies to supplement OLT.

According to current concepts of liver regeneration, the liver has three levels of cells that can respond to liver injury and loss of hepatocytes (Sell 2001). First, mature hepatocytes, which are numerous, respond to mild liver injury by 1 to 2 cell cycles. Second, intra-organ ductal progenitor cells, which are less numerous, respond by longer and limited proliferation. Third, stem cells entering from the circulation participate in liver regeneration. These cells, in part of bone marrow origin, enter the liver in a seemingly random distribution, as isolated cells, or in a portal and periportal distribution when there is marked injury (Petersen et al. 1999). In this latter mode, responding to severe injury, they enter first as an intermediate cell population, which then mature into hepatocytes.

New strategies for generating a viable source of healthy hepatocytes for treating hepatic disorders are under investigation. Potential sources include the expansion of existing hepatocytes, ES cells, progenitor/stem cells in the liver, and bone marrow stem cells. Direct transplantation of hepatocytes has already been successfully tried in a small number of patients as a bridge to OLT or as a therapeutic alternative but there remain several limitations, which mainly include limited hepatocyte amplification, the fact that replacement of less than 10% of the liver mass is achieved, the short duration of the functioning replacement cells and the need for immunosuppression (Ohashi et al. 2001; Najimi and Sokal 2005). Additionally, the number of patients that can be treated in this manner is severely limited due to the paucity of healthy donors. This limited success fuelled the interest in haematopoietic stem cells for hepatic disorders.

3.1
Generation of Hepatocytes by Haematopoietic Stem Cells

In vitro studies have demonstrated the potential of bone marrow stem cells to differentiate towards the hepatic lineage. Oh et al. (2000) and Miyazaki et al. (2002) have shown that rat bone marrow contains a subpopulation (3%) of cells co-expressing haematopoietic stem cell markers (CD34, c-kit, Thy-1), α-fetoprotein (AFP) and c-met. They have also demonstrated the expression of albumin, a marker of terminally differentiated hepatocytes, after culturing crude bone marrow in the presence of hepatocyte growth factor (HGF) and epidermal growth factor (EGF). Similar observations were made by Okumoto et al. (2003), where rat bone marrow cells enriched for Sca-1 began expressing liver-enriched transcription factor HNF1α and cytokeratin 8 (CK8) after being cultured in the presence of HGF. Moreover, Ratajczak et al. (2004) demonstrated that a subpopulation of murine mononuclear bone marrow cells isolated by chemotaxis in response to the α-chemokine stromal-derived factor-1 (SDF-1) expressed messenger RNA (mRNA) for AFP and a population enriched for Sca-1 expressed mRNA for AFP, c-met and CK19. Purified murine HSC were able to differentiate into liver-like cells expressing (AFP, GATA4, HNF4, HNF3β, HNF1α) and mature hepatocyte markers (CK18, albumin, transferrin) when co-cultured with injured liver tissue, suggesting that micro-environmental cues are responsible for conversion (Jang et al. 2004).

Studies on human bone marrow confirmed observations from rodent models and showed that human bone marrow cells have the potential to differentiate into liver-like cells. When cultured on collagen matrix and in the presence of

HGF they expressed liver-specific genes, albumin and CK19 (Fiegel et al. 2003). Furthermore, when selected by SDF-1 chemotaxis they appear to be multipotent and express AFP (Kucia et al. 2004). Numerous cytokines and growth factors have been shown to have potent effects on hepatic growth and differentiation under *in vitro* culture conditions (Heng et al. 2005). HGF, EGF, transforming growth factor (TGF), and acid fibroblast growth factor (aFGF) are the most commonly used in the majority of studies and have been shown to promote hepatic differentiation *in vitro* (Block et al. 1996; Michalopoulos et al. 2003).

Like HSC, a subpopulation of human MSC known as multipotent adult progenitor cells (MAPC) has the potential to differentiate into hepatocyte-like cells in the presence of HGF and fibroblast growth factor-4 (FGF-4) (Schwartz et al. 2002). However, the MAPC culture is fastidious, with a substantial delay between the isolation and the appearance of hepatocyte-like cells, which calls into question their use in the clinic. *In vitro* hepatic differentiation of MSC in the presence of HGF and oncostatin M was confirmed by another group (Lee et al. 2004). Again, the differentiation process was lengthy but they demonstrated the expression of liver-specific genes in differentiated cells and other characteristics of liver cells, including albumin production, glycogen storage, urea secretion, uptake of low-density lipoprotein, and phenobarbital-inducible cytochrome P450 activity.

3.2
Haematopoietic Stem Cell Therapy for Liver Diseases

3.2.1
Animal Studies

Liver has long been known to exhibit considerable regenerative potential, but it has been only recently that we began to understand the implication of stem/progenitor cells in this process. The extrahepatic stem cells such as HSC are of particular interest since they are easily accessible.

Petersen et al. (1999) were first to show that bone marrow stem cells could be a potential source of hepatic oval cells. The liver injury was induced with 2-AAF-CCl$_4$, followed by cross-sex bone marrow transplantation. Regenerated hepatic cells were shown to be of bone marrow origin, demonstrated by using markers for the Y chromosome, dipeptidyl peptidase IV enzyme, and L21-6 antigen to identify donor-derived cells.

The other demonstration of hepatocyte regeneration from bone marrow cells is the extensive repopulation of damaged livers of fumarylacetoacetate hydrolase (FAH) mice, an inducible animal model of tyrosinemia type 1 (Grompe et al. 1995), a lethal hereditary liver disease (Lagasse et al. 2000). Intravenous injection of adult bone marrow cells in the FAH$^{(-/-)}$ mouse rescued the mouse and restored the biochemical function of its liver. The liver repopulation by bone marrow cells was slow, although significant. The first hepatocytes of bone marrow origin appeared 7 weeks after transplantation. However, later on, at 22 weeks, one-third of the liver comprised bone marrow-derived cells. This suggested that bone marrow stem cells contribute to hepatocyte generation in the presence of injury, where the regenerative potential of hepatocytes is impaired. Furthermore, recent data have demonstrated that bone marrow stem cells injected during liver injury can reduce the resulting liver fibrosis (Sakaida et al. 2004). The exact mechanism of this therapeutic effect is not fully understood yet, but it may be facilitated by the matrix metalloproteases, which enable degradation of hepatic scars and are expressed by bone marrow stem cells. More studies have shown that HSC engraft, repopulate and have survival advantage when transplanted into injured liver. Mallet et al. (2002) used JO_2 antibody, the murine anti-Fas agonist, to induce hepatic apoptosis. Unfractionated bone marrow cells expressing Bcl-2 under the control of a liver-specific promoter were transplanted into normal mice. Some mice received repeated weekly injections of JO_2 antibody to induce liver injury. Bone marrow-derived hepatocytes expressing Bcl-2 were only seen in the liver of the mice, which received JO_2 antibody injections. Moreover, in mice with induced liver cirrhosis, 25% of the recipient liver was repopulated in 4 weeks by bone marrow-derived hepatocytes (Terai et al. 2003). Jang et al. (2004) reported that murine HSC converted into viable hepatocytes with increasing liver injury. They noticed that liver function was restored 2–7 days after transplantation of murine $Fr25lin^-$ cells into irradiated and CCl_4 injured murine liver, suggesting that HSC contribute to the regeneration of injured liver by differentiating into functional hepatocytes. Misawa et al. (2006) used desialylated bone marrow cells in order to increase their accumulation in rat livers in a rodent model of human hepatic Wilson's disease. They demonstrated that direct accumulation of desialylated bone marrow cells into liver increased the proportion of functional bone marrow cell-derived hepatocytes. However, some studies have shown that bone marrow stem cells can repopulate liver even in the absence of liver injury. In one study, Theise et al. (2000a) transplanted unfractionated male bone marrow or $CD34^+lin^-$ cells into irradiated female mice and looked for bone marrow-derived hepatocytes. They identified up to 2.2% Y-positive hepatocytes at 7 days and 2 months or longer post-transplantation. Moreover, positive fluorescent in situ hybridization (FISH) for the Y-chromosome and albumin mRNA confirmed male-derived cells were mature hepatocytes, suggesting that hepatocytes can be derived from bone marrow cells in the absence of severe acute injury. Krause et al. (2001) injected single male HSC into irradiated mice and obtained engraftment in several organs, including liver, gastrointestinal tract, bronchus and skin of recipient animals. In support of the findings, Wang et al. (2003) found albumin-expressing hepatocyte-like cells in the livers of mice transplanted with highly purified HSC.

In contrast, several other studies failed to show trans-differentiation of bone marrow stem cells into liver parenchyma. Wagers et al. (2002) transplanted single HSC into lethally irradiated mice and showed that HSC reconstituted blood leukocytes of irradiated mice, but did not contribute to non-haematopoietic tissues, including liver, brain, kidney, gut, and muscle. Kanazawa and Verma (2003) used various liver injury models to assess hepatic regeneration by bone marrow stem cells. By using three different models—CCl_4 treatment, albumin-urokinase transgenic mouse [TgN(Alb1Plau)], or hepatitis B transgenic mouse [TgN(Alb1HBV)]—they concluded there was little or no contribution of bone marrow cells to the replacement of injured livers. Similarly, Dahlke et al. (2003) concluded that bone marrow cell infusion was not able to enhance liver regeneration in a retrorsine/CCl_4 model of acute liver injury. It was shown that there is only a small contribution of bone marrow stem cell to liver regeneration after chronic liver injury. Vig et al. (2006) have found only 4.7% bone marrow-derived hepatocytes at 3 months after transplantation and even less (1.6%) at 6 months after bone marrow transplantation.

Although views concerning the contribution of HSC to hepatocyte lineages *in vivo* still remain divided, the differences between the studies may in part reflect the types of cells used, different injury models used and the method used to detect engrafted stem cells.

3.2.2
Human Studies

Several studies have also shown the presence of cells of bone marrow origin in the human liver. Alison et al. (2000) showed that adult human liver cells can be derived from the stem cells originated in bone marrow. They examined livers derived from female patients who received bone marrow transplantation from a male donor and have found Y-chromosome and CK8-positive hepatocytes, suggesting that extrahepatic stem cells can colonise the liver. Similarly, Theise et al. (2000b) investigated archival autopsy and biopsy liver specimens from recipients of sex-mismatched therapeutic bone marrow transplantation and orthotopic liver transplantations. They identified hepatocytes and cholangiocytes of bone marrow origin by immunocytochemistry staining for CK8, CK18 and CK19 and FISH analysis for the Y-chromosome. Using double staining analysis, they found a large number of engrafted hepatocytes (4%–43%) and cholangiocytes (4%–38%), showing that they can be derived and differentiated from bone marrow and replenish hepatic parenchymal cells. Although Korbling et al. (2002) confirmed bone marrow-derived hepatocytes in liver biopsies of sex-mismatched bone marrow transplantation, they found less significant numbers (4%–7%) than Theise et al. (2000b). Similarly, Ng et al. (2003) showed that most of the recipient-derived cells in the liver allografts were macrophages/Kupffer cells and only a small proportion of hepatocytes (1.6%) was recipient derived. Two other studies of liver transplant patients did not detect bone marrow-derived hepatocytes at all (Fogt et al. 2002; Wu et al. 2003).

The difference and inconsistent results of the published studies could be due to use of different techniques to identify recipient-derived hepatocytes in transplanted patients. Also, various markers can be used for hepatocyte identification, and the accuracy of the methods used for identification is variable.

Although studies have shown that bone marrow stem cells could give rise to hepatocytes, the use of bone marrow stem cells as therapeutic agents is still in its infancy. So far, only three studies have published the results from the clinical use of bone marrow stem cells for liver insufficiency. The first clinical study was performed by am Esch et al. (2005) who infused 3 patients with autologous $CD133^+$ cells subsequent to portal vein embolisation of right liver segments. Computerised tomography scan volumetry showed a 2.5-fold increase in the growth rate of the left lateral segments compared to the control group of 3 patients that had been subjected to portal vein embolisation without bone marrow stem cell administration. Despite the small number of patients and the lack of an adequately sized randomised control group, these data suggested that cell therapy enhances and accelerates liver regeneration. The second clinical study using adult haematopoietic stem cells in treatment of liver diseases was carried out by our group (Gordon et al. 2006). We have performed a phase I clinical trial in which five patients with liver insufficiency were given granulocyte colony-stimulating factor (G-CSF) to mobilise their stem cells for collection by leukapheresis. Between 1×10^6 and 2×10^8 $CD34^+$ cells were injected into either the portal vein or hepatic artery. Of the 5 patients, 3 showed improvement in serum bilirubin and 4 in serum albumin. Clinically, the procedure was well tolerated with no observed procedure-related complications. The data clearly suggested the contribution of stem cells to the regeneration of liver damage and are encouraging for the future development of stem cell therapy for liver diseases. A more recent study confirmed the observations from the previous two studies. Terai et al. (2006) treated nine liver cirrhosis patients with autologous bone marrow. They infused $5.20+/-0.63\times10^9$ mononuclear cells ($CD34^+$, $CD45^+$, c-kit$^+$) via the peripheral vein and followed the patients for 24 weeks. Significant improvements in serum albumin levels, total protein and improved Child-Pugh scores were observed at 24 weeks after therapy, suggesting that bone marrow cell therapy should be considered as a novel treatment for liver cirrhosis patients.

4
Diabetes

Type 1 diabetes accounts for only 5%–10% of all diabetes cases worldwide but its incidence is increasing with current estimates of those affected at approximately 10^5 in the UK and 10^6 in the USA (Burns et al. 2004; Daneman 2006). Type 1 (insulin-dependent) diabetes is a chronic disease affecting genetically predisposed individuals in which insulin-secreting β-cells within pancreatic islets of Langerhans are selectively and irreversibly destroyed by autoimmune assault. Once activated the continued destruction of β-cells leads to a progressive loss of insulin, then to clinical diabetes, and finally—in almost all affected—to a state of absolute insulin deficiency (Devendra et al. 2004). Type 2 diabetes was once considered a disease of wealthy nations but now it is a global affliction. The incidence of type 2 (adult onset or insulin resistant) diabetes is increasing globally and has reached pandemic proportions (Petersen and Shulman 2006). Estimates by the International Diabetes Federation (IDF) anticipate that the worldwide incidence of diabetes among those 20–79 years old will increase by around 70% in the next 20 years (International Diabetes Federation 2006). Although the aetiology of type 2 diabetes remains obscure, obesity and a sedentary lifestyle are the most common epidemiologic factors associated with development of the disease.

The common denominator in both types 1 and 2 diabetes is a decrease in β-cell mass resulting in an absolute or relative state of insulin insufficiency, respectively. Long-term normalisation of glucose metabolism is a prerequisite for prevention of secondary complications and to date has only been achieved with transplantation of the whole organ or with a reasonable number of islets. Recent success with the Edmonton protocol in 2000 (hepatic portal vein infusion of organ donor islets of Langerhans) provided proof that the restoration of a sufficient functional β-cell mass re-establishes euglycaemia (Ryan et al. 2002). However, the chief limitation to transplantation, either whole gland or islets, is the paucity of donors. Current protocols for islet transplantation require up to 1×10^6 primary human islets per recipient the approximate equivalent of $2-4\times10^6$ β-cells (Burns et al. 2004). Thus, such treatment can be offered to only an estimated 0.5% of needy recipients (Lechner and Habener 2003). Additionally, differentiated β-cells cannot be expanded efficiently *in vitro* and senesce rapidly (Halvorsen et al. 2000).

Multiple approaches are now being investigated to generate insulin-producing cells *in vitro* either by genetic engineering of β-cells or by utilising various β-cell precursor cells and stem/progenitor cells with the ability to grow *in vitro* and to differentiate into insulin producing cells and ultimately into β-cells (Hess et al. 2003). ES cell usage has been the subject of both ethical and scientific debate (Frankel 2000). The demonstration of plasticity of haematopoietic stem cells has been published and has led to their study in the treatment of diabetes (Hess et al. 2003). Insulin secreting cells (β-cells or otherwise) could be transplanted into patients to help maintain blood glucose homeostasis, reduce the burden of diabetes-related complications, overcome the limitation of donor organs and provide benefit to millions of diabetics.

4.1
Haematopoietic Stem Cells for β-Cell Generation

In diabetics, 90% of insulin-secreting β-cells of the pancreatic islets of Langerhans are destroyed. Thus, the replacement of insulin-secreting β-cells would be an ideal cure. Studies of the growth, development and differentiation of pancreatic insulin-secreting cells have explored both embryonic and adult-derived stem cells as candidates for β-cell differentiation. Little is known about the cellular and molecular mechanisms of β-cell differentiation, but recent work has begun to pose speculative hypotheses on pancreatic cell development.

Several investigators have isolated multipotent bone marrow-derived cell populations which have been described as capable of trans-differentiating and expressing a pancreatic β-cell phenotype under differentiation culture conditions. Such populations include marrow-isolated adult multilineage-inducible (MIAMI) cells (D'Ippolito et al. 2004) and bone marrow-derived stem (BMDS) cells (Lee and Stoffel 2003; Tang et al. 2004), as well as a subpopulation of peripheral blood cells of monocytic origin which have been induced to express characteristics of pancreatic cells (Ruhnke et al. 2005). In addition, a subpopulation of multipotent human mesenchymal stem cells (hMSC) has been identified which constitutively expresses pancreatic islet transcription factors (Moriscot et al. 2005). Below, we look at these cell populations derived from the adult bone marrow, which have been successfully induced *in vitro* to genetically and phenotypically resemble pancreatic islet β-cells.

MIAMI cells, derived from whole bone marrow after 14 days in culture, are described as a homogeneous population expressing a unique set of markers. Interestingly, these cells are negative for the haematopoietic stem cell marker CD34, but express the hallmark embryonic markers Oct-4 and Rex-1, as well as markers for cells of mesodermal, ectodermal and endodermal lineages. With manipulation of culture conditions, MIAMI cells are described as being capable of undergoing neural, osteogenic, chondrogenic, adipogenic and endodermal differentiation. To promote the expression of pancreatic islet genes, D'Ippolito et al. (2004) treated MIAMI cells under conditions previously known to promote stem cell commitment to a pancreatic lineage which includes high glucose culture medium supplemented with basic fibroblast growth factor (bFGF), followed by exposure to butylated hydroxyanisole (BHA) and exendin-4. Nicotinamide, exendin-4 and activin-A were then added as factors known to induce pancreatic islet differentiation. RT-PCR analysis detected the expression of the β-cell transcription factors Beta2/NeuroD, Nkx6.1 and Isl1 in differentiation-induced cells as compared to untreated MIAMI cells. This suggests that MIAMI cells may have the potential for pancreatic β-cell differentiation and potential use *in vivo*.

As with MIAMI cells, manipulation of BMDS cell culture medium with high glucose, followed by treatments with nicotinamide and exendin, has been shown to induce cells into expressing features of pancreatic islet cells (Tang et al. 2004). Genetic analysis showed the upregulation of pancreatic β-cell genes such as *insulin I* and *II*, *Glut-2*, *nestin*, *Pdx-1*, *glucose kinase* and *Pax-6* in differentiation-treated cells. Immunocytochemical analysis also showed insulin and c-peptide synthesis in up to 20% of differentiation-treated cells. Induced cells also responded to glucose challenge and secreted insulin according to glucose concentration. These data imply that BMDS cells are capable of differentiating into pancreatic islet cells as evidenced by genetic and phenotypic expression, as well as insulin-secreting function.

Monocytes derived from peripheral blood, known as programmable cells of monocytic origin, have also been shown to express characteristics of pancreatic β-cells in the presence of islet cell-conditioned medium, which contains EGF, HGF and nicotinamide (Ruhnke et al. 2005). After exposure to islet cell-conditioned medium, these cells expressed transcription factors involved in β-cell differentiation such as Ngn3, Nkx6.1 and NeuroD, as well as transcription factors involved in the regulation of the insulin gene. In terms of functionality, monocytic cells which were differentiated into pancreatic neo-islet cells were responsive to glucose challenge, as determined by insulin release pre- and post-differentiation induction. Programmable cells of monocytic origin were found to be genetically and functionally similar to pancreatic β-cells *in vitro*. However, *in vivo* experiments have yet to validate cell engraftment and insulin production in a diabetic model.

Moriscot et al. (2005) have identified plastic adherent hMSCs which have an Oct-4$^+$ pluripotent phenotype and which constitutively express Nkx6.1, an islet-associated transcription factor. hMSCs were plated with adenoviruses encoding Ipf1, HlxB9 or FoxA2, and were then cultured alone, cultured with islet-conditioned medium or co-cultured with human islets. Various culture conditions revealed that infection with adenovirus expressing HlxB9 and FoxA2, and co-culture with human islets led to insulin secretion as well as upregulation of NeuroD and insulin I. Although the described combination of adenoviral infection and co-culture led to insulin secretion, not all transcription factors described in β-cell development were upregulated, suggesting that either these hMSCs have not terminally differentiated into pancreatic cells, or that culture conditions need to be further optimised for proper β-cell differentiation. hMSCs have the capacity of at least partial differentiation into β-cells and may be a potential target for therapeutic purposes, as they are easily isolated from adults.

In vitro work proves that adult BMDS cells are capable of differentiation into pancreatic islet β-cells and allows for manipulation cell growth and development. Several investigators have been able to induce bone marrow-derived cells into cells which phenotypically express pancreatic β-cell characteristics and respond to glucose challenge. However, cells manufactured *in vitro* have not yet demonstrated insulin secretion to the same degree as endogenous β-cells. Insulin-secreting cells generated *in vitro* have been found to secrete only about 1% of the level insulin produced by endogenous β-cells (Bonner-Weir and Weir 2005). This information supports the notion that *in vitro* cell culture may not adequately mimic the physiological microenvironment where endogenous cells develop and regenerate in the body. Importantly, assessing the engraftment and functionality of cultured cells *in vivo* is an important part of the follow-up to *ex vivo* studies.

4.2
Haematopoietic Stem Cells for Diabetes

4.2.1
Animal Studies

Promising data regarding the potential of BMDS cells to reconstitute the β-cell endocrine portion of the pancreas have been produced by Ianus and co-workers (2003). Irradiated female wild-type mice were injected with BMDS cells expressing enhanced green fluorescent protein (EGFP) via a Cre-LoxP system controlled by the active transcription of the murine insulin gene. All donors were male mice sharing the same C57BL/6 background with recipient animals. At 4–6 weeks post-transplantation, pancreatic islet tissue was harvested and EGFP-positive cells were isolated by means of fluorescence-activated cell-sorting (FACS) or manual picking under a fluorescence microscope for further analysis. Up to 3% of total cells per islet were found to express EGFP in transplanted animals with no evident expression observed in peripheral blood and bone marrow samples. RT-PCR analysis of sorted cells revealed besides insulin I and insulin II, the expression of other β-cell markers including GLUT2, IPF-1, HNF1α, HNF1β, HNF3β and Pax-6. At the same time, cells lacked the common haematopoietic/leukocyte marker CD45. Finally, demonstration of glucose-dependent and incretin-enhanced insulin secretion was reported as proof of functionality.

Using the same experimental strategy as Ianus et al. (2003), another group was unable to confirm their findings (Taneera et al. 2006). Whole BMDS cells from transgenic mice in which green fluorescent protein (GFP) expression was under the control of the murine insulin promoter failed to demonstrate any signs of GFP expression within the pancreatic parenchyma of transplanted animals. As this was not informative per se of pancreatic engraftment, another series of experiments was conducted using BMDS cells expressing GFP under the control β-actin promoter. Despite the large degree of engraftment, none of the GFP-positive cells co-expressed insulin or the β-cell transcription factors Pdx-1 or Nkx6.1, while more than 99.9% expressed the pan-haematopoietic marker CD45 as well as myeloid antigens. The authors concluded that although BMDS cells demonstrated efficient pancreatic engraftment, a haematopoietic cell fate was almost exclusively adopted.

In order to study the regenerative potential of BMDS cells in diabetes mellitus, Hess et al. (2003) utilised a murine model of streptozotocin-induced pancreatic damage leading to hyperglycaemia. Hyperglycaemic mice were injected intravenously with BMDS cells from GFP transgenic donor mice. Irradiation status and c-kit expression served as intergroup comparison criteria. One of the most interesting observations made was that pancreatic injury was a prerequisite for BMDS cells taking on an insulin-producing phenotype within the pancreatic parenchyma, an observation also made by other researchers

(Mathews et al. 2004). No improvement was witnessed in the c-kit⁻ transplanted subgroup in comparison to the c-kit⁺ group or the subgroup using whole bone marrow, which points to the c-kit⁺ stem cell population as the potent subpopulation. The lack of any clinical improvement in the subgroup transplanted with irradiated whole bone marrow revealed the unlikelihood of paracrine factors being responsible for any observed improvement in the non-irradiated groups. Even though 2.5% of insulin-positive cells from islets of streptozotocin-treated animals were positive for GFP, there was no expression of Pdx-1 in these cells. In addition to their low frequency, these cells were noticed to be absent during the onset of hyperglycaemic reduction. Finally, a large proportion of donor cells documented in ductal or islet regions were of endothelial lineage, thus associating the regenerative process with various endothelial interactions. Lechner et al. (2004) transplanted BMDS cells from sex-mismatched GFP-transgenic mice into animals treated with streptozotocin in one experiment and animals that had undergone partial pancreatectomy in another. Both GFP immunostaining and Y chromosome FISH failed to detect significant evidence of donor BMDS cell trans-differentiation into β-cells.

Murine BMDS cells have been shown to undergo differentiation *in vitro* into cells that express molecular markers of pancreatic β-cells and stain positive for c-peptide and insulin (Tang et al. 2004). *In vivo* transplantation of these cells into the left renal capsule and the distal tip of the spleen of streptozotocin-mediated diabetic mice not only reversed the existing hyperglycaemia, but also restored the animals' ability to respond to *in vivo* glucose challenges. Another interesting approach is that of multiple in contrast to single injections of BMDS cells in order to restore normoglycaemia (Banerjee et al. 2005). Banerjee et al. (2005) demonstrated improved glycaemic control in addition to histopathologic evidence of pancreatic islet regeneration following multiple intravenous injections of BMDS cells in streptozotocin-treated mice. The fact that BMDS cells were harvested from experimental-diabetic mice is evidence of the retained regenerative potential of the bone marrow.

It is quite striking that a number of groups have not been able to reproduce the results of Ianus et al. (2003). Most authors noted a number of technical considerations in an attempt to explain this. Many groups have reversed the process of identifying pancreatic-specific markers in engrafted cells by trying to identify cellular populations expressing such markers prior to any *in vitro* manipulation. Human mononuclear umbilical cord blood cells have been reported to express a vast number of cell markers and transcription factors necessary for β-cell differentiation (Pessina et al. 2004). When transplanted into non-obese diabetic mice with autoimmune type 1 diabetes, human umbilical cord blood mononuclear cells were able to reduce blood glucose levels and improve survival compared to untreated animals (Ende et al. 2004). The argument regarding the true nature and fate of candidate 'insulin-producing' cells remains unresolved (Sipione et al. 2004), as they may not be true β-cell precursors and furthermore may be unsuitable for clinical transplantation.

Pre-transplantation *in vitro* manipulation (Oh et al. 2004; Tang et al. 2004) may also provide a useful tool for delivering large numbers of predefined cells, thus avoiding complications such as suboptimal delivery rates and cell differentiation down non-pancreatic pathways.

Recent data have also documented the capacity for self-renewal of post-natal murine β-cells (Dor et al. 2004), thus casting a shadow of uncertainty over scientists' expectation of multipotent stem cells. The existence of unknown molecular and cellular pathways is highly likely. Endothelial signals have been linked with induction of pancreatic differentiation (Lammert et al. 2001), an association that may underlie observations in some of the above studies (Choi et al. 2003; Hess et al. 2003; Mathews et al. 2004). Many studies have failed to demonstrate expression of insulin or Pdx-1 in donor bone marrow-derived cells seen in peri-ductal or peri-islet locations (Choi et al. 2003; Hess et al. 2003; Mathews et al. 2004; Taneera et al. 2006), whilst many of these cells express endothelial markers (Choi et al. 2003; Hess et al. 2003; Mathews et al. 2004). Further to identifying actual potent subpopulations (Hess et al. 2003), a few technical considerations may come in useful. A dose-dependent effect has been noted (Ende et al. 2004), suggesting possible population expansion to be required prior to clinical transplantation. Finally, the demonstration of multi-step cell delivery being more efficient long-term (Banerjee et al. 2005) also underlines the possibility of introducing this principle in experimental and possibly clinical work.

4.2.2
Human Studies

The first reported data on cellular therapy for type 1 and 2 diabetes were presented by Fernandez Vina et al. (2006a, b). They have treated 23 patients with type 1 diabetes with $CD34^+CD38^-$ cells isolated from bone marrow and followed them up for 90 days. After 90 days from cell transplantation, the blood sugar decreased by 9.7% and c-peptide significantly increased by 55%. It was also observed that the exogenous insulin, taken daily by the patients, decreased by 17% suggesting that autologous bone marrow stem cells could improve pancreatic function. Similar results were obtained for type 2 diabetes patients where autologous $CD34^+CD38^-$ cells were transplanted via spleen artery into 16 patients. The blood sugar significantly decreased by 27% 90-days post-transplantation, while c-peptide and insulin increased by 26% and 19%,respectively. Even more impressive is the fact that 90 days post-transplantation, 84% of treated patients did not need any more anti-diabetic drugs or insulin.

5
Conclusions

Over the past 5 years attention has been focussed sharply on stem cells and the extraordinary potential that they offer in treating a number of currently intractable human diseases. However, for liver diseases and diabetes, the stem cell therapy treatment approach is still in its infancy and further work is necessary. Many technical questions have yet to be answered before stem cell therapy can be applied to its fullest potential in the clinic. In order to achieve the goal of cell therapy, the source and types of cells, generation of cells in sufficient numbers, maintenance of the differentiated phenotype and cell engraftment and homing when transplanted in damaged tissue must be considered. Ideally, cells should expand extensively *in vitro*, have minimal immunogenicity and be able to reconstitute tissue when transplanted in damaged tissue. Defining which patient groups are suitable for this therapy and which stem cell types are the most effective given the underlying pathology is also important. The optimum timing and method of delivery need to be determined as they may have a significant influence on the outcome of cell transplantation. Long-term side-effects of treatment are unknown as most of the clinical studies are very recent. In addition, there is growing evidence that transplanted cells, being multipotent, do not simply replace missing tissue but also trigger local mechanisms to initiate a repair response. Paracrine effects and immune regulation of the transplanted cells may also play a role in functional restoration of the tissue (Pluchino et al. 2003).

As more scientific knowledge is gained in this field, hopefully some of the technical concerns will be answered and we will soon see stem cell therapy in more clinical applications.

References

Alison MR, Poulsom R, Jeffery R, Dhillon AP, Quaglia A, Jacob J, Novelli M, Prentice G, Williamson J, Wright NA (2000) Hepatocytes from non-hepatic adult stem cells. Nature 406:257

am Esch JS 2nd, Knoefel WT, Klein M, Ghodsizad A, Fuerst G, Poll LW, Piechaczek C, Burchardt ER, Feifel N, Stoldt V, Stockschlader M, Stoecklein N, Tustas RY, Eisenberger CF, Peiper M, Haussinger D, Hosch SB (2005) Portal application of autologous $CD133^+$ bone marrow cells to the liver: a novel concept to support hepatic regeneration. Stem Cells 23:463–470

Assmus B, Schachinger V, Teupe C, Britten M, Lehmann R, Dobert N, Grunwald F, Aicher A, Urbich C, Martin H, Hoelzer D, Dimmeler S, Zeiher AM (2002) Transplantation of progenitor cells and regeneration enhancement in acute myocardial infarction (TOPCARE-AMI). Circulation 106:3009–3017

Banerjee M, Kumar A, Bhonde RR (2005) Reversal of experimental diabetes by multiple bone marrow transplantation. Biochem Biophys Res Commun 328:318–325

Bianco P, Riminucci M, Gronthos S, Robey PG (2001) Bone marrow stromal stem cells: nature, biology, and potential applications. Stem Cells 19:180–192

Block GD, Locker J, Bowen WC, Petersen BE, Katyal S, Strom SC, Riley T, Howard TA, Michalopoulos GK (1996) Population expansion, clonal growth, and specific differentiation patterns in primary cultures of hepatocytes induced by HGF/SF, EGF and TGF alpha in a chemically defined (HGM) medium. J Cell Biol 132:1133–1149

Bonner-Weir S, Weir GC (2005) New sources of pancreatic beta-cells. Nat Biotechnol 23:857–861

Burns CJ, Persaud SJ, Jones PM (2004) Stem cell therapy for diabetes: do we need to make beta cells? J Endocrinol 183:437–443

Choi JB, Uchino H, Azuma K, Iwashita N, Tanaka Y, Mochizuki H, Migita M, Shimada T, Kawamori R, Watada H (2003) Little evidence of transdifferentiation of bone marrow-derived cells into pancreatic beta cells. Diabetologia 46:1366–1374

D'Ippolito G, Diabira S, Howard GA, Menei P, Roos BA, Schiller PC (2004) Marrow-isolated adult multilineage inducible (MIAMI) cells, a unique population of postnatal young and old human cells with extensive expansion and differentiation potential. J Cell Sci 117:2971–2981

Dahlke MH, Popp FC, Bahlmann FH, Aselmann H, Jager MD, Neipp M, Piso P, Klempnauer J, Schlitt HJ (2003) Liver regeneration in a retrorsine/CCl4-induced acute liver failure model: do bone marrow-derived cells contribute? J Hepatol 39:365–373

Daneman D (2006) Type 1 diabetes. Lancet 367:847–858

Devendra D, Liu E, Eisenbarth GS (2004) Type 1 diabetes: recent developments. BMJ 328:750–754

Dor Y, Brown J, Martinez OI, Melton DA (2004) Adult pancreatic beta-cells are formed by self-duplication rather than stem-cell differentiation. Nature 429:41–46

Ende N, Chen R, Reddi AS (2004) Effect of human umbilical cord blood cells on glycemia and insulitis in type 1 diabetic mice. Biochem Biophys Res Commun 325:665–669

Fallowfield JA, Iredale JP (2004) Targeted treatments for cirrhosis. Expert Opin Ther Targets 8:423–435

Fang TC, Alison MR, Wright NA, Poulsom R (2004) Adult stem cell plasticity: will engineered tissues be rejected? Int J Exp Pathol 85:115–124

Fernandez Vina R, Andrin O, Saslavsky J, Ferreyra de Silva J, Vrsalovick F, Camozzi L, Ferreyra O, D Adamo C, Foressi F, Fernandez Vina R, Clasen A (2006a) Increase of 'c' peptide level in type 1 diabetics patients after direct pancreas implant by endovascular way of autologous adult mononuclear CD34+CD38(−) cells (Teceldiab 2 study). In: 4th International Society for Stem Cell Research Annual Meeting, Toronto

Fernandez Vina R, Saslavsky J, Andrin O, Vrsalovick F, Ferreyra de Silva J, Ferreyra O, Camozzi L, Foressi F, D Adamo C, Fernandez Vina R (2006b) First word reported data from Argentina of implant and cellular therapy with autologous adult stem cells in type 2 diabetic patients (Teceldiar study 1). In: 4th International Society for Stem Cell Research Annual Meeting, Toronto

Fiegel HC, Lioznov MV, Cortes-Dericks L, Lange C, Kluth D, Fehse B, Zander AR (2003) Liver-specific gene expression in cultured human hematopoietic stem cells. Stem Cells 21:98–104

Fodor WL (2003) Tissue engineering and cell based therapies, from the bench to the clinic: the potential to replace, repair and regenerate. Reprod Biol Endocrinol 1:102

Fogt F, Beyser KH, Poremba C, Zimmerman RL, Khettry U, Ruschoff J (2002) Recipient-derived hepatocytes in liver transplants: a rare event in sex-mismatched transplants. Hepatology 36:173–176

Frankel MS (2000) In search of stem cell policy. Science 287:1397

Gordon MY, Levičar N, Pai M, Bachellier P, Dimarakis I, Al-Allaf F, M'Hamdi H, Thalji T, Welsh JP, Marley SB, Davis J, Dazzi F, Marelli-Berg F, Tait P, Playford R, Jiao L, Jensen S, Nicholls JP, Ayav A, Nohandani M, Farzaneh F, Gaken J, Dodge R, Alison M, Apperley JF, Lechler R, Habib NA (2006) Characterisation and clinical application of human CD34$^+$ stem/progenitor cell populations mobilised into the blood by G-CSF. Stem Cells 24:1822–1830

Grompe M, Lindstedt S, al-Dhalimy M, Kennaway NG, Papaconstantinou J, Torres-Ramos CA, Ou CN, Finegold M (1995) Pharmacological correction of neonatal lethal hepatic dysfunction in a murine model of hereditary tyrosinaemia type I. Nat Genet 10:453–460

Gussoni E, Soneoka Y, Strickland CD, Buzney EA, Khan MK, Flint AF, Kunkel LM, Mulligan RC (1999) Dystrophin expression in the mdx mouse restored by stem cell transplantation. Nature 401:390–394

Halvorsen TL, Beattie GM, Lopez AD, Hayek A, Levine F (2000) Accelerated telomere shortening and senescence in human pancreatic islet cells stimulated to divide *in vitro*. J Endocrinol 166:103–109

Heng BC, Yu H, Yin Y, Lim SG, Cao T (2005) Factors influencing stem cell differentiation into the hepatic lineage *in vitro*. J Gastroenterol Hepatol 20:975–987

Hess D, Li L, Martin M, Sakano S, Hill D, Strutt B, Thyssen S, Gray DA, Bhatia M (2003) Bone marrow-derived stem cells initiate pancreatic regeneration. Nat Biotechnol 21:763–770

Ianus A, Holz GG, Theise ND, Hussain MA (2003) *In vivo* derivation of glucose-competent pancreatic endocrine cells from bone marrow without evidence of cell fusion. J Clin Invest 111:843–850

International Diabetes Federation (2006) Diabetes atlas. In: http://www.eatlas.idf.org/. Cited 23 Dec 2009

Jang YY, Collector MI, Baylin SB, Diehl AM, Sharkis SJ (2004) Hematopoietic stem cells convert into liver cells within days without fusion. Nat Cell Biol 6:532–539

Kanazawa Y, Verma IM (2003) Little evidence of bone marrow-derived hepatocytes in the replacement of injured liver. Proc Natl Acad Sci USA 100 [Suppl 1]:11850–11853

Korbling M, Katz RL, Khanna A, Ruifrok AC, Rondon G, Albitar M, Champlin RE, Estrov Z (2002) Hepatocytes and epithelial cells of donor origin in recipients of peripheral-blood stem cells. N Engl J Med 346:738–746

Krause DS, Theise ND, Collector MI, Henegariu O, Hwang S, Gardner R, Neutzel S, Sharkis SJ (2001) Multi-organ, multi-lineage engraftment by a single bone marrow-derived stem cell. Cell 105:369–377

Kucia M, Ratajczak J, Reca R, Janowska-Wieczorek A, Ratajczak MZ (2004) Tissue-specific muscle, neural and liver stem/progenitor cells reside in the bone marrow, respond to an SDF-1 gradient and are mobilized into peripheral blood during stress and tissue injury. Blood Cells Mol Dis 32:52–57

Lagasse E, Connors H, Al-Dhalimy M, Reitsma M, Dohse M, Osborne L, Wang X, Finegold M, Weissman IL, Grompe M (2000) Purified hematopoietic stem cells can differentiate into hepatocytes *in vivo*. Nat Med 6:1229–1234

Lammert E, Cleaver O, Melton D (2001) Induction of pancreatic differentiation by signals from blood vessels. Science 294:564–567

Lechner A, Habener JF (2003) Stem/progenitor cells derived from adult tissues: potential for the treatment of diabetes mellitus. Am J Physiol Endocrinol Metab 284:E259–E266

Lechner A, Yang YG, Blacken RA, Wang L, Nolan AL, Habener JF (2004) No evidence for significant transdifferentiation of bone marrow into pancreatic beta-cells *in vivo*. Diabetes 53:616–623

Lee KD, Kuo TK, Whang-Peng J, Chung YF, Lin CT, Chou SH, Chen JR, Chen YP, Lee OK (2004) *In vitro* hepatic differentiation of human mesenchymal stem cells. Hepatology 40:1275–1284

Lee VM, Stoffel M (2003) Bone marrow: an extra-pancreatic hideout for the elusive pancreatic stem cell? J Clin Invest 111:799–801

Li L, Xie T (2005) Stem cell niche: structure and function. Annu Rev Cell Dev Biol 21:605–631

Mallet VO, Mitchell C, Mezey E, Fabre M, Guidotti JE, Renia L, Coulombel L, Kahn A, Gilgenkrantz H (2002) Bone marrow transplantation in mice leads to a minor population of hepatocytes that can be selectively amplified *in vivo*. Hepatology 35:799–804

Mathews V, Hanson PT, Ford E, Fujita J, Polonsky KS, Graubert TA (2004) Recruitment of bone marrow-derived endothelial cells to sites of pancreatic beta-cell injury. Diabetes 53:91–98

Mezey E, Chandross KJ, Harta G, Maki RA, McKercher SR (2000) Turning blood into brain: cells bearing neuronal antigens generated *in vivo* from bone marrow. Science 290:1779–1782

Michalopoulos GK, Bowen WC, Mule K, Luo J (2003) HGF-, EGF-, and dexamethasone-induced gene expression patterns during formation of tissue in hepatic organoid cultures. Gene Expr 11:55–75

Ministry of Health (2001) Annual report of the chief medical officer. In: Ministry of Health Report, UK, London

Misawa R, Ise H, Takahashi M, Morimoto H, Kobayashi E, Miyagawa S, Ikeda U (2006) Development of liver regenerative therapy using glycoside-modified bone marrow cells. Biochem Biophys Res Commun 342:434–440

Miyazaki M, Akiyama I, Sakaguchi M, Nakashima E, Okada M, Kataoka K, Huh NH (2002) Improved conditions to induce hepatocytes from rat bone marrow cells in culture. Biochem Biophys Res Commun 298:24–30

Moriscot C, de Fraipont F, Richard MJ, Marchand M, Savatier P, Bosco D, Favrot M, Benhamou PY (2005) Human bone marrow mesenchymal stem cells can express insulin and key transcription factors of the endocrine pancreas developmental pathway upon genetic and/or microenvironmental manipulation *in vitro*. Stem Cells 23:594–603

Najimi M, Sokal E (2005) Liver cell transplantation. Minerva Pediatr 57:243–257

Ng IO, Chan KL, Shek WH, Lee JM, Fong DY, Lo CM, Fan ST (2003) High frequency of chimerism in transplanted livers. Hepatology 38:989–998

Oh SH, Miyazaki M, Kouchi H, Inoue Y, Sakaguchi M, Tsuji T, Shima N, Higashio K, Namba M (2000) Hepatocyte growth factor induces differentiation of adult rat bone marrow cells into a hepatocyte lineage *in vitro*. Biochem Biophys Res Commun 279:500–504

Oh SH, Muzzonigro TM, Bae SH, LaPlante JM, Hatch HM, Petersen BE (2004) Adult bone marrow-derived cells trans-differentiating into insulin-producing cells for the treatment of type I diabetes. Lab Invest 84:607–617

Ohashi K, Park F, Kay MA (2001) Hepatocyte transplantation: clinical and experimental application. J Mol Med 79:617–630

Okumoto K, Saito T, Hattori E, Ito JI, Adachi T, Takeda T, Sugahara K, Watanabe H, Saito K, Togashi H, Kawata S (2003) Differentiation of bone marrow cells into cells that express liver-specific genes *in vitro*: implication of the Notch signals in differentiation. Biochem Biophys Res Commun 304:691–695

Orlic D, Kajstura J, Chimenti S, Jakoniuk I, Anderson SM, Li B, Pickel J, McKay R, Nadal-Ginard B, Bodine DM, Leri A, Anversa P (2001) Bone marrow cells regenerate infarcted myocardium. Nature 410:701–705

Pessina A, Eletti B, Croera C, Savalli N, Diodovich C, Gribaldo L (2004) Pancreas developing markers expressed on human mononucleated umbilical cord blood cells. Biochem Biophys Res Commun 323:315–322

Petersen BE, Bowen WC, Patrene KD, Mars WM, Sullivan AK, Murase N, Boggs SS, Greenberger JS, Goff JP (1999) Bone marrow as a potential source of hepatic oval cells. Science 284:1168–1170

Petersen KF, Shulman GI (2006) Etiology of insulin resistance. Am J Med 119:S10–S16

Pittenger MF, Mackay AM, Beck SC, Jaiswal RK, Douglas R, Mosca JD, Moorman MA, Simonetti DW, Craig S, Marshak DR (1999) Multilineage potential of adult human mesenchymal stem cells. Science 284:143–147

Pluchino S, Quattrini A, Brambilla E, Gritti A, Salani G, Dina G, Galli R, Del Carro U, Amadio S, Bergami A, Furlan R, Comi G, Vescovi AL, Martino G (2003) Injection of adult neurospheres induces recovery in a chronic model of multiple sclerosis. Nature 422:688–694

Preston SL, Alison MR, Forbes SJ, Direkze NC, Poulsom R, Wright NA (2003) The new stem cell biology: something for everyone. Mol Pathol 56:86–96

Ratajczak MZ, Kucia M, Reca R, Majka M, Janowska-Wieczorek A, Ratajczak J (2004) Stem cell plasticity revisited: CXCR4-positive cells expressing mRNA for early muscle, liver and neural cells 'hide out' in the bone marrow. Leukemia 18:29–40

Reubinoff BE, Pera MF, Fong CY, Trounson A, Bongso A (2000) Embryonic stem cell lines from human blastocysts: somatic differentiation *in vitro*. Nat Biotechnol 18:399–404

Ruhnke M, Ungefroren H, Nussler A, Martin F, Brulport M, Schormann W, Hengstler JG, Klapper W, Ulrichs K, Hutchinson JA, Soria B, Parwaresch RM, Heeckt P, Kremer B, Fandrich F (2005) Differentiation of *in vitro*-modified human peripheral blood monocytes into hepatocyte-like and pancreatic islet-like cells. Gastroenterology 128:1774–1786

Ryan EA, Lakey JR, Paty BW, Imes S, Korbutt GS, Kneteman NM, Bigam D, Rajotte RV, Shapiro AM (2002) Successful islet transplantation: continued insulin reserve provides long-term glycemic control. Diabetes 51:2148–2157

Sakaida I, Terai S, Yamamoto N, Aoyama K, Ishikawa T, Nishina H, Okita K (2004) Transplantation of bone marrow cells reduces CCl4-induced liver fibrosis in mice. Hepatology 40:1304–1311

Schwartz RE, Reyes M, Koodie L, Jiang Y, Blackstad M, Lund T, Lenvik T, Johnson S, Hu WS, Verfaillie CM (2002) Multipotent adult progenitor cells from bone marrow differentiate into functional hepatocyte-like cells. J Clin Invest 109:1291–1302

Sell S (2001) Heterogeneity and plasticity of hepatocyte lineage cells. Hepatology 33:738–750

Sipione S, Eshpeter A, Lyon JG, Korbutt GS, Bleackley RC (2004) Insulin expressing cells from differentiated embryonic stem cells are not beta cells. Diabetologia 47:499–508

Taneera J, Rosengren A, Renstrom E, Nygren JM, Serup P, Rorsman P, Jacobsen SE (2006) Failure of transplanted bone marrow cells to adopt a pancreatic beta-cell fate. Diabetes 55:290–296

Tang DQ, Cao LZ, Burkhardt BR, Xia CQ, Litherland SA, Atkinson MA, Yang LJ (2004) *In vivo* and *in vitro* characterization of insulin-producing cells obtained from murine bone marrow. Diabetes 53:1721–1732

Terai S, Sakaida I, Yamamoto N, Omori K, Watanabe T, Ohata S, Katada T, Miyamoto K, Shinoda K, Nishina H, Okita K (2003) An *in vivo* model for monitoring trans-differentiation of bone marrow cells into functional hepatocytes. J Biochem (Tokyo) 134:551–558

Terai S, Ishikawa T, Omori K, Aoyama K, Marumoto Y, Urata Y, Yokoyama Y, Uchida K, Yamasaki T, Fujii Y, Okita K, Sakaida I (2006) Improved liver function in liver cirrhosis patients after autologous bone marrow cell infusion therapy. Stem Cells 24:2292–2298

Theise ND, Badve S, Saxena R, Henegariu O, Sell S, Crawford JM, Krause DS (2000a) Derivation of hepatocytes from bone marrow cells in mice after radiation-induced myeloablation. Hepatology 31:235–240

Theise ND, Nimmakayalu M, Gardner R, Illei PB, Morgan G, Teperman L, Henegariu O, Krause DS (2000b) Liver from bone marrow in humans. Hepatology 32:11–16

Till JE, McCulloch E (1961) A direct measurement of the radiation sensitivity of normal mouse bone marrow cells. Radiat Res 14:213–222

Vig P, Russo FP, Edwards RJ, Tadrous PJ, Wright NA, Thomas HC, Alison MR, Forbes SJ (2006) The sources of parenchymal regeneration after chronic hepatocellular liver injury in mice. Hepatology 43:316–324

Wagers AJ, Sherwood RI, Christensen JL, Weissman IL (2002) Little evidence for developmental plasticity of adult hematopoietic stem cells. Science 297:2256–2259

Wang X, Ge S, McNamara G, Hao QL, Crooks GM, Nolta JA (2003) Albumin-expressing hepatocyte-like cells develop in the livers of immune-deficient mice that received transplants of highly purified human hematopoietic stem cells. Blood 101:4201–4208

Wollert KC, Meyer GP, Lotz J, Ringes-Lichtenberg S, Lippolt P, Breidenbach C, Fichtner S, Korte T, Hornig B, Messinger D, Arseniev L, Hertenstein B, Ganser A, Drexler H (2004) Intracoronary autologous bone-marrow cell transfer after myocardial infarction: the BOOST randomised controlled clinical trial. Lancet 364:141–148

Wu T, Cieply K, Nalesnik MA, Randhawa PS, Sonzogni A, Bellamy C, Abu-Elmagd K, Michalopolous GK, Jaffe R, Kormos RL, Gridelli B, Fung JJ, Demetris AJ (2003) Minimal evidence of transdifferentiation from recipient bone marrow to parenchymal cells in regenerating and long-surviving human allografts. Am J Transplant 3:1173–1181

The Participation of Mesenchymal Stem Cells in Tumor Stroma Formation and Their Application as Targeted-Gene Delivery Vehicles

B. Hall[1] · M. Andreeff[2] · F. Marini[2] (✉)

[1] Center for Childhood Cancer, Columbus Children's Research Institute,
700 Children's Drive, Columbus OH, 43205, USA

[2] Department of Blood and Marrow Transplantation, Unit 81,
The University of Texas M. D. Anderson Cancer Center,
1515 Holcombe Blvd., Houston TX, 77030, USA
fmarini@mdanderson.org

1	Introduction	263
2	Understanding the Tumor Microenvironment	265
3	Role of Tumor–Stroma Interactions in Tumor Progression	268
4	The Participation of MSC in Tumors and in Wound Healing	269
5	Rationale for Using MSC as Cellular Delivery Vehicles	270
6	Experimental Support for Use of MSC as Therapeutic Agents	273
7	Future Directions in Designing MSC-Based Delivery Strategies	274
8	Conclusions	275
	References	276

Abstract Recent evidence suggests that mesenchymal stem cells (MSC) selectively proliferate to tumors and contribute to the formation of tumor-associated stroma. The biological rationale for tumor recruitment of MSC remains unclear but may represent an effort of the host to blunt tumor cell growth and improve survival. There is mounting experimental evidence that normal stromal cells can revert malignant cell behavior, and separate studies have demonstrated that stromal cells can enhance tumor progression after acquisition of tumor-like genetic lesions. Together, these observations support the rationale for modifying normal MSC to deliver therapeutic proteins directly into the tumor microenvironment. Modified MSC can produce high concentrations of antitumor proteins directly within the tumor mass, which have been shown to blunt tumor growth kinetics in experimental animal model systems. In this chapter we will address the biological properties of MSC within the tumor microenvironment and discuss the potential use of MSC and other bone marrow-derived cell populations as delivery vehicles for antitumor proteins.

Keywords Mesenchymal Stem Cell (MSC) · Tumor Microenvironment · Tumor-associated Fibroblast (TAF) · Stromagenesis · Myofibroblast · Desmoplasia · Tumor Stroma

1
Introduction

Solid tumors are composed of tumor cells and supportive nontumor components known as tumor stroma. Tumor stroma can be divided into four main elements: (1) tumor vasculature, (2) cells of the immune system, (3) extracellular matrix (ECM), and (4) fibroblastic stromal cells—also known as tumor-associated fibroblasts (TAF) (Kunz-Schughart and Knuechel 2002a, b), carcinoma-associated fibroblasts (CAF) (Orimo et al. 2005), and reactive stroma (Rowley 1998). Most if not all solid tumors have some degree of tumor stroma, and the presence of reactive stroma is often an indicator of poor prognosis (De Wever and Mareel 2003; Hasebe et al. 2000; Kenny and Bissell 2003; Kurosumi et al. 2003).

Fibroblastic stromal cells have been linked to several activities that promote cancer metastasis and growth including angiogenesis (Orimo et al. 2005), epithelial to mesenchymal transition (Radisky et al. 2005), and progressive genetic instability (Kurose et al. 2001; Moinfar et al. 2000). Additionally, fibroblastic stromal cells can dysregulate antitumor immune responses as exemplified by experiments demonstrating that allogeneic murine tumor cells, when coinjected with fibroblastic stromal cells, can engraft across immunologic barriers (Djouad et al. 2003). Together, these studies suggest that tissue-specific fibroblasts are influential players in progression of metastatic cancer. However, with the exception of promoting epithelial to mesenchymal transition (Radisky et al. 2005), the direct biological impact on cancer cells themselves has been difficult to distinguish from indirect mechanisms such as enhanced support for angiogenesis (Orimo et al. 2005) or recruitment of inflammatory cells (Jin et al. 2006; Silzle et al. 2003). Given these data, the role of fibroblasts in cancer progression may, at first glance, appear to benefit tumor growth and decrease overall patient survival. However, there is mounting experimental evidence that healthy tumor microenvironments suppress tumor growth, and it is only after acquisition of tumor-like genetic lesions that fibroblasts appear to promote tumor progression (Kenny and Bissell 2003; Kurose et al. 2001; Hill et al. 2005; McCullough et al. 1998).

While some stromal fibroblasts are recruited to the expanding tumor mass from local tissue fibroblasts (Ronnov-Jessen et al. 1995), recent experimental evidence supports the contention that additional TAF can be recruited from peripheral fibroblast pools, such as bone marrow (BM)-derived mesenchymal stem cells (MSC) (Studeny et al. 2002, 2004), fibrocytes (Direkze et al. 2004), and Tie2-positive cell populations (De Palma et al. 2005). In addition, it was recently demonstrated that BM-derived cells may also be involved in cancer metastasis by pre-establishing a metastatic niche preceding the arrival of cancer cells themselves at secondary sites of tumor growth (Kaplan et al. 2005).

Recent work from our group demonstrated that BM-derived MSC contribute to stroma formation in tumors and provided proof of principle for using MSC as delivery vehicles for antitumor agents (Studeny et al. 2002, 2004; Nakamizo et al. 2005). Some studies have begun to evaluate the feasibility, safety, and practicality of the therapeutic use of MSC in tissue regeneration or cell replacement therapies (Le Blanc and Pittenger 2005; Garcia-Olmo et al. 2005; Bang et al. 2005; Lazarus et al. 2005; Kan et al. 2005; Gojo and Umezawa

2003; Chen et al. 2004; Horwitz et al. 1999), but overall, our understanding of the biological role of MSC within the tumor microenvironment is limited and needs to be improved to better understand the nature of MSC–tumor cell interactions and to ensure high efficacy of MSC-based anticancer strategies. This report will outline our current understanding of the biological impact of MSC as tumor stromal cells and MSC persistence within the tumor microenvironment. We finish our discussion on the use of MSC as cellular delivery vehicles, particularly to deliver anticancer agents directly into the tumor microenvironment, a site where MSC selectively engraft and participate in tumor stroma development. This "Trojan horse" approach uses tumor recruitment of fibroblasts, such as MSC, as a therapeutic tool to deliver MSC that produce potent anticancer agents *in situ*.

2
Understanding the Tumor Microenvironment

Solid tumors are composed of multiple cell types, and tumor viability is dependent on the nonmalignant cells of the tumor microenvironment. The tumor microenvironment is often referred to as "tumor stroma" and includes fibroblasts, vasculature and perivasculature cells, and ECM (Fig. 1; Le Blanc et al. 2004; Tlsty and Hein 2001; Bissell and Radisky 2001; Coussens and Werb 2002; Mueller and Fusenig 2004). Stromal cells provide physical architecture (ECM), growth factors, cytokines, blood supply, and remove metabolic and biological waste (Philip et al. 2004).

Even small tumors (>1–2 mm) require new access to blood vessels for survival (angiogenesis: the sprouting of new blood vessels from existing blood vessels) (Folkman 2003). This process requires complex cellular and molecular cooperation between multiple cell types, including endothelial cells, fibroblasts, and macrophages. Under abnormal conditions, such as wound healing or tumorigenesis, endothelial cells can be activated to divide and form new blood vessels. Frequently detected growth factors that activate endothelial cells include vascular endothelial growth factor (VEGF) and basic fibroblast growth factor (bFGF). Tumor cells, tumor-associated macrophages, and TAF produce VEGF and are involved in tumor-induced angiogenesis (Polverini and Leibovich 1984; Mantovani et al. 2002; Silzle et al. 2004; Dong et al. 2004). Both tumor-associated macrophages and TAF are abundantly detected in tumors and often support tumor growth (Kurose et al. 2001; Richter et al. 1993). The cellular origin of TAF remains unclear, but accumulating evidence suggests that TAF originally derive from resident organ fibroblasts (Kammertoens et al. 2005; Kiaris et al. 2005). While studies have indicated that organ fibroblasts in the proximity of the developing tumor became TAF (Sivridis et al. 2005; Yang et al. 2005), these data do not preclude the possibility that circulating MSC or mesenchymal progenitor (stem) cells directly contribute to the heterogeneous

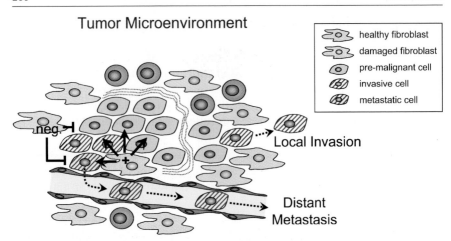

Fig. 1 Tumor–stroma interactions are critical for the survival of the tumor. Stromal cells provide structural support for malignant cells, modulate the tumor microenvironment, and influence biological behavior including the aggressiveness of the malignancy. In response, the tumor provides growth factors, cytokines, and cellular signals that continually initiate new stromal reactions and recruit new cells into the microenvironment to further support tumor growth. It is not fully understood how stroma influences the neoplastic cells, but experimental evidence suggests that healthy tissue microenvironments can suppress tumor growth, and in contrast damaged or altered tumor microenvironments can enhance tumor growth and aggression

organ-specific fibroblast population or the TAF cell pool (Fig. 2; Studeny et al. 2002, 2004; Nakamizo et al. 2005; Prockop 1997; Emura et al. 2000; Ishii et al. 2003; Direkze et al. 2003).

Some TAF are tissue-resident matrix-synthesizing or matrix-degrading cells, whereas others are contractile cells (myofibroblasts), circulating precursor cells (fibrocytes), or blood vessel-associated pericytes. TAF biologically impact the tumor microenvironment through the production of growth factors, cytokines, chemokines, matrix-degrading enzymes, and immunomodulatory mechanisms (Silzle et al. 2004). The BM is a unique and accessible source of multiple lineages of cells with therapeutic value, especially progenitor and stem cells. The adherent fraction of BM cells contains differentiated mesenchymal stromal cells and pluripotent MSC, which give rise to differentiated cells belonging to the osteogenic, chondrogenic, adipogenic, myogenic, and fibroblastic lineages (Dong et al. 2004). Although MSC are routinely recovered from BM, they have also been isolated from a number of other tissues such as muscle, synovium, umbilical cord, and adipose tissue.

In addition to production of numerous growth factors and their support of angiogenesis, TAF provide organization to the tumor stroma by producing ECM components. Thus, several critical events required for tumor growth are not sufficient unless the tumor cells attract and stimulate fibroblasts. Based

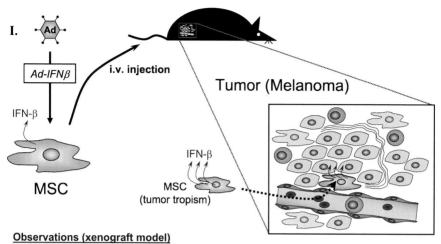

Fig. 2 Engraftment of MSC within human tumor xenografts. We previously demonstrated that human MSC had a strong propensity to engraft and persist within human tumor xenografts. In that study, subcutaneous tumor xenografts were established prior to intravenous injection of human MSC, which were subsequently found within the xenografts tumor stroma. We also established that the therapeutic potential of delivering interferon β (*IFN-β*) to tumors via adenovirus-modified MSC was effective in as little as one MSC-expressing IFN-β per 100 tumor cells. (Studeny et al. 2002)

on our understanding of tumor stroma development (Richter et al. 1993), we propose that stromagenesis is a multistep process involving the concomitant recruitment of local tissue fibroblasts and circulating BM-derived stem cells or MSC from the BM into the tumor, followed by intratumoral proliferation of these cells, and finally, acquisition of several biological TAF characteristics (Emura et al. 2000). Once in the tumor microenvironment, MSC convert into activated myofibroblasts (i.e., TAF) and may also differentiate into fibrocytes, which produce ECM components, and perivascular or vascular structures. Little is known about the dynamics of fibroblast involvement and the molecules that regulate it. As a side note, the reactive stroma associated with many solid tumors exhibits several of the same biological markers seen in tissue stroma at sites of wound repair (Garcia-Olmo et al. 2005). Biological processes in wound repair, including stromal cell acquisition of the myofibroblast phenotype, deposition of type I collagen, and induction of angiogenesis are observed in reactive stroma during cancer progression.

3
Role of Tumor–Stroma Interactions in Tumor Progression

Carcinomas are the most frequent human tumors and arise from epithelial cells that line the inner surfaces of organs. The conversion of normal epithelial cells to metastatic cancer cells is accepted as a multi-stage process that requires progressive genetic alterations within the epithelial tumor cell and has been the focus of intense investigation (reviewed in Tlsty and Hein 2001; Hanahan and Weinberg 2000; Ronnov-Jessen et al. 1996). At the same time, the many other cell types in the tumor microenvironment are increasingly appreciated as components of a complex biological network, akin to an organ system, that are critical for tumor progression (Tlsty and Hein 2001; Hanahan and Weinberg 2000; Ronnov-Jessen et al. 1996). The interactions between cancer cells and endothelial cells that lead to tumor vascularization are one example of tumor-microenvironment communication that has received considerable attention over the past decade and is the first example demonstrating that targeting noncancer cells in the microenvironment can have benefit to cancer patients (Folkman 2004).

While it has been appreciated that the molecular dialog between cancer cells and other cells in the microenvironment is necessary for tumor development, the cancer cells were often thought to primarily initiate and orchestrate the conversations with other cell types (Rowley 1998; Haffen et al. 1987). Thus, the microenvironment could be considered a reactive participant in tumor progression, responding to signals initiated by an altered genetic program in the epithelial cell. This view is at odds with what has been learned from studying epithelial cells, especially mammary epithelial cells, during embryonic development, where it is clear that mesenchymal cells, such as fibroblasts and adipocytes, play an active and instructive role in programming epithelial cell structure and function (Cunha and Hom 1996; Gallego et al. 2001; Sakakura et al. 1976; Wiesen et al. 1999). The bidirectional molecular dialog between mesenchymal and epithelial cell types is critical for normal organ development and function (Cunha and Hom 1996; Gallego et al. 2001; Sakakura et al. 1976; Wiesen et al. 1999), and it is highly probable that such interactions are equally important during tumor progression.

There is a rapidly expanding literature documenting changes in the mammary tumor stromal fibroblasts during cancer progression, including the possible role of stromal interactions with tumor stem cells (reviewed in Ronnov-Jessen et al. 1996; Bissell and Labarge 2005). For example, oncogene-primed epithelial cells form tumors more rapidly when transplanted into stroma that has been exposed to X-ray radiation before transplantation than when transplanted into untreated stroma (Barcellos-Hoff and Ravani 2000). Another example is work demonstrating that directed expression of a self-activating form of the metalloprotease MMP3/stromelysin 1, a gene product normally expressed in stromal cells during development, to epithelial cells can lead to extensive remodeling of the stroma, genetic changes in the epithelial cells, and tumorigenesis (Radisky et al. 2005; Sternlicht et al. 1999, 2000). Recently, human genetic evidence supports this active role of tumor stroma in breast cancer progression by demonstrating somatic mutations in TP53 and PTEN genes in both breast neoplastic epithelium and tumor stroma (Kurose et al. 2002). Taken together, these results indicate that the action of genes in multiple cell types is required for normal development and for tumor progression.

Although damaged tissue microenvironments may serve to promote tumor growth and progression, there is strong evidence that healthy tissue microenvironments effectively repress tumor growth. This dominant influence was first reported over 30 years ago from experiments using teratocarcinoma cell-injected blastocysts to produce tumor-free, viable mice (Mintz and Illmensee 1975). The malignant potential of the teratocarcinoma cells were suppressed as suggested by the generation of healthy mice, with many tissues originated from the teratocarcinoma cells (also reviewed in: Kenny and Bissell 2003). Another example of the protective effect of a healthy tissue microenvironment is observed in the prostate cancer mouse model ($TgAPT_{121}$), where a truncated form of the SV40 large T antigen is expressed in prostate epithelial cells to dysregulate cell cycle control while leaving p53 function intact (Hill et al. 2005). When $TgAPT_{121}$ mice were generated on wildtype, hemizygous, or nullizygous p53 backgrounds (i.e., damaged microenvironments), increased rates of disease progression and death were observed corresponding to progressive p53 loss (e.g., $p53^{-/-} > p53^{+/-} > p53^{+/+}$) (Hill et al. 2005). Finally, the reduced ability of aged microenvironments to repress tumor growth was demonstrated in an elegant study by McCullough et al. (1998). Young (3 months old) and old (19 months old) Fischer 344 rats were challenged with the congenic hepatocarcinoma cell line, BAG2-GN6TF, and while the old rats rapidly developed aggressive liver carcinomas, young rats suppressed tumor growth through apparent promotion of tumor cell differentiation. Interestingly, suppressed hepatocarcinoma cells maintained their malignant potential within young rat livers, because when these cells were re-isolated from young rat livers and transplanted into aged rats, they rapidly formed aggressive, undifferentiated tumors (McCullough et al. 1998). Taken together, these data suggest that healthy tissue microenvironments serve to prevent tumor outgrowth, and it is only when tumor stroma becomes damaged (by age, carcinogens, or tumor-derived pressure) that tumor stroma supports tumor progression.

4
The Participation of MSC in Tumors and in Wound Healing

Homing of MSC after systemic or local infusion has been studied in animal models in a variety of experimental settings (Allers et al. 2004; Almeida-Porada et al. 2004; Ortiz et al. 2003). A study by Erices et al. (2003) reported the systemic infusion of MSC in irradiated syngeneic mice. After 1 month, 8% of bone cells and 5% of lung cells in the recipient mice were positive for the transplanted cells. In a separate study, baboons received labeled MSC and were detected in the BM more than 500 days after transplantation (Deans and Moseley 2000). Apparently, the native capacity of MSC to adhere to matrix components favors their preferential homing to bone, lungs, and cartilage when injected intravenously. However, conditioning regimens prior to cell transplantation (such as irradiation or chemotherapy treatment) may greatly influence the efficiency and sites of MSC homing. In this regard, a growing number of studies of various pathological conditions have demonstrated that MSC selectively home to sites of injury, irrespective of the tissue or organ (Lange et al. 2005; Rojas et al. 2005; Phinney and Isakova 2005; Sato et al. 2005; Natsu et al. 2004; Silva et al. 2005). This ability of MSC has been demonstrated in brain injury (Kurozumi et al. 2004; Dai et al. 2005), wound healing, and tissue regeneration (Almeida-Porada et al. 2004; Mansilla et al. 2005; Satoh et al. 2004). Evidence suggests that tumors can be considered sites of tissue damage, or "wounds that never heal" (Dvorak 1986), as well as sites of potential inflammatory cytokine and chemokine production. Thus, these properties may enable MSC to home and deliver therapeutic agents to tumors.

A number of factors have been implicated in the homing of BM cells to sites of injury, yet it is still unknown what processes and factors underlie MSC homing to tumors (Fig. 3). In wound repair, as in cancer, cells that usually divide infrequently are induced to proliferate rapidly, the ECM is invaded, connective tissues are remodeled, epithelial and stromal cells migrate, and new blood vessels are formed. We speculate that many of the same factors involved in wound healing are upregulated in the tumor microenvironment to initiate the homing process.

5
Rationale for Using MSC as Cellular Delivery Vehicles

The BM is a unique and accessible source of multiple lineages of cells with potential therapeutic value, especially progenitor and stem cells. The adherent fraction of BM cells contains differentiated mesenchymal BM stromal cells and pluripotent MSC, which give rise to differentiated cells belonging to the osteogenic, chondrogenic, adipogenic, myogenic, and fibroblastic lineages (Prockop 1997). As mentioned above, MSC are routinely recovered from BM; however, they have also been isolated from a number of other tissues such as muscle, synovium, umbilical cord, and adipose tissue (Bianchi et al. 2001; Pittenger et al. 1999; Gronthos et al. 2003; Reyes et al. 2001). There has been increasing interest in recent years in the BM stromal cell system in various fields of cell therapy, such as hematopoietic stem cell transplantation and connective tissue engineering. In these studies (Lazarus et al. 2005; Baksh et al. 2004; Devine

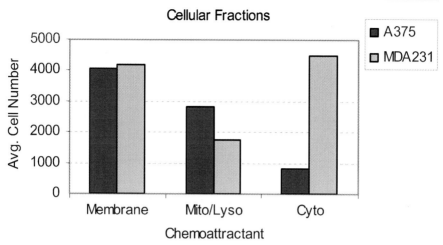

Fig. 3 MSC tropism for tumor cell components. In order to assess what attracts MSC toward tumor xenografts, an *in vitro* migration assay was performed. Tumor cells (either A375 melanoma or MDA-MB-231 breast carcinoma) were harvested, lysed, and fractionated into cytosolic or membrane fractions. These fractions were placed in the lower chamber of the Chemicon cell migration assay well. Culture inserts containing known amounts of preseeded MSC were placed into these wells and MSC were allowed to migrate toward cell fractions for 8 h. Migration was assessed by staining the containment well of the upper chamber; MSC were briefly stained with crystal violet and visually counted under a 40× microscope. The data shown enumerate the average number of migrated cells from three wells. These data suggest that isolated membrane fractions of tumor cells appear to contain potent MSC attractants, more so than cytoplasmic fractions from the same cells. As described in the text, characterization of the exact chemoattractants is under investigation

et al. 2003; Fukuda 2003), MSC have been mainly used to support hematopoietic cell engraftment, repair tissue defects, or trigger regeneration of various mesenchymal tissues (Koc et al. 2002; Ballas et al. 2002; McNiece et al. 2004).

Many observations support the rationale for using MSC as delivery vehicles for antitumor agents. We and others have observed dynamic interactions between BM-derived MSC and tumor cell lines *in vitro* (Fig. 4; Studeny et al. 2002, 2004; Nakamizo et al. 2005; Zhu et al. 2006; Houghton et al. 2004; Prindull and Zipori 2004; Hombauer and Minguell 2000), and we have shown *in vivo* that human MSC engraft and persist within existing tumor microenvironments (Studeny et al. 2002, 2004; De Palma et al. 2003, 2005; Nakamizo et al. 2005). In addition, MSC phenotypically resemble TAF in the presence of transforming growth factor (TGF)-β1 or in coculture with tumor cells (Wright et al. 2003; Forbes et al. 2004). Finally, two independent studies have shown that BM fibroblasts contribute to the development of stromal cell populations in tumors in mice (Direkze et al. 2004; Ishii et al. 2005).

Elegant work by Ishii et al. strongly implicates BM-derived cells in the development of tumors (Ishii et al. 2003). In their study, severe combined immunod-

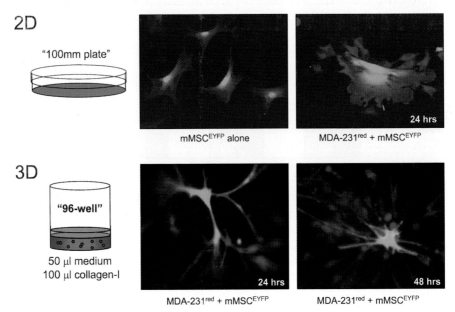

Fig. 4 Dynamic 2D and 3D interactions between murine MSC and human tumor cells *in vitro*. Enhanced yellow fluorescent protein (EYFP)-labeled murine mesenchymal stem cells ($mMSC^{EYFP}$) were cocultured with dsRed-labeled tumor cells (MDA-231^{red}) under two-dimensional (*2D*) and three-dimensional (*3D*) conditions. In the *top two panels* murine MSC were grown in traditional 2D culture conditions alone or with MDA-MB-231 (10:1 tumor:MSC ratio). Under 3D culture conditions, the interplay between MSC and human breast cancer cells appears more robust (3D matrix = 1 mg/ml bovine collagen I). We have observed great variability between individual tumor cells lines in their interactions with MSC

eficient (SCID) mice received BM cells from β-gal$^+$ and recombination activating gene-1 (RAG-1)-deficient mice carrying a unique major histocompatibility complex molecule (H2Kb). Once engraftment was observed in the recipient mice, pancreatic tumors were implanted and allowed to progress. The tumors were subsequently removed, and the contribution of β-gal$^+$ cells to the tumor microenvironment was assessed. The authors found that a small proportion (13%) of H2Kb$^+$α-SMA$^+$ double-positive myofibroblasts were present around the developing malignancy 2 weeks after tumor xenotransplantation, suggesting that a small number of BM-derived myofibroblasts are incorporated into the cancer stroma during the early stages of tumor growth. By 4 weeks after tumor implantation, approximately 40% of the myofibroblasts in the cancer-induced stroma were of BM origin. Immunohistochemical staining of serial sections revealed that non-BM-derived myofibroblasts were adjacent to the cancer nest, whereas BM-derived myofibroblasts were outside of the non-BM-derived myofibroblasts, indicating that most BM-derived myofibroblasts incorporate into the cancer-induced stroma mainly in the late stage of tumor development.

Taking these data into consideration, it has been suggested that BM-derived myofibroblasts contribute to the pathogenesis of cancer-induced desmoplastic reaction by altering the tumor microenvironment. The findings of Ishii et al. also suggest that the local tissue reaction starts at the circumference of a neoplasm in the early stage of tumor development. However, it is unclear from these data whether the tumor-resident β-gal$^+$ myofibroblasts are generated from a circulating stem cell population originating from an adherent BM population (such as MSC or endothelial precursors) or from a circulating hematopoietic population. Nevertheless, the findings demonstrate a specific contribution of BM-derived cells to the developing tumor stroma, which strongly suggests that BM-derived cells target tumors as a result of physiological requests by the expanding tumor (Sangai et al. 2005; Sugimoto et al. 2005). The study by Ishii et al. also support our initial observations that tumors recruit cells from the circulation and that this population of stromal precursor cells could be used to target novel therapies to growing tumors. A number of other groups have since reported similar findings in ovarian, rhabdomyosarcoma, and breast tumor models (Roni et al. 2003; Jankowski et al. 2003; Yoneda and Hiraga 2005).

Direct targeting of anticancer agents into the tumor microenvironment would increase their efficacy (Burns and Weiss 2003; Dennis et al. 2004). There are several advantages to using BM-derived MSC as cellular delivery vehicles: (1) Most invasive cancers are likely to induce a desmoplastic reaction to some extent, thereby providing a common target for the treatment of many types of cancers; (2) MSC have a low intrinsic mutation rate and are therefore less likely than the genetically unstable cancer cells to acquire a drug-resistant phenotype; (3) MSC are simple to isolate and culture (Kassem 2004); (4) they can engraft after *ex vivo* expansion (Schoeberlein et al. 2005; Ye et al. 2006); (5) they have high metabolic activity and efficient machinery for the secretion of therapeutic proteins (Chan et al. 2005); (6) MSC can be expanded for more than 50 population doublings in culture without the loss of their phenotype or multilineage potential, which is in striking contrast to terminally differentiated cells such as endothelial or muscle cells (In 't Anker et al. 2004).

6
Experimental Support for Use of MSC as Therapeutic Agents

Practical attempts to use MSC as cellular delivery vehicles have focused mainly on delivery of therapeutic gene products (such as interleukin-3, growth hormone, and factor IX) via the systemic circulation (Chan et al. 2005; Hurwitz et al. 1997; Evans et al. 2005). Recent studies from our group (Studeny et al. 2002, 2004; Nakamizo et al. 2005) have reported that exogenously administered MSC migrate and preferentially survive and proliferate within tumor masses. Once in the tumor microenvironment, MSC incorporate into the tumor architecture and serve as precursors for stromal elements, predominantly fibroblasts. The preferential distribution of MSC in lung tumor nodules, but not in lung parenchyma, was demonstrated after systemic administration of MSC in mice bearing melanoma xenografts. In addition, when MSC were systemically injected into mice with subcutaneously established tumors, MSC-derived fibroblasts were consistently identified in tumors but not in healthy organs (Pittenger et al. 1999).

We also examined the therapeutic potential of MSC to deliver interferon β (IFN-β) into the tumor microenvironment. Our data showed that IFN-β acts through local paracrine effects after it is delivered into tumors by cellular *in vitro*-modified MSC "mini-pumps," thus emphasizing the importance of tumor-targeted delivery of a cytokine. No survival benefits were seen when the same MSC were made to produce systemic IFN-β at sites distant from tumors or when recombinant IFN-β protein was delivered systemically without a carrier (Studeny et al. 2004). MSC may also have similar capabilities to specifically home to and selectively engraft into established gliomas *in vivo*. Data from our group demonstrated that MSC injected into the carotid artery of mice can specifically target U87 gliomas and that the MSC can migrate from either the contralateral or the ipsilateral carotid artery, suggesting that blood flow or a perfusion effect is not directly responsible for getting the MSC to the tumor (Nakamizo et al. 2005). When those MSC were armed to secrete IFN-β, we observed statistically significant decreases in tumor size, suggesting that MSC producing IFN-β were able to control tumor growth kinetics. Additional work by Hamada et al. has shown that genetically modified MSC expressing interleukin-2 can also control tumor development when injected intratumorally into established gliomas, again suggesting that intratumoral production of this immunostimulatory cytokine is responsible for controlling tumor kinetics (Kurozumi et al. 2005; Honma et al. 2006). Thus, the cultured BM-adherent cell population contains cells with extraordinarily high proliferative capacity that can contribute to the maintenance of both tumor stroma and connective tissue in organs remote from the BM (Hamada et al. 2005). At the same time, the successful engraftment of MSC in tissues would most likely take place only in a state of increased cell turnover triggered by tissue damage or tumor growth. This property makes MSC excellent candidates for the cell-based delivery of therapeutics to tumor sites.

7
Future Directions in Designing MSC-Based Delivery Strategies

Potential concerns in using MSC as delivery vehicles stem from how little we understand about the homeostatic maintenance of this cell population *in vivo* (Baksh et al. 2004; Caplan and Bruder 2001; Feldmann et al. 2005) and the possibility that MSC themselves might enhance or initiate tumor growth (Parham 2001; Iacobuzio-Donahue et al. 2002). However, unlike homeostatic MSC niches, tumor microenvironments are pathologically altered tissues (see discussion above) that resemble unresolved wounds (Rowley 1998; Dvorak 1986; Robinson and Coussens 2005), and the studies that reported tumorigenic properties of MSC used extensively passaged and/or genetically altered MSC, consistent with the notion that damaged tumor stroma can enhance tumor progression (Hill et al. 2005; McCullough et al. 1998; Serakinci et al. 2004; Rubio et al. 2005). Additional studies have demonstrated that genetic alterations must accumulate in tumor stromal cells, presumably over time, before they can enhance tumor growth (Burns et al. 2005; Fierro et al. 2004; Cunha et al. 2003; Chung et al. 2005). We and others have shown that lower-passage MSC do not form tumors *in vivo* (Ohlsson et al. 2003; Xia et al. 2004; De Kok et al. 2005; Chen et al. 2005).

Although the use of stem cells for cancer gene therapy usually provides some degree of tumor selectivity, strategies to improve tumor homing may greatly increase the applicability and success of tumor-targeted cell delivery vehicles (Harrington et al. 2002). One can envision two major approaches for increased targeting. First, other cell populations with specific properties can be identified and tested, such as more primitive stem cells with higher replicative or migratory potential, such as the "RS" phenotype (Prockop et al. 2001). For instance, isolation methods could be developed to specifically select cell types with innate specific targeting or other phenotypic properties that can be exploited to enhance tumor targeting or incorporation into the tumor microenvironment (Sekiya et al. 2002). Second, cell homing may be manipulated by using specific culture agents or medium treatments to alter the expression of cell surface receptors (Fiedler et al. 2004; Kraitchman et al. 2005). It might be possible to manipulate cell targeting or homing properties by choosing specific pretreatment or isolation protocols. These two approaches represent another area of investigation to enhance the efficacy of cellular vehicle strategies.

8
Conclusions

In this review, we have described recent progress in the development of BM-derived stem cell populations, particularly MSC, for use as cellular delivery vehicles with specific tropisms for solid tumors and their respective tumor microenvironments. This tumor tropism is based primarily on the innate physiological ability of MSC to home to sites of inflammation and tissue repair. The most obvious advantage offered by MSC is the local delivery and intratumoral release of therapeutic agents, especially proteins with poor pharmacokinetic profiles. An added benefit of cell-based treatments is the possibility of *ex vivo* manipulation to maximize cell loading, as shown by our direct comparison of therapeutic effects provided by cell-based or systemic injection of recombinant proteins. In addition to delivering cancer-related protein substances, cellular vehicles could also serve as protective "coatings" for the delivery of payloads such as replicating oncoselective viruses [e.g., Onyx-15 (Forte et al. 2006) or Delta 24 (Jiang et al. 2003)]. Not only would the cells protect these viruses from the immune system, but they would also allow the amplification of the initial viral load with the possibility of increasing viral spread through cell-to-cell contact specifically at tumor sites.

Much work remains to be done before the potential therapeutic benefits of MSC-based approaches can be fully exploited. The development of optimally targeted cellular vectors will require advances in cell science and virology and will certainly benefit from continued innovative application of knowledge gained from basic biology, tissue stem cell engineering, and gene therapy.

References

Allers C, Sierralta WD, Neubauer S, Rivera F, Minguell JJ, Conget PA (2004) Dynamic of distribution of human bone marrow-derived mesenchymal stem cells after transplantation into adult unconditioned mice. Transplantation 78:503–508

Almeida-Porada G, Porada C, Zanjani ED (2004) Plasticity of human stem cells in the fetal sheep model of human stem cell transplantation. Int J Hematol 79:1–6

Baksh D, Song L, Tuan RS (2004) Adult mesenchymal stem cells: characterization, differentiation, and application in cell and gene therapy. J Cell Mol Med 8:301–316

Ballas CB, Zielske SP, Gerson SL (2002) Adult bone marrow stem cells for cell and gene therapies: implications for greater use. J Cell Biochem Suppl 38:20–28

Bang OY, Lee JS, Lee PH, Lee G (2005) Autologous mesenchymal stem cell transplantation in stroke patients. Ann Neurol 57:874–882

Barcellos-Hoff MH, Ravani SA (2000) Irradiated mammary gland stroma promotes the expression of tumorigenic potential by unirradiated epithelial cells. Cancer Res 60:1254–1260

Bianchi G, Muraglia A, Daga A, Corte G, Cancedda R, Quarto R (2001) Microenvironment and stem properties of bone marrow-derived mesenchymal cells. Wound Repair Regen 9:460–466

Bissell MJ, Labarge MA (2005) Context, tissue plasticity, and cancer: are tumor stem cells also regulated by the microenvironment? Cancer Cell 7:17–23

Bissell MJ, Radisky D (2001) Putting tumours in context. Nat Rev Cancer 1:46–54

Burns JS, Abdallah BM, Guldberg P, Rygaard J, Schroder HD, Kassem M (2005) Tumorigenic heterogeneity in cancer stem cells evolved from long-term cultures of telomerase-immortalized human mesenchymal stem cells. Cancer Res 65:3126–3135

Burns MJ, Weiss W (2003) Targeted therapy of brain tumors utilizing neural stem and progenitor cells. Front Biosci 8:e228–234

Caplan AI, Bruder SP (2001) Mesenchymal stem cells: building blocks for molecular medicine in the 21st century. Trends Mol Med 7:259–264

Chan J, O'Donoghue K, de la Fuente J, et al (2005) Human fetal mesenchymal stem cells as vehicles for gene delivery. Stem Cells 23:93–102

Chen J, Wang C, Lu S, et al (2005) In vivo chondrogenesis of adult bone-marrow-derived autologous mesenchymal stem cells. Cell Tissue Res 319:429–438

Chen SL, Fang WW, Ye F, et al (2004) Effect on left ventricular function of intracoronary transplantation of autologous bone marrow mesenchymal stem cell in patients with acute myocardial infarction. Am J Cardiol 94:92–95

Chung LW, Baseman A, Assikis V, Zhau HE (2005) Molecular insights into prostate cancer progression: the missing link of tumor microenvironment. J Urol 173:10–20

Coussens LM, Werb Z (2002) Inflammation and cancer. Nature 420:860–867

Cunha GR, Hom YK (1996) Role of mesenchymal-epithelial interactions in mammary gland development. J Mammary Gland Biol Neoplasia 1:21–35

Cunha GR, Hayward SW, Wang YZ, Ricke WA (2003) Role of the stromal microenvironment in carcinogenesis of the prostate. Int J Cancer 107:1–10

Dai W, Hale SL, Martin BJ, et al (2005) Allogeneic mesenchymal stem cell transplantation in postinfarcted rat myocardium: short- and long-term effects. Circulation 112:214–223

De Kok IJ, Drapeau SJ, Young R, Cooper LF (2005) Evaluation of mesenchymal stem cells following implantation in alveolar sockets: a canine safety study. Int J Oral Maxillofac Implants 20:511–518

De Palma M, Venneri MA, Roca C, Naldini L (2003) Targeting exogenous genes to tumor angiogenesis by transplantation of genetically modified hematopoietic stem cells. Nat Med 9:789–795

De Palma M, Venneri MA, Galli R, et al (2005) Tie2 identifies a hematopoietic lineage of proangiogenic monocytes required for tumor vessel formation and a mesenchymal population of pericyte progenitors. Cancer Cell 8:211–226

De Wever O, Mareel M (2003) Role of tissue stroma in cancer cell invasion. J Pathol 200:429–447

Deans RJ, Moseley AB (2000) Mesenchymal stem cells: biology and potential clinical uses. Exp Hematol 28:875–884

Dennis JE, Cohen N, Goldberg VM, Caplan AI (2004) Targeted delivery of progenitor cells for cartilage repair. J Orthop Res 22:735–741

Devine SM, Cobbs C, Jennings M, Bartholomew A, Hoffman R (2003) Mesenchymal stem cells distribute to a wide range of tissues following systemic infusion into nonhuman primates. Blood 101:2999–3001

Direkze NC, Forbes SJ, Brittan M, et al (2003) Multiple organ engraftment by bone-marrow-derived myofibroblasts and fibroblasts in bone-marrow-transplanted mice. Stem Cells 21:514–520

Direkze NC, Hodivala-Dilke K, Jeffery R, et al (2004) Bone marrow contribution to tumor-associated myofibroblasts and fibroblasts. Cancer Res 64:8492–8495

Djouad F, Plence P, Bony C, et al (2003) Immunosuppressive effect of mesenchymal stem cells favors tumor growth in allogeneic animals. Blood 102:3837–3844

Dong J, Grunstein J, Tejada M, et al (2004) VEGF-null cells require PDGFR alpha signaling-mediated stromal fibroblast recruitment for tumorigenesis. EMBO J 23:2800–2810

Dvorak HF (1986) Tumors: wounds that do not heal. Similarities between tumor stroma generation and wound healing. N Engl J Med 315:1650–1659

Emura M, Ochiai A, Horino M, Arndt W, Kamino K, Hirohashi S (2000) Development of myofibroblasts from human bone marrow mesenchymal stem cells cocultured with human colon carcinoma cells and TGF beta 1. In Vitro Cell Dev Biol Anim 36:77–80

Erices AA, Allers CI, Conget PA, Rojas CV, Minguell JJ (2003) Human cord blood-derived mesenchymal stem cells home and survive in the marrow of immunodeficient mice after systemic infusion. Cell Transplant 12:555–561

Evans CH, Robbins PD, Ghivizzani SC, et al (2005) Gene transfer to human joints: progress toward a gene therapy of arthritis. Proc Natl Acad Sci U S A 102:8698–8703

Feldmann RE Jr, Bieback K, Maurer MH, et al (2005) Stem cell proteomes: a profile of human mesenchymal stem cells derived from umbilical cord blood. Electrophoresis 26:2749–2758

Fiedler J, Etzel N, Brenner RE (2004) To go or not to go: migration of human mesenchymal progenitor cells stimulated by isoforms of PDGF. J Cell Biochem 93:990–998

Fierro FA, Sierralta WD, Epunan MJ, Minguell JJ (2004) Marrow-derived mesenchymal stem cells: role in epithelial tumor cell determination. Clin Exp Metastasis 21:313–319

Folkman J (2003) Fundamental concepts of the angiogenic process. Curr Mol Med 3:643–651

Folkman J (2004) Endogenous angiogenesis inhibitors. Apmis 112:496–507

Forbes SJ, Russo FP, Rey V, et al (2004) A significant proportion of myofibroblasts are of bone marrow origin in human liver fibrosis. Gastroenterology 126:955–963

Forte G, Minieri M, Cossa P, et al (2006) Hepatocyte growth factor effects on mesenchymal stem cells: proliferation, migration, and differentiation. Stem Cells 24:23–33

Fukuda K (2003) Use of adult marrow mesenchymal stem cells for regeneration of cardiomyocytes. Bone Marrow Transplant 32 [Suppl 1]:S25–27

Gallego MI, Binart N, Robinson GW, et al (2001) Prolactin, growth hormone, and epidermal growth factor activate Stat5 in different compartments of mammary tissue and exert different and overlapping developmental effects. Dev Biol 229:163–175

Garcia-Olmo D, Garcia-Arranz M, Herreros D, Pascual I, Peiro C, Rodriguez-Montes JA (2005) A phase I clinical trial of the treatment of Crohn's fistula by adipose mesenchymal stem cell transplantation. Dis Colon Rectum 48:1416–1423

Gojo S, Umezawa A (2003) Plasticity of mesenchymal stem cells—regenerative medicine for diseased hearts. Hum Cell 16:23–30

Gronthos S, Zannettino AC, Hay SJ, et al (2003) Molecular and cellular characterisation of highly purified stromal stem cells derived from human bone marrow. J Cell Sci 116:1827–1835

Haffen K, Kedinger M, Simon-Assmann P (1987) Mesenchyme-dependent differentiation of epithelial progenitor cells in the gut. J Pediatr Gastroenterol Nutr 6:14–23

Hamada H, Kobune M, Nakamura K, et al (2005) Mesenchymal stem cells (MSC) as therapeutic cytoreagents for gene therapy. Cancer Sci 96:149–156

Hanahan D, Weinberg RA (2000) The hallmarks of cancer. Cell 100:57–70

Harrington K, Alvarez-Vallina L, Crittenden M, et al (2002) Cells as vehicles for cancer gene therapy: the missing link between targeted vectors and systemic delivery? Hum Gene Ther 13:1263–1280

Hasebe T, Mukai K, Tsuda H, Ochiai A (2000) New prognostic histological parameter of invasive ductal carcinoma of the breast: clinicopathological significance of fibrotic focus. Pathol Int 50:263–272

Hill R, Song Y, Cardiff RD, Van Dyke T (2005) Selective evolution of stromal mesenchyme with p53 loss in response to epithelial tumorigenesis. Cell 123:1001–1011

Hombauer H, Minguell JJ (2000) Selective interactions between epithelial tumour cells and bone marrow mesenchymal stem cells. Br J Cancer 82:1290–1296

Honma T, Honmou O, Iihoshi S, et al (2006) Intravenous infusion of immortalized human mesenchymal stem cells protects against injury in a cerebral ischemia model in adult rat. Exp Neurol 199:56–66

Horwitz EM, Prockop DJ, Fitzpatrick LA, et al (1999) Transplantability and therapeutic effects of bone marrow-derived mesenchymal cells in children with osteogenesis imperfecta. Nat Med 5:309–313

Houghton J, Stoicov C, Nomura S, et al (2004) Gastric cancer originating from bone marrow-derived cells. Science 306:1568–1571

Hurwitz DR, Kirchgesser M, Merrill W, et al (1997) Systemic delivery of human growth hormone or human factor IX in dogs by reintroduced genetically modified autologous bone marrow stromal cells. Hum Gene Ther 8:137–156

Iacobuzio-Donahue CA, Argani P, Hempen PM, Jones J, Kern SE (2002) The desmoplastic response to infiltrating breast carcinoma: gene expression at the site of primary invasion and implications for comparisons between tumor types. Cancer Res 62:5351–5357

In 't Anker PS, Scherjon SA, Kleijburg-van der Keur C, et al (2004) Isolation of mesenchymal stem cells of fetal or maternal origin from human placenta. Stem Cells 22:1338–1345

Ishii G, Sangai T, Oda T, et al (2003) Bone-marrow-derived myofibroblasts contribute to the cancer-induced stromal reaction. Biochem Biophys Res Commun 309:232–240

Ishii G, Sangai T, Ito T, et al (2005) *In vivo* and *in vitro* characterization of human fibroblasts recruited selectively into human cancer stroma. Int J Cancer 117:212–220

Jankowski K, Kucia M, Wysoczynski M, et al (2003) Both hepatocyte growth factor (HGF) and stromal-derived factor-1 regulate the metastatic behavior of human rhabdomyosarcoma cells, but only HGF enhances their resistance to radiochemotherapy. Cancer Res 63:7926–7935

Jiang H, Conrad C, Fueyo J, Gomez-Manzano C, Liu TJ (2003) Oncolytic adenoviruses for malignant glioma therapy. Front Biosci 8:d577–588

Jin H, Su J, Garmy-Susini B, Kleeman J, Varner J (2006) Integrin alpha4beta1 promotes monocyte trafficking and angiogenesis in tumors. Cancer Res 66:2146–2152

Kammertoens T, Schuler T, Blankenstein T (2005) Immunotherapy: target the stroma to hit the tumor. Trends Mol Med 11:225–231

Kan I, Melamed E, Offen D (2005) Integral therapeutic potential of bone marrow mesenchymal stem cells. Curr Drug Targets 6:31–41

Kaplan RN, Riba RD, Zacharoulis S, et al (2005) VEGFR1-positive haematopoietic bone marrow progenitors initiate the pre-metastatic niche. Nature 438:820–827

Kassem M (2004) Mesenchymal stem cells: biological characteristics and potential clinical applications. Cloning Stem Cells 6:369–374

Kenny PA, Bissell MJ (2003) Tumor reversion: correction of malignant behavior by microenvironmental cues. Int J Cancer 107:688–695

Kiaris H, Chatzistamou I, Trimis G, Frangou-Plemmenou M, Pafiti-Kondi A, Kalofoutis A (2005) Evidence for nonautonomous effect of p53 tumor suppressor in carcinogenesis. Cancer Res 65:1627–1630

Koc ON, Day J, Nieder M, Gerson SL, Lazarus HM, Krivit W (2002) Allogeneic mesenchymal stem cell infusion for treatment of metachromatic leukodystrophy (MLD) and Hurler syndrome (MPS-IH). Bone Marrow Transplant 30:215–222

Kraitchman DL, Tatsumi M, Gilson WD, et al (2005) Dynamic imaging of allogeneic mesenchymal stem cells trafficking to myocardial infarction. Circulation 112:1451–1461

Kunz-Schughart LA, Knuechel R (2002a) Tumor-associated fibroblasts (part I): active stromal participants in tumor development and progression? Histol Histopathol 17:599–621

Kunz-Schughart LA, Knuechel R (2002b) Tumor-associated fibroblasts (part II): functional impact on tumor tissue. Histol Histopathol 17:623–637

Kurose K, Hoshaw-Woodard S, Adeyinka A, Lemeshow S, Watson PH, Eng C (2001) Genetic model of multi-step breast carcinogenesis involving the epithelium and stroma: clues to tumour-microenvironment interactions. Hum Mol Genet 10:1907–1913

Kurose K, Gilley K, Matsumoto S, Watson PH, Zhou XP, Eng C (2002) Frequent somatic mutations in PTEN and TP53 are mutually exclusive in the stroma of breast carcinomas. Nat Genet 32:355–357

Kurosumi M, Tabei T, Inoue K, et al (2003) Prognostic significance of scoring system based on histological heterogeneity of invasive ductal carcinoma for node-negative breast cancer patients. Oncol Rep 10:833–837

Kurozumi K, Nakamura K, Tamiya T, et al (2004) BDNF gene-modified mesenchymal stem cells promote functional recovery and reduce infarct size in the rat middle cerebral artery occlusion model. Mol Ther 9:189–197

Kurozumi K, Nakamura K, Tamiya T, et al (2005) Mesenchymal stem cells that produce neurotrophic factors reduce ischemic damage in the rat middle cerebral artery occlusion model. Mol Ther 11:96–104

Lange C, Togel F, Ittrich H, et al (2005) Administered mesenchymal stem cells enhance recovery from ischemia/reperfusion-induced acute renal failure in rats. Kidney Int 68:1613–1617

Lazarus HM, Koc ON, Devine SM, et al (2005) Cotransplantation of HLA-identical sibling culture-expanded mesenchymal stem cells and hematopoietic stem cells in hematologic malignancy patients. Biol Blood Marrow Transplant 11:389–398

Le Blanc K, Pittenger M (2005) Mesenchymal stem cells: progress toward promise. Cytotherapy 7:36–45

Le Blanc K, Rasmusson I, Sundberg B, et al (2004) Treatment of severe acute graft-versus-host disease with third party haploidentical mesenchymal stem cells. Lancet 363:1439–1441

Mansilla E, Marin GH, Sturla F, et al (2005) Human mesenchymal stem cells are tolerized by mice and improve skin and spinal cord injuries. Transplant Proc 37:292–294

Mantovani A, Sozzani S, Locati M, Allavena P, Sica A (2002) Macrophage polarization: tumor-associated macrophages as a paradigm for polarized M2 mononuclear phagocytes. Trends Immunol 23:549–555

McCullough KD, Coleman WB, Ricketts SL, Wilson JW, Smith GJ, Grisham JW (1998) Plasticity of the neoplastic phenotype *in vivo* is regulated by epigenetic factors. Proc Natl Acad Sci U S A 95:15333–15338

McNiece I, Harrington J, Turney J, Kellner J, Shpall EJ (2004) *Ex vivo* expansion of cord blood mononuclear cells on mesenchymal stem cells. Cytotherapy 6:311–317

Mintz B, Illmensee K (1975) Normal genetically mosaic mice produced from malignant teratocarcinoma cells. Proc Natl Acad Sci U S A 72:3585–3589

Moinfar F, Man YG, Arnould L, Bratthauer GL, Ratschek M, Tavassoli FA (2000) Concurrent and independent genetic alterations in the stromal and epithelial cells of mammary carcinoma: implications for tumorigenesis. Cancer Res 60:2562–2566

Mueller MM, Fusenig NE (2004) Friends or foes—bipolar effects of the tumour stroma in cancer. Nat Rev Cancer 4:839–849

Nakamizo A, Marini F, Amano T, et al (2005) Human bone marrow-derived mesenchymal stem cells in the treatment of gliomas. Cancer Res 65:3307–3318

Natsu K, Ochi M, Mochizuki Y, Hachisuka H, Yanada S, Yasunaga Y (2004) Allogeneic bone marrow-derived mesenchymal stromal cells promote the regeneration of injured skeletal muscle without differentiation into myofibers. Tissue Eng 10:1093–1112

Ohlsson LB, Varas L, Kjellman C, Edvardsen K, Lindvall M (2003) Mesenchymal progenitor cell-mediated inhibition of tumor growth *in vivo* and *in vitro* in gelatin matrix. Exp Mol Pathol 75:248–255

Orimo A, Gupta PB, Sgroi DC, et al (2005) Stromal fibroblasts present in invasive human breast carcinomas promote tumor growth and angiogenesis through elevated SDF-1/CXCL12 secretion. Cell 121:335–348

Ortiz LA, Gambelli F, McBride C, et al (2003) Mesenchymal stem cell engraftment in lung is enhanced in response to bleomycin exposure and ameliorates its fibrotic effects. Proc Natl Acad Sci U S A 100:8407–8411

Parham DM (2001) Pathologic classification of rhabdomyosarcomas and correlations with molecular studies. Mod Pathol 14:506–514

Philip M, Rowley DA, Schreiber H (2004) Inflammation as a tumor promoter in cancer induction. Semin Cancer Biol 14:433–439

Phinney DG, Isakova I (2005) Plasticity and therapeutic potential of mesenchymal stem cells in the nervous system. Curr Pharm Des 11:1255–1265

Pittenger MF, Mackay AM, Beck SC, et al (1999) Multilineage potential of adult human mesenchymal stem cells. Science 284:143–147

Polverini PJ, Leibovich SJ (1984) Induction of neovascularization *in vivo* and endothelial proliferation *in vitro* by tumor-associated macrophages. Lab Invest 51:635–642

Prindull G, Zipori D (2004) Environmental guidance of normal and tumor cell plasticity: epithelial mesenchymal transitions as a paradigm. Blood 103:2892–2899

Prockop DJ (1997) Marrow stromal cells as stem cells for nonhematopoietic tissues. Science 276:71–74

Prockop DJ, Sekiya I, Colter DC (2001) Isolation and characterization of rapidly self-renewing stem cells from cultures of human marrow stromal cells. Cytotherapy 3:393–396

Radisky DC, Levy DD, Littlepage LE, et al (2005) Rac1b and reactive oxygen species mediate MMP-3-induced EMT and genomic instability. Nature 436:123–127

Reyes M, Lund T, Lenvik T, Aguiar D, Koodie L, Verfaillie CM (2001) Purification and *ex vivo* expansion of postnatal human marrow mesodermal progenitor cells. Blood 98:2615–2625

Richter G, Kruger-Krasagakes S, Hein G, et al (1993) Interleukin 10 transfected into Chinese hamster ovary cells prevents tumor growth and macrophage infiltration. Cancer Res 53:4134–4137

Robinson SC, Coussens LM (2005) Soluble mediators of inflammation during tumor development. Adv Cancer Res 93:159–187

Rojas M, Xu J, Woods CR, et al (2005) Bone marrow-derived mesenchymal stem cells in repair of the injured lung. Am J Respir Cell Mol Biol 33:145–152

Roni V, Habeler W, Parenti A, et al (2003) Recruitment of human umbilical vein endothelial cells and human primary fibroblasts into experimental tumors growing in SCID mice. Exp Cell Res 287:28–38

Ronnov-Jessen L, Petersen OW, Koteliansky VE, Bissell MJ (1995) The origin of the myofibroblasts in breast cancer. Recapitulation of tumor environment in culture unravels diversity and implicates converted fibroblasts and recruited smooth muscle cells. J Clin Invest 95:859–873

Ronnov-Jessen L, Petersen OW, Bissell MJ (1996) Cellular changes involved in conversion of normal to malignant breast: importance of the stromal reaction. Physiol Rev 76:69–125

Rowley DR (1998) What might a stromal response mean to prostate cancer progression? Cancer Metastasis Rev 17:411–419

Rubio D, Garcia-Castro J, Martin MC, et al (2005) Spontaneous human adult stem cell transformation. Cancer Res 65:3035–3039

Sakakura T, Nishizuka Y, Dawe CJ (1976) Mesenchyme-dependent morphogenesis and epithelium-specific cytodifferentiation in mouse mammary gland. Science 194:1439–1441

Sangai T, Ishii G, Kodama K, et al (2005) Effect of differences in cancer cells and tumor growth sites on recruiting bone marrow-derived endothelial cells and myofibroblasts in cancer-induced stroma. Int J Cancer 115:885–892

Sato Y, Araki H, Kato J, et al (2005) Human mesenchymal stem cells xenografted directly to rat liver are differentiated into human hepatocytes without fusion. Blood 106:756–763

Satoh H, Kishi K, Tanaka T, et al (2004) Transplanted mesenchymal stem cells are effective for skin regeneration in acute cutaneous wounds. Cell Transplant 13:405–412

Schoeberlein A, Holzgreve W, Dudler L, Hahn S, Surbek DV (2005) Tissue-specific engraftment after in utero transplantation of allogeneic mesenchymal stem cells into sheep fetuses. Am J Obstet Gynecol 192:1044–1052

Sekiya I, Larson BL, Smith JR, Pochampally R, Cui JG, Prockop DJ (2002) Expansion of human adult stem cells from bone marrow stroma: conditions that maximize the yields of early progenitors and evaluate their quality. Stem Cells 20:530–541

Serakinci N, Guldberg P, Burns JS, et al (2004) Adult human mesenchymal stem cell as a target for neoplastic transformation. Oncogene 23:5095–5098

Silva GV, Litovsky S, Assad JA, et al (2005) Mesenchymal stem cells differentiate into an endothelial phenotype, enhance vascular density, and improve heart function in a canine chronic ischemia model. Circulation 111:150–156

Silzle T, Kreutz M, Dobler MA, Brockhoff G, Knuechel R, Kunz-Schughart LA (2003) Tumor-associated fibroblasts recruit blood monocytes into tumor tissue. Eur J Immunol 33:1311–1320

Silzle T, Randolph GJ, Kreutz M, Kunz-Schughart LA (2004) The fibroblast: sentinel cell and local immune modulator in tumor tissue. Int J Cancer 108:173–180

Sivridis E, Giatromanolaki A, Koukourakis MI (2005) Proliferating fibroblasts at the invading tumour edge of colorectal adenocarcinomas are associated with endogenous markers of hypoxia, acidity, and oxidative stress. J Clin Pathol 58:1033–1038

Sternlicht MD, Lochter A, Sympson CJ, et al (1999) The stromal proteinase MMP3/stromelysin-1 promotes mammary carcinogenesis. Cell 98:137–146

Sternlicht MD, Bissell MJ, Werb Z (2000) The matrix metalloproteinase stromelysin-1 acts as a natural mammary tumor promoter. Oncogene 19:1102–1113

Studeny M, Marini FC, Champlin RE, Zompetta C, Fidler IJ, Andreeff M (2002) Bone marrow-derived mesenchymal stem cells as vehicles for interferon-beta delivery into tumors. Cancer Res 62:3603–3608

Studeny M, Marini FC, Dembinski JL, et al (2004) Mesenchymal stem cells: potential precursors for tumor stroma and targeted-delivery vehicles for anticancer agents. J Natl Cancer Inst 96:1593–1603

Sugimoto T, Takiguchi Y, Kurosu K, et al (2005) Growth factor-mediated interaction between tumor cells and stromal fibroblasts in an experimental model of human small-cell lung cancer. Oncol Rep 14:823–830

Tlsty TD, Hein PW (2001) Know thy neighbor: stromal cells can contribute oncogenic signals. Curr Opin Genet Dev 11:54–59

Wiesen JF, Young P, Werb Z, Cunha GR (1999) Signaling through the stromal epidermal growth factor receptor is necessary for mammary ductal development. Development 126:335–344

Wright N, de Lera TL, Garcia-Moruja C, et al (2003) Transforming growth factor-beta1 down-regulates expression of chemokine stromal cell-derived factor-1: functional consequences in cell migration and adhesion. Blood 102:1978–1984

Xia Z, Ye H, Choong C, et al (2004) Macrophagic response to human mesenchymal stem cell and poly(epsilon-caprolactone) implantation in nonobese diabetic/severe combined immunodeficient mice. J Biomed Mater Res A 71:538–548

Yang F, Tuxhorn JA, Ressler SJ, McAlhany SJ, Dang TD, Rowley DR (2005) Stromal expression of connective tissue growth factor promotes angiogenesis and prostate cancer tumorigenesis. Cancer Res 65:8887–8895

Ye J, Yao K, Kim JC (2006) Mesenchymal stem cell transplantation in a rabbit corneal alkali burn model: engraftment and involvement in wound healing. Eye 20:482–490

Yoneda T, Hiraga T (2005) Crosstalk between cancer cells and bone microenvironment in bone metastasis. Biochem Biophys Res Commun 328:679–687

Zhu W, Xu W, Jiang R, et al (2006) Mesenchymal stem cells derived from bone marrow favor tumor cell growth *in vivo*. Exp Mol Pathol 80:267–274

Subject Index

6-OHDA, 228, 230

ACTH, 7, 12
activated astrocytes, 233
activated microglia, 233
adhesion, 77
adipocyte, 136
adrenergic nerves, 15
adult, 137
age, 73
AGM, 95
akinesia, 132
Akt, 130
allogeneic, 130
α_1-antichymotrypsin, 13
α_1-antitrypsin, 13
Alzheimer's disease, 223, 224, 227, 231, 233
AMD3100, 11, 18
amiodarone, 127
amyotrophic lateral sclerosis, 223, 224, 227, 229, 231, 234
angina, 136
angiogenesis, 101
angiopoietin-1, 7, 9, 19
antagonists, 7, 151
anterior, 128
antioxidant, 80
aorta-gonad-mesonephros, 94
apical, 128
arrhythmia, 133
arrhythmogenesis, 127
artery, 137
assessment, 135
atherosclerosis, 83
autologous, 123
autotoxicity, 234

B lymphoid progenitor, 11

Bacterial endotoxins, 7
BDNF, 230–232
bedside, 147
bench, 147
biomarker, 153
biorepository, 148
BL-CFCs, 96
blast colony-forming cell, 95
blood, 134
blood islands, 92
BMDPC, 39–43
bone marrow-derived progenitor cells, 39
bone marrow-derived mononuclear cells, 121
border, 128

c-KIT, 12
c-Mpl, 22
Ca^{2+}, 9
cancer, 82
cardiomyocyte, 134
cardiomyoplasty, 151
cathepsin G, 12
catheter, 153
CCL3, 7
CD133, 20
CD146, 20
CD150, 10
CD31, 10, 20
CD34, 20
CD45, 20
CD90, 24
cell, 121
cell fusion, 106
cell therapy, 121, 222
center, 128
CFU-fibroblasts, 25
chemo, 82

chemokines, 6
choline acetyltransferase, 228, 231
chondrocyte, 136
cirrhosis, 247
comparison, 150
concentration, 129
connexin-43, 127
construct, 130
contractility, 122, 145
cord blood, 79
coronary artery bypass grafting, 123
culture expansion, 225
CXCL12, 7–9, 11, 12, 14, 22
CXCL2, 7
CXCL8, 7
CXCR4, 7, 11, 21, 42, 43
CXCR4 antagonists, 11
CY, 7
cyclin-dependent kinase inhibitor, 80
cyclooxygenase-2, 234
cyclophosphamide, 6
cytokines, 149

damage, 135, 146
database, 147
death, 133
Defibrotide, 7
degree, 137
delivery, 153
density, 135
desmoplastic reaction, 276
diastolic, 128, 146
dietary restriction, 74
differentiation, 149
DNA damage, 74, 80
dobutamine, 132
dopamine, 228
dopamine β hydroxylase, 14
downregulation, 138
drugs, 154
dysfunction, 151

E-selectin, 10
EDNO, 40–43
effect of granulocyte-colony stimulating factor, 39
efficacy, 127
embryonic dopaminergic neurons, 226
endosteum, 6
endothelial dysfunction, 39, 40, 43

endothelial HSC niche, 10
endothelial nitric oxide, 39
endothelial nitric oxide synthase, 40
endothelial progenitor, 39
endothelial progenitor cells, 19
endotoxin, 12
engraftment, 76, 122, 154
eNOS, 22, 39–43
eNOS-KO, 41–43
EPC, 19, 20, 23, 40–42
erythropoietin, 7, 19, 22
example, 135
exercise, 145
expansion, 124
expression, 137

factor, 134
fibroblast, 136
fibroblast contribution, 271
fibroblastic stromal cells, 266
filgrastim, 146
Flt-3 ligand, 7
5′-fluorouracil, 6
5-FU, 6, 7
Fucoidans, 7
function, 134

G-CSF, 6, 7, 19, 22, 26, 27, 42, 43
G-SCF, 43
GDNF, 230–232
gestation, 25
glucose, 130
glutamine, 129
glutathione peroxidases, 79
GM-CSF, 7, 17, 19, 22, 42
granulocyte, 146
granulocyte macrophage-colony stimulating factor, 42
granulocyte-colony stimulating factor, 42
growth hormone, 7

heart failure, 120
hemangioblast, 92, 94–96, 99
hemangiocytes, 21–23
hematopoiesis, 92–94
hematopoietic, 134
hematopoietic cells, 108, 109
hematopoietic stem cells, 6
homing, 135
hospitalization, 156

Subject Index   287

HSC exhaustion, 73
hyaluronic, 9
hypertension, 151
hypoxia, 27

IGF, 231
IL-1, 17
IL-3, 7, 17
IL-6, 17
IL-7, 11
immunomodulation, 235
immunosuppression, 123
implantation, 152
improvement, 124
infarct size, 137
inflammation, 135
inflammatory process, 233
inflammatory stress, 233
integrin α4β1, 11
integrity, 135
interferon and stromal-derived factor 1, 130
interferon β, 277
interleukin, 130
intervention, 153
interventional, 145
ischemia, 19, 102, 129, 136

Jagged-1, 9

KIT ligand, 7

L-DOPA, 228
lateral, 128
lifespan, 81
limiting dilution, 76
lineages, 146
LSK cells, 77
LV function, 122

MACE, 132
macrophages, 105
management, 148
marker, 135
marrow stroma, 136
matrix metalloproteinase-9, 12
maturation, 135
mature, 146
mediator, 136
mesenchymal Stem Cells, 25

mesenchymal stem cells, 24, 225
mesodermal, 136
microarray, 75
microenvironment, 76
microenvironmental dialog, 271
milieu only, 145
minocycline, 234
MMP-9, 23
$MMP\text{-}9^{-/-}$, 22
mobilization, 5, 17, 19, 77, 138
molecular, 128
monocytes, 100, 101, 105, 108, 130
Mozobil, 18
MPTP, 223, 228
MSC homing, 273
multipotent adult progenitor cells, 225, 226, 229
mycophenolate, 132
myocardial infarction, 120
myogenesis, 123

N-acetyl-cysteine, 80
N-cadherin, 9, 127
neonatal, 137
network, 153
neurodegeneration, 222
neurogenesis, 223, 232
neutrophil elastase, 12
NF-κB, 130
NGF, 231, 232
niches, 6, 8, 11
nitric oxide, 22
nitric oxide synthase enzyme, 40
nitric oxide synthases, 40
nomenclature, 152
nonrevascularizable, 123
nonsteroidal antiinflammatory drugs, 234
$Nos3^{-/-}$, 22
Notch1, 9
NTN, 230

origin, 136
osteoblast, 136
osteoblasts, 6, 8, 14
osteopontin, 9
oxygen, 130
oxygen tension, 78

P-selectin, 10
p38 mitogen-activated protein kinase, 81

paraaortic splanchnopleura, 94
paracrine, 122
parathyroid hormone receptor 1, 8
Parkinson's disease, 223, 224, 226, 228, 230, 233
PAS, 95
pathway, 137
patient-to-patient variability, 17
PECAM-1, 10
pegylated G-CSF, 17, 27
perfusion, 135
pericyte, 103, 107–109
Pertussis toxin, 7
pharmacological, 145
pharmacology, 156
placebo, 132
plaque, 130
polyanions, 12
preconditioning, 151
precursor, 133
product, 136
progenipoietin-1, 17
progenitor, 133
progenitor cells, 41

radiation, 82
ranolazine, 138
re-vascularization, 21, 23
reactive oxygen species (ROS), 79
receptors, 155
reduction, 133
regeneration, 121
registry, 148
rejection, 136
remodeling, 121
repair, 135
reproducibility, 154
resistant, 127
resonance imaging, 129
restenosis, 138
revascularization, 151

safety evaluation, 155
scar, 124
SCF, 7, 12, 17, 27, 42, 43
SDF-1, 9, 42
senescence, 70, 82
serpina1, 13
serpina3, 13
SH2, 24

SH3, 24
SH4, 24
SHH, 230
skeletal myoblasts, 121
soluble SCF, 22
ST-elevation, 145
stem cell factor, 39, 42
stem cells, 75, 134
– adult, 223
– embryonic, 95, 222
– hematopoietic, 94, 97, 99, 224
– mesenchymal, 99
– neural, 223
stress, 132
STRO-1, 24
stromal cell-derived factor-1, 40
stromal function, 266
sulfated polysaccharides, 12
superoxide, 79
survival, 129
sympathetic nerves, 15
systolic, 123

tacrolimus, 132
targeted delivery, 230
taxonomy, 152
telomerase, 72
telomere, 71
telomere loss, 225
thickness, 135
thrombopoietin, 7, 22
Thy-1, 24
Tie-2, 9, 21
Tie-2 expressing monocytes, 23
Tie-2-expressing monocytes, 21, 22
Tie-2^+, 8
tissue, 130
toxicity, 129
TPO, 27
transmembrane c-KIT ligand, 12
transmembrane stem cell factor, 9
transplant, 75
transplantation, 73, 128
treatment, 124
trial, 133
tumor, 102, 104–107
type 1 diabetes, 252
type 2 diabetes, 253
tyrosine hydroxylase, 227–229

umbilical, 146
umbilical cord blood, 25
uridine diphosphate-galactose ceramide
  galactosyltransferase, 14

vascular endothelial growth factor, 19, 40
vasculature, 84, 136
vasculogenesis, 102, 135

VCAM-1, 8, 11, 12
VE-cadherin, 20
VEGF, 7, 40, 42, 149
VEGFR1, 21
VEGFR2, 20
VLA-4, 11

wall stress, 122